Controlled Stochastic Processes

I. I. Gihman
A. V. Skorohod

Controlled Stochastic Processes

Translated by Samuel Kotz

Springer-Verlag
New York Heidelberg Berlin

Iosif Il'ich Gihman
Academy of Sciences of the Ukranian SSR
Institute of Applied Mathematics
and Mechanics
Donetsk
USSR

Anatoliĭ Vladimirovich Skorohod
Academy of Sciences of the Ukranian SSR
Institute of Mathematics
Kiev
USSR

Translator:

Samuel Kotz
Department of Management Science and Statistics
University of Maryland
College Park, MD 20742
USA

AMS Subject Classification (1980): 34H05, 49E99, 60H99, 93Exx

Library of Congress Cataloging in Publication Data

Gihman, Iosif Il'ich.
Controlled stochastic processes.

Translation of Upravli͡aemye sluchaĭnye prot͡sessy.
Bibliography: p.
Includes index.
1. Stochastic processes. 2. Control theory.
I. Skorohod, Anatoliĭ Vladimirovich, joint
author. II. Title.
QA274.G5613 519.2 79-4107

Title of the Russian Original Edition: Upravliaemye sluchaĭnye protsessy.
Publisher: Nauka, Moscow, 1977

Softcover reprint of the hardcover 1st edition 1979

9 8 7 6 5 4 3 2 1

ISBN-13: 978-1-4612-6204-6 e-ISBN-13: 978-1-4612-6202-2
DOI: 10.1007/978-1-4612-6202-2

Preface

The theory of controlled processes is one of the most recent mathematical theories to show very important applications in modern engineering, particularly for constructing automatic control systems, as well as for problems of economic control. However, actual systems subject to control do not admit a strictly deterministic analysis in view of random factors of various kinds which influence their behavior. Such factors include, for example, random noise occurring in the electrical system, variations in the supply and demand of commodities, fluctuations in the labor force in economics, and random failures of components on an automated line. The theory of controlled processes takes the random nature of the behavior of a system into account. In such cases it is natural, when choosing a control strategy, to proceed from the average expected result, taking note of all the possible variants of the behavior of a controlled system.

An extensive literature is devoted to various economic and engineering systems of control (some of these works are listed in the Bibliography). However, as of now there is no text which adequately covers the general mathematical theory of controlled processes. The authors of this monograph have attempted to fill this gap.

In this volume the general theory of discrete-parameter (time) controlled processes (Chapter 1) and those with continuous-time (Chapter 2), as well as the theory of controlled stochastic differential equations (Chapter 3), are presented. In this book, the notion of a controlled stochastic object serves as a departing basic concept; this allows us to substantially avoid the difficulties associated with continuous-parameter processes that are so familiar to specialists in the field. The traditional problems of optimal stopping rules for processes and the derivation of Bellman's equations and their application to the construction of optimal equations for controlled Markov processes are also examined in this volume. The authors have attempted to minimize the use of advanced mathematical methods; however, a knowledge of the basic notions of the theory of stochastic processes and the rudiments of measure theory and functional analysis are presupposed.

The authors would appreciate receiving comments and suggestions from the readers.

I. I. Gihman
A. V. Skorohod

Contents

Discrete-Parameter Controlled Stochastic Processes 1

1 Definitions

Let two sets X and U with σ-algebras of measurable subsets $\mathfrak{A}$ and $\mathfrak{B}$ respectively, i.e. two measurable spaces $(X, \mathfrak{A})$ and $(U, \mathfrak{B})$ be given. The first space is called the *phase space of the basic process* and the second *the phase space of control*. Let N be the set of non-negative integers. In this Chapter all the processes are defined on the set N. To define a controlled process it is necessary to define the probability distribution of a random process with values in X provided a sequence of controls at each instant of time is given and also to define a rule according to which these controls are selected. We shall now describe the components of a controlled process in a more precise manner.

It is natural to assume that the distribution of the random variable x_n—representing the values of the basic process at time n—is completely determined provided the values of the basic process $x_0, \ldots, x_{n-1}$ at the preceding instants of time and the values of the controls $u_0, \ldots, u_{n-1}$ at these times are known. Let

$$p_n(dx_n/x_0, \ldots, x_{n-1}; u_0, \ldots, u_{n-1}) \tag{1.1}$$

define the conditional distribution of the variable x_n given $x_0, \ldots, x_{n-1}$; $u_0, \ldots, u_{n-1}$. We postulate that the collection of functions $\{p_n(\cdot/\cdot), n = 0, 1, \ldots\}$ defines *the controlled object*. In order that the functions (1.1) will serve as distributions of the sequence of variables $\{x_n, n = 0, 1, \ldots\}$ in $(X, \mathfrak{A})$ it is necessary and sufficient that the following two conditions be satisfied:

1. $p_n(\cdot/\cdot)$ be a measure on $\mathfrak{A}$ with respect to the first argument.

2. $p_n(A_n/x_0, \ldots, x_{n-1}; \cdot)$ be measurable in $x_0, \ldots, x_{n-1}$ relative to $\mathfrak{A}$ for all $A \in \mathfrak{A}$ and $u_0, \ldots, u_{n-1} \in U$. These conditions are assumed to be satisfied for all the controlled objects under consideration.

If a controlled object $\{p_n(\cdot/\cdot), n = 0, 1, \ldots\}$ is defined, then using it one can construct a family of distributions in X^N which depends on a point of the space U^N as on a parameter. We denote by $\mathbf{x}$ and $\mathbf{u}$ the points in X^N and U^N respectively:

$$\mathbf{x} = \{x_0, x_1, \ldots\}, \qquad \mathbf{u} = \{u_0, u_1, \ldots\}.$$

Let $\mathfrak{A}^N$ and $\mathfrak{B}^N$ be σ-algebras in X^N and U^N generated by cylinders, and $\mathfrak{A}_n$ and $\mathfrak{B}_n$ be σ-algebras of cylinders in X^N and U^N with bases over $\{0, 1, \ldots, n\}$. We define a family of measures $\mu(\cdot/u)$ on $\mathfrak{A}^N$ by relation

$$\mu(C/\mathbf{u}) = \int_{C_0} p(dx_0) \int_{C_1} p_1(dx_1/x_0; u_0) \cdots \int_{C_n} p(dx_n/x_0, \ldots, x_{n-1}; u_0, \ldots, u_{n-1}); \tag{1.2}$$

here C is a cylinder in $\mathfrak{A}_n$ of the form:

$$C = \{\mathbf{x} : x_0 \in C_0, \ldots, x_n \in C_n\},$$

$C_k \in \mathfrak{A}$. Relation (1.2) uniquely defines a measure on $\mathfrak{A}^N$. It is easy to verify that the family of measures $\mu(\cdot/\mathbf{u})$ possesses the following property: if $V \in \mathfrak{A}_n$ then $\mu(V/\mathbf{u})$ depends only on $u_0, \ldots, u_{n-1}$. This property may be more succinctly stated if we assume instead of condition 2 the following condition:

3. $p_n(A/x_0, \ldots, x_{n-1}; u_0, \ldots, u_{n-1})$ is measurable jointly in the variables relative to σ-algebra $\mathfrak{A}^n \times \mathfrak{B}^n$. It is natural to impose this condition on the controlled object in order to be able to utilize random controls.

If condition 3 is fulfilled then $\mu(V/\mathbf{u})$ is $\mathfrak{B}^N$-measurable for $V \in \mathfrak{A}^N$ and $\mathfrak{B}_{n-1}$-measurable for $V \in \mathfrak{A}_n$. Now let a family of measures $\mu(V/\mathbf{u})$ be defined on $\mathfrak{A}^N$ satisfying the condition:

4. $\mu(V/\mathbf{u})$ is a measurable function for $V \in \mathfrak{A}_n$. Consider the set $A^{(n)} = \{\mathbf{x} : x_n \in A\}$. Let $\mu(A^{(n)}/\mathfrak{A}_n/\mathbf{u})$ be the conditional probability of $A^{(n)}$ relative to σ-algebra $\mathfrak{A}_{n-1}$ calculated on the probability space $\{X^N, \mathfrak{A}^N, \mu(\cdot/\mathbf{u})\}$. Since $A^{(n)} \in \mathfrak{A}_n$ we have

$$\mu(A^{(n)}/\mathfrak{A}_{n-1}/\mathbf{u}) = p_n(A/x_0, \ldots, x_{n-1}; u_0, \ldots, u_{n-1}) \tag{1.3}$$

for almost all $\mathbf{x}$ in measure $\mu(\cdot/\mathbf{u})$.

The only thing that can be said about the function appearing on the r.h.s. is that it is $\mathfrak{A}_{n-1}$-measurable for fixed A and $u_0, \ldots, u_{n-1}$. However, under quite general conditions this function can be defined in such a manner that condition 1 is satisfied for this function. From theorem 3 in Section 3 of

Chapter 1 of [20] it follows that this condition in particular can be assumed to be fulfilled if X is a complete metric separable space and $\mathfrak{A}$ is a σ-algebra of its Borel sets. In this case one can select a variant of the conditional probability in the r.h.s. of (1.3) such that condition 3 will be satisfied as well. First we shall establish the following auxiliary assertion.

Lemma 1.1. *Let X be a separable complete metric space. There exists a Borel one-to-one mapping $f: X \to [0, 1]$ such that $f(A)$ is a Borel set on $[0, 1]$ for all $A \in \mathfrak{A}$.*

PROOF. Let $r(x, y)$ be a metric in X. Without loss of generality it may be assumed that $r(x, y) \leq 1$ for $x, y \in X$ (for example, one may consider an equivalent initial metric $1 - \exp\{-r(x, y)\}$). Let $\{x_k, k = 1, 2, \ldots\}$ be a dense set in X. Set $r_k(x) = r(x, x_k)$. It is easy to see that

$$\sup_k |r_k(x) - r_k(y)| = r(x, y).$$

Therefore $g: x \to (r_1(x), r_2(x), \ldots)$ isometrically maps X into l_∞—the Banach space of numerical sequences $(r_1, r_2, \ldots)$ with the norm

$$\|(r_1, r_2, \ldots)\| = \sup_k |r_k|.$$

Denote by $g(X)$ the image of X; $g(X)$ is a closed set (since X is dense). Let B be the closed linear hull of $g(X)$. Since for a separable X, B is a separable subspace of l_∞ it follows that B is a separable Banach space. In view of the well known theorem concerning the universality of $C_{[0,1]}$ (see, e.g., [1] p. 264) B is isometric to a subspace of the space $C_{[0,1]}$. Denote this mapping of B into $C_{[0,1]}$ by g_1. Furthermore the mapping $g_2: x(\cdot) \to x(\cdot)$ is a continuous mapping of $C_{[0,1]}$ into $L_2[0, 1]$, and moreover $g_2(C_{[0,1]})$ is the Borel set:

$$g_2(C_{[0,1]}) = \bigcap_{m=1}^{\infty} \bigcup_{N=1}^{\infty} \bigcap_{r=N}^{\infty} \bigcap_{n=1}^{\infty} \bigcap_{|k-l| \leq n/r}^{\infty} \left\{x(\cdot): n^2 \left| \int_{k/n}^{k+1/n} \int_{l/n}^{l+1/n} (x(t) - x(s))\, dt\, ds \right| \leq \frac{1}{m}\right\}.$$

Therefore the image of any Borel set in $C_{[0,1]}$ will be a Borel set. Denote by g_3 the isometric mapping of $L_2[0, 1]$ into l_2 and by g_4 the natural imbedding of l_2 into l_∞. g_4 is continuous and measurable with respect to the Borel σ-algebra in l_2 and the σ-algebra $\mathfrak{B}_\infty$ generated by coordinate functionals in l_∞. The image of l_2 under this mapping is a $\mathfrak{B}_\infty$-measurable set. Therefore the image of any Borel set in l_2 will also be $\mathfrak{B}_\infty$-measurable. Now construct a measurable mapping of $(l_\infty, \mathfrak{B}_\infty)$ into $([0, 1], \mathfrak{A}_{[0,1]})$; $\mathfrak{A}_{[0,1]}$ is a Borel σ-algebra on $[0, 1]$ for which the image of l_∞ is a Borel set. Let $l_k(y)$ be the k-th coordinate functional:

$$y = (l_1(y), l_2(y), \ldots).$$

Set

$$\alpha_{nk}(y) = \text{Ent}\left[\left(\frac{1}{2} + \frac{1}{\pi}\operatorname{arctg} l_k(y) - \sum_{j=1}^{n-1} \alpha_{jk}(y)2^{-j}\right)2^n\right]$$

(Ent[t] is the integral part of t). $\alpha_{nk}(y)$ maps measurably $(l_\infty, \mathfrak{B}_\infty)$ into $([0, 1], \mathfrak{A}_{[0, 1]})$. Therefore the function

$$g_5(y) = \sum_{k=1}^{\infty} \sum_{n=1}^{\infty} \alpha_{nk}(y)2^{-2^{n}3^{k}}$$

possesses the same property. It is easy to see that g_5 maps l_∞ into [0, 1] in a one-to-one manner. The image of l_∞ differs from the perfect set of points $t \in [0, 1]$ of the form $t = \sum_{k,n} \alpha_{nk} 2^{-2^n 3^k}$ where α_{nk} takes on values 0 or 1—only on a certain countable set. The existence of the mapping $g_5 g_4 g_3 g_2 g_1 g$ was asserted in lemma 1.1. □

Remark 1.1. One can correspond to every controlled object with a complete separable metric phase space X a controlled object with the phase space [0, 1] by setting for $A \in \mathfrak{A}$, $x_0, \ldots, x_{n-1}$

$$\begin{aligned}\hat{p}_n(f(A)/f(x_0), \ldots, f(x_{n-1});\quad u_0, \ldots, u_{n-1})& \\ = p_n(A/x_0, \ldots, x_{n-1};\quad u_0, \ldots, u_{n-1}).&\end{aligned} \tag{1.4}$$

If at least one of the points $t_0, t_1, \ldots, t_{n-1}$ does not belong to $f(X)$ we set

$$\hat{p}_n(A/t_0, t_1, \ldots, t_{n-1};\quad u_0, u_1, \ldots, u_{n-1}) = \chi_A(0).$$

Let $\hat{\mu}(A/\mathbf{u})$ be a controlled object constructed by means of $\hat{p}_n$ in $[0, 1]^N$ and f_N be the mapping of X^N into $[0, 1]$; $f_N(x) = (f(x_0), f(x_1), \ldots)$. Then for $A \in \mathfrak{A}^N$

$$\hat{\mu}(f_N(A)/\mathbf{u}) = \mu(A/\mathbf{u}).$$

Therefore if for the controlled object $\hat{\mu}(\cdot/\mathbf{u})$ there exist functions $\hat{p}_n$ satisfying conditions 1 and 3 such that $\hat{\mu}$ and $\hat{p}_n$ are connected by the expression (1.2) then the functions $p_n(\cdot/\cdot, \cdot)$ also exist for the controlled object $\mu(\cdot/\cdot)$ (these are expressed in terms of $\hat{p}_n$ via equation (1.4)). Consequently, relation (1.2) is fulfilled and p_n satisfies conditions 1 and 3.

Theorem 1.1. *Let $(X, \mathfrak{A})$ be a complete separable metric space with a σ-algebra of Borel sets. If the family of measures $\mu(V/\mathbf{u})$ satisfies condition* 4 *then there exists a collection of functions $\{p_n(dx_n/x_0, \ldots, x_{n-1}, u_0, \ldots, u_{n-1})$, $n = 0, 1, \ldots\}$ satisfying conditions* 1 *and* 3 *such that formula* (1.2) *is valid.*

PROOF. As it follows from lemma 1.1 and Remark 1.1 it is sufficient to prove the theorem for the case when X is [0, 1]. Let $\mathfrak{A}^{(k)}$ denote the σ-algebra generated by intervals

$$\left[0, \frac{1}{2^k}\right), \left[\frac{1}{2^k}, \frac{2}{2^k}\right), \ldots, \left[\frac{2^k - 1}{2^k}, 1\right]$$

and $\mathfrak{A}_{n-1}^{(k)}$ denotes the σ-algebra in $\mathfrak{A}_{[0,1]}^{N}$ generated by sets of the form $\{x : x_0 \in A_0, \ldots, x_{n-1} \in A_{n-1}\}$ where $A_i \in \mathfrak{A}^{(k)}$. Clearly $\mathfrak{A}_{n-1}^{(k)} \subset \mathfrak{A}_{n-1}^{(k+1)}$ and $\mathfrak{A}_{n-1}$ coincides with the σ-closure of $\bigcup_k \mathfrak{A}_{n-1}^{(k)}$. Therefore for $A \in \mathfrak{A}_{[0,1]}^{N}$

$$\mu(A/\mathfrak{A}_{n-1}, \mathbf{u}) = \lim_{k\to\infty} \mu(A/\mathfrak{A}_{n-1}^{(k)}, \mathbf{u})$$

for almost all x in measure $\mu(dx/\mathbf{u})$, ($\mu(A/\mathfrak{L}, \mathbf{u})$ denotes the conditional distribution with respect to σ-algebra $\mathfrak{L}$ relative to measure $\mu(\cdot/\mathbf{u})$). Hence the conditional distribution function

$$F_n(z/x_0, \ldots, x_{n-1}, \mathbf{u}) = \mu(\{x : x_n < z\}/\mathfrak{A}_{n-1}, \mathbf{u})$$

for almost all $x_0, \ldots, x_{n-1}$, in measure $\mu(\cdot/u)$, is a limit of the functions

$$F_n^{(k)}(z/x_0, \ldots, x_{n-1}, u_0, \ldots, u_{n-1}) = \sum_i \frac{\mu(\{x : x_n < z\} \cap V_i/\mathbf{u})}{\mu(V_i/\mathbf{u})} \chi V_i(x)$$

where $V_1, \ldots, V_{2^{kn}}$ are the sets of X^N generating $\mathfrak{A}_{n-1}^{(k)}$. Set

$$\tilde{F}_n(z/x_0, \ldots, x_{n-1}, u_0, \ldots, u_{n-1}) = \lim_{k\to\infty} F_n^{(k)}(z/x_0, \ldots, x_{n-1}, u_0, \ldots, u_{n-1}) \tag{1.5}$$

whenever the limit exists. For all $z \in [0, 1]$ the function $\tilde{F}(z/\cdot)$ is determined on a measurable set $S_n(z) \subset X^n \times U^n$ and is measurable in $x_0, \ldots, x_{n-1}, u_0, \ldots, u_{n-1}$ with respect to $\mathfrak{A}_{[0,1]}^{n} \times \mathfrak{L}^n$ and moreover for every $\mathbf{u}$

$$\mu(\{x : (x_0, \ldots, x_{n-1}, u_0, \ldots, u_{n-1}) \in S_n(z)\}/\mathbf{u}) = 1. \tag{1.6}$$

Let

$$S_n - \bigcap_m \bigcap_{k \le 2^m} S_n\left(\frac{k}{2^m}\right),$$

and $P(A)$ be an arbitrary measure on $\mathfrak{A}_{[0,1]}$. Set

$$F_n(z/\cdot) = \sup_{k/2^m < z} \tilde{F}_n((k/2^m)/\cdot)$$

$$P_n(A/x_0, \ldots, x_{n-1}, u_0, \ldots, u_{n-1}) = \begin{cases} \int_A dF_n(z/x_0, \ldots, x_{n-1}, u_0, \ldots, u_{n-1}) & \\ \qquad \text{if } (x_0, \ldots, x_{n-1}, u_0, \ldots, u_{n-1}) \in S_n, \\ P(A) \quad \text{if } (x_0, \ldots, x_{n-1}, u_0, \ldots, u_{n-1}) \notin S_n. \end{cases}$$

It is easy to verify that for every continuous function $\zeta(z)$ on $[0, 1]$

$$\int \zeta(z) P_n(dz) x_0, \ldots, x_{n-1}, u_0, \ldots, u_{n-1}) = \int \zeta(z) \mu(dz/\mathfrak{A}_{n-1}, u) \tag{1.7}$$

for almost all $(x_0, \ldots, x_{n-1}, u_0, \ldots, u_{n-1})$ in measure $\mu(\cdot/u)$. Hence (1.7) holds for all bounded measurable functions. Formula (1.2) now follows from (1.7) and properties of iterated conditional expectations. □

Thus if a phase space is a complete separable metric space with a σ-algebra of Borel sets we have two equivalent definitions of a controlled object. In the first place, it is a family of functions (1.1) satisfying conditions 1–3. In the second, it is a family of measures $\mu(\cdot/\mathbf{u})$ on $(X^N, \mathfrak{B}^N)$ depending on $\mathbf{u} \in \mathfrak{B}^N$ as on a parameter and satisfying the following consistency condition: if $V \in \mathfrak{A}_n$ then $\mu(V/\mathbf{u})$ is a $\mathfrak{B}_{n-1}$-measurable function of $\mathbf{u}$. The latter definition is more convenient, it is shorter and may be carried over to the case of continuous processes. This fact will be utilized below.

We now proceed to define a *sequence of controls* or a *strategy*. It is natural to suppose that when choosing a control at time n we know the value of the basic process up to that time inclusively and we also know the value of the control at the preceding (moment of) time. The actual value of the control u_n at time n is assumed to be random.

Let

$$q_n(du_n/x_0, \ldots, x_n; u_0, \ldots, u_{n-1}) \tag{1.8}$$

be the conditional distribution of the variable u_n given that the values of the basic process is $x_0, \ldots, x_n$ and that of the controls in the preceding times is $u_0, \ldots, u_{n-1}$. The function (1.8) is defined for $n = 0, 1, \ldots$ (for $n = 0$ the function is $q_0(du_0/x_0)$ and satisfies the following conditions:

5. $q_n(\cdot/x_0, \ldots, x_n; \;\; u_0, \ldots, u_{n-1})$ is a probability measure in the first variable for all $x_i \in X$ and $u_i \in U$;
6. for $B \in \mathfrak{B}$ $q_n(B/x_0, \ldots, x_n; \; u_0, \ldots, u_{n-1})$ is a $\mathfrak{A}^{n+1} \times \mathfrak{B}^n$-measurable function in x_i and u_i.

By means of functions $q_n(\cdot/\cdot)$ we can construct a family of measures $\nu(\cdot/\mathbf{x})$ on $(U^N, \mathfrak{B}^N)$ depending on $\mathbf{x} \in X^N$ as on a parameter. Let D be a cylinder in $\mathfrak{B}_n$ such that

$$D = \{\mathbf{u} : u_0 \in D_0, \ldots, u_n \in D_n\}.$$

Then

$$\nu(D/\mathbf{x}) = \int_{D_0} q_0(du_0/x_0) \int_{D_1} q_1(du_1/x_0, x_1; u_0) \cdots \int_{D_n} q_n(du_n/x_0, \ldots, x_n; u_0, \ldots, u_{n-1}). \tag{1.9}$$

Formula (1.9) for each $\mathbf{x}$ determines a consistent family of finite-dimensional distributions and hence there exists a unique measure on $(U^N, \mathfrak{B}^N)$ for which (1.8) is fulfilled. The family of measures $\nu(D/x)$ satisfies the following consistency condition:

7. if $W \in \mathfrak{B}_n$ then $\nu(W/\mathbf{x})$ is a $\mathfrak{A}_n$-measurable function of $\mathbf{x}$.

It follows from theorem 1.1 that in the case when U is a complete separable metric space with the σ-algebra of Borel sets, then for any family of measures $\nu(\cdot/\mathbf{x})$ satisfying condition 7 one can construct a family of functions $q_n(\cdot/\cdot)$ satisfying conditions 5 and 6 and connected with $\nu(\cdot/\cdot)$ by relation (1.9). Such a family of measures $\nu(\cdot/\cdot)$ will be called a *strategy* (or sometimes a *control*).

If a controlled object $\mu(\cdot/\mathbf{u})$ and a control $\nu(\cdot/\mathbf{x})$ are given, one can construct a random sequence (ξ_n, η_n) with values in $X \times U$ (more precisely one can construct a distribution on $X^N \times U^N$) such that

$$\left.\begin{aligned} &\mathbf{P}\{\xi_n \in A_n/\xi_0, \eta_0, \ldots, \xi_{n-1}, \eta_{n-1}\} \\ &\qquad = p_n(A_n/\xi_0, \ldots, \xi_{n-1}; \eta_0, \ldots, \eta_{n-1}), \\ &\mathbf{P}\{\eta_n \in B_n/\xi_0, \eta_0, \ldots, \xi_{n-1}, \eta_{n-1}, \xi_n\} \\ &\qquad = q_n(B_n/\xi_0, \ldots, \xi_n, \eta_0, \ldots, \eta_{n-1}), \end{aligned}\right\} \tag{1.10}$$

where the measures $p_n(\cdot/\cdot)$ and $q_n(\cdot/\cdot)$ are determined in terms of $\mu(\cdot/\cdot)$ and $\nu(\cdot/\cdot)$. Finite-dimensional distributions of the sequence (ξ_n, η_n) are defined by the equality

$$\begin{aligned} &\mathbf{P}\{\xi_0 \in A_0, \eta_0 \in B_0, \ldots, \xi_n \in A_n, \eta_n \in B_n\} \\ &= \int_{A_0} p_0(dx_0) \int_{B_0} q_1(du_0/x_0), \ldots, \int_{A_n} p_n(dx_n/x_0, \ldots, x_{n-1}; u_0, \ldots, u_{n-1}) \\ &\qquad \times \int_{B_n} q_n(du_n/x_0, \ldots, x_n; u_0, \ldots, u_{n-1}). \end{aligned} \tag{1.11}$$

A sequence (ξ_n, η_n) in $X \times U$ for which the condition (1.10) is satisfied is called a *controlled random sequence* (process) with the controlled object $\mu(\cdot/\cdot)$ and the control $\nu(\cdot/\cdot)$. Clearly, for any controlled object and control there exists a controlled process and its distribution is uniquely determined by a control and a controlled object. The sequence $\{\xi_n; n = 0, 1, \ldots\}$ is called the *basic* or *controlled process* and the sequence $\{\eta_n; n = 0, 1, \ldots\}$ is called *control*.

If a controlled object and control are given, then the controlled process "performs" in the following manner: we choose η_0 from ξ_0; these two variables define the state of the basic process ξ_1 at time 1 (more precisely the distribution of the process at time 1); by means of ξ_0, η_0 and ξ_1 we construct control η_1 at time 1; $\xi_0, \eta_0, \xi_1, \eta_1$ then determine ξ_2 and so on.

Any sequence (ξ_n, η_n) with values in $X \times U$ can be considered as a controlled process with a controlled object and a control. The latter can be constructed by means of functions $p_n(\cdot/\cdot)$ and $q_n(\cdot/\cdot)$ which are determined with the aid of conditional probabilities appearing in the l.h.s. of equation (1.5).

Non-randomized controls constitute an important subclass of controls; these are controls for which the measure $\nu(\cdot/\mathbf{x})$ is concentrated at one point of the space U^N for any $\mathbf{x} \in X$. For a non-randomized control there exist a sequence of functions $\varphi_n(x_0, \ldots, x_n)$ with values in U measurable with respect to $\mathfrak{A}_n$ such that

$$p_n(B_n/x_0, x_1, \ldots, x_n; u_0, u_1, \ldots, u_{n-1}) = \chi_{B_n}(\varphi_n(x_0, \ldots, x_n))$$

(here $\chi_B(\cdot)$ is the indicator of B). Clearly the class of non-randomized controls is substantially smaller than the class of all controls. As will be shown below it is sufficient in many cases to consider only controlled processes with a non-randomized control.

2 Optimization Problem

The basic problem in the theory of controlled processes is to choose an optimal control (this choice is made by a given controlled object). Let a controlled object $\mu(\cdot/\cdot)$ and a class of admissible controls (strategies) $\mathfrak{N}$ be given. Moreover, it is assumed that a functional $F(\mathbf{x}, \mathbf{u})$ on $X^N \times U^N$ measurable with respect to $\mathfrak{A}^N \times \mathfrak{B}^N$ is defined. This functional is called a *cost of controlling* (or *control cost*). It describes the expenditures required for controlling the given controlled object provided a sequence of controls $\mathbf{u} = (u_0, u_1, \ldots)$ is selected and the basic process takes on the sequence of values $\mathbf{x} = (x_0, x_1, \ldots)$. Assume that a certain strategy $\nu(\cdot/\cdot) \in \mathfrak{N}$ is chosen. By means of the controlled object $\mu(\cdot/\cdot)$ and strategy $\nu(\cdot/\cdot)$ one can construct the controlled process (ξ, η), $\xi = (\xi_0, \xi_1, \ldots)$, $\eta = (\eta_0, \eta_1, \ldots)$ as it was described in Section 1. Denote by $\mathbf{E}_\nu$ the mean value with respect to a measure corresponding to (ξ, η) in $X^N \times U^N$ provided the control $\nu(\cdot/\cdot)$ is selected (clearly this measure is determined by $\mu(\cdot/\cdot)$ and $\nu(\cdot/\cdot)$, however $\mu(\cdot/\cdot)$ is considered to be fixed and therefore the dependence of the mean value on μ is not indicated). Then the mean cost of controlling when strategy $\nu(\cdot/\cdot)$ is used is determined by the expression

$$S(\nu) = \mathbf{E}_\nu F(\xi, \eta).$$

An optimization problem consists in determining a strategy for which $S(\nu)$ is minimal, i.e. a control with minimal expenditures. To solve this problem one should naturally first determine the optimal control cost:

$$S = \inf_{\nu \in \mathfrak{N}} S(\nu).$$

After that we search for controls (or at least one control) $\bar{\nu}$ such that $S = S(\bar{\nu})$. It may turn out that no such $\bar{\nu}$ exists. In that case controls ν_ε for which $S(\nu_\varepsilon) \leq S + \varepsilon$ will be of interest. These are called *ε-optimal controls.* 0-optimal controls are called *optimal.*

Thus the basic problem of the theory of controlled processes is stated as follows: *For a given controlled object $\mu(\cdot/\cdot)$, cost of control $F(\cdot, \cdot)$ and a class of admissible controls $\mathfrak{N}$, determine an optimal control, and if such a control does not exist, determine an ε-optimal control for all $\varepsilon > 0$.*

We shall assume that X and U are complete separable metric spaces. We shall now describe the class $\mathfrak{N}$ of controls with constraints for which both optimal and ε-optimal controls can be found among non-randomized controls.

A class $\mathfrak{N}$ of controls is called *a class of controls with constraints* if it contains all the controls for which the following 2 conditions are satisfied: (1) the functions $q(\cdot/x_0, \ldots, x_n; u_0, \ldots, u_{n-1})$ constructed by means of control $v(\cdot/\cdot)$ are measurable with respect to $\tilde{\mathfrak{A}}_n \times \mathfrak{B}_{n-1}$ where $\tilde{\mathfrak{A}}_n \subset \mathfrak{A}_n$ is a fixed monotone sequence of σ-algebras; (2) let $(\xi_k; \eta_k)$ be a controlled process constructed from the controlled object $\mu(\cdot/\cdot)$ and a control $v(\cdot/\cdot)$, then $P\{(\eta_0, \eta_1, \ldots, \eta_n) \in \Gamma_n\} = 1$ for all n, where Γ_n is a sequence of Borel sets in U^{n+1}.

Theorem 1.2. *If $\mathfrak{N}$ is a class of controls with constraints, then for any $v \in \mathfrak{N}$ there exists a non-randomized control $\bar{v} \in \mathfrak{N}$ such that*

$$S(\bar{v}) \leq S(v).$$

The proof of this theorem is based on the following lemma.

Lemma 1.2. *Let μ_s be a family of measures on $(X, \mathfrak{A})$, $s \in (S, \mathscr{L})$ ($(S, \mathscr{L})$ is a measurable space). Denote by $\mathfrak{B}_{[0,1]}$ the σ-algebra of Borel sets on $[0, 1]$ and by m the Lebesgue measure on $[0, 1]$. If $\mu_s(E)$ is $\mathscr{L}$-measurable in s for all $E \in \mathfrak{A}$, then there exists a function $f(t, s)$ on $[0, 1] \times S$ measurable with respect to $\mathfrak{B}_{[0,1]} \times \mathscr{L}$ with values in X such that*

$$m(\{t : f(t, s) \in E\}) = \mu_s(E)$$

for all $E \in \mathfrak{A}$.

Proof. In view of lemma 1.1 for any complete separable metric space X there exists a measurable one-to-one mapping $\lambda(x)$ of $(X, \mathfrak{A})$ into $([0, 1], \mathfrak{B}_{[0,1]})$. Therefore one can assume without loss of generality that X coincides with $[0, 1]$. Let

$$\Phi(s, x) = \mu_s([0, x]), \; x \in [0, 1].$$

The function $\Phi(s, x)$ is measurable jointly in the variables, is monotone and continuous from the right in x. For all x_0 such that $\Phi(s, x) > \Phi(s, x_0)$ provided $x > x_0$, we set

$$f(\Phi(s, x_0), s) = x_0.$$

Thus $f(t, x)$ is defined on the domain of values of the function $\Phi(s, x)$. If t does not belong to this domain, then either $t \in [0, \Phi(s, 0)]$ in which case we set $f(t, s) = 0$, or there exists an x such that $t \in [\Phi(x - 0, s), \Phi(x, s)]$ in which case we set $f(t, s) = x$. The function $f(t, s)$ is measurable jointly in the variables and is monotone in t:

$$\{t : f(t, s) \leq t_0\} = [0, \Phi(s, t_0)].$$

Therefore

$$m(\{t : f(t, s) \in [0, x_0]\}) = \Phi(s, x) = \mu_s([0, x]).$$

This implies the validity of the lemma's assertion for any Borel set $E \in \mathfrak{A}_{[0, 1]}$. □

Proof of theorem 1.2. Denote by $f_k(t, x_0, \ldots, x_{k-1}; u_0, \ldots, u_{k-1})$ a $\mathfrak{B}_{[0, 1]} \times \mathfrak{A}_k \times \mathfrak{B}_k$-measurable function on $[0, 1] \times X^k \times U^k$ taking on values in X, such that for a fixed $x_0, \ldots, x_{k-1}, u_0, \ldots, u_{k-1}$ the equality

$$m(\{t : f_k(t, x_0, \ldots, x_{k-1}, u_0, \ldots, u_{k-1}) \in A\}) \\ = p_k(A/x_0, \ldots, x_{k-1}; u_0, \ldots, u_{k-1}) \tag{1.12}$$

is valid (here p_k are determined by the controlled object $\mu(\cdot/\cdot)$). Next denote by $g_k(t, x_0, \ldots, x_k; u_0, \ldots, u_{k-1})$ a function with values in U possessing the same measurability properties such that for a fixed $x_0, \ldots, x_k, u_0, \ldots, u_{k-1}$ the equality

$$m(\{t : g_k(t, x_0, \ldots, x_k, u_0, \ldots, u_{k-1}) \in B\}) \\ = q_k(B/x_0, \ldots, x_k; u_0, \ldots, u_{k-1}), \tag{1.13}$$

is valid (here $q_k(\cdot/\cdot)$ is constructed from the given control $v \in \mathfrak{N}$.) Moreover $g_k(t, x_0, \ldots, x_k, u_0, \ldots, u_{k-1})$ can be assumed to be a $\mathfrak{A}_{[0, 1]} \times \tilde{\mathfrak{A}}_k \times \mathfrak{B}_{k-1}$-measurable function for which $(u_0, \ldots, u_{k-1}, g_k(t, x_0, \ldots, u_{k-1}) \in \Gamma_k$. The existence of functions f_k and g_k with the required properties follow from Lemma 1.2.

Let $\zeta_0, \zeta_1, \ldots, \theta_0, \theta_1, \ldots$ be a sequence of mutually independent random variables and jointly uniformly distributed on [0, 1]. Set

$$\xi_0 = f_0(\zeta_0), \qquad \eta_0 = g_0(\theta_0, \xi_0), \qquad \xi_1 = f_1(\zeta_1, \xi_0, \eta_0),$$
$$\eta_1 = g_1(\theta_1, \xi_0, \xi_1, \eta_0), \ldots,$$
$$\xi_n = f_n(\zeta_n, \xi_0, \ldots, \xi_{n-1}, \eta_0, \ldots, \eta_{n-1}),$$
$$\eta_n = g_n(\theta_n, \xi_0, \ldots, \xi_n, \eta_0, \ldots, \eta_{n-1}).$$

Formulas (1.12) and (1.13) imply that a sequence $\{(\xi_n, \eta_n); n = 0, 1, \ldots,\}$ forms a controlled Markov process with controlled object $\mu(\cdot/\cdot)$ and control $v(\cdot/\cdot)$. Denote by $\mathbf{E}(\cdot/\theta)$ the conditional mathematical expectation with respect to a σ-algebra generated by the variables $(\theta_0, \theta_1, \ldots)$. Let

$$F_v(\theta) = \mathbf{E}_v(F(\xi, \eta)/\theta).$$

This function is a $\mathfrak{A}_{[0, 1]}^N$-measurable function of θ on $[0, 1]^N$. Let m^N be a countable product of Lebesgue measures on [0, 1]. The measure m^N is defined on $\mathfrak{A}_{[0, 1]}^N$. Since

$$\int F_v(\theta) m^N(d\theta) = S(v),$$

there exists θ^0 such that

$$F(\theta^0) \le S(\nu).$$

Observe that

$$\xi = f(\zeta, \theta), \qquad \eta = g(\zeta, \theta),$$

where $\zeta = (\zeta_0, \zeta_1, \ldots) \in [0, 1]^N$, and $f(\zeta, \theta)$ and $g(\zeta, \theta)$ are measurable in their arguments. Since ζ and θ are independent, it follows that

$$\mathbf{E}_\nu(F(\xi, \eta)/\theta)\Big|_{\theta=\theta^0} = \mathbf{E}F(f(\zeta, \theta^0), g(\zeta, \theta^0)).$$

However $f(\zeta, \theta^0) = \xi^0$, $g(\zeta, \theta^0) = \eta^0$, where

$$\xi_n^0 = f_n(\zeta_n, \xi_0^0, \ldots, \xi_{n-1}^0, \eta_0^0, \ldots, \eta_{n-1}^0),$$
$$\eta_n^0 = g_n(\theta_n^0, \xi_0^0, \ldots, \xi_n^0, \eta_0^0, \ldots, \eta_{n-1}^0).$$

The process (ξ_n^0, η_n^0) is a controlled process (for the same controlled object $\mu(\cdot/\cdot)$) with a *non-randomized control* for which

$$u_n = g_n(\theta_n^0, x_0, \ldots, x_n, u_0, \ldots, u_{n-1}).$$

To express u_n in terms of $x_0, \ldots, x_n$ we ought to substitute into this formula the values of u_k for $k \le n - 1$. Denote by ν^0 such a non-randomized control. This control belongs to $\mathfrak{N}$. Moreover

$$S(\nu^0) = \mathbf{E}_{\nu^0} F(\xi^0, \eta^0) = F(\theta^0) \le S(\nu). \qquad \square$$

We shall now determine a general condition under which there exists an optimal control. For this purpose some properties of semi-continuous functions will be required.

Recall that a function $f(x)$ defined on a metric space X is called *lower semi-continuous* if for all $x \in X$

$$\varliminf_{y \to x} f(y) \ge f(x).$$

A lower semi-continuous function attains its minimum on every compact set. Moreover if $f(x)$ is lower semi-continuous, then the set $\{x : f(x) \le c\}$ is closed for all c.

Let a sequence of finite measures μ_n be defined on X converging weakly to a measure μ. This means that for any bounded continuous function $\varphi(x)$ on X $(\varphi \in C_X)$

$$\lim_{n \to \infty} \int \varphi(x)\mu_n(dx) = \int \varphi(x)\mu(dx)$$

(cf. [20] Vol 1, Chapter VI, §1. p. 362). It was proved in [20], Vol 1, p. 367 that the weak convergence of a sequence of measures μ_n to a measure μ for any closed $K \subset X$ implies the inequality

$$\varlimsup \mu_n(K) \le \mu(K). \tag{1.14}$$

Lemma 1.3. *If f is bounded from below, is lower semi-continuous and μ_n weakly converges to μ, then*

$$\varliminf_{n\to\infty} \int f(x)\mu_n(dx) \geq \int f(x)\mu(dx).$$

PROOF. Assume first that f is bounded. Let

$$c_0 \leq \inf f(x) < c_1 < \cdots < c_N \geq \sup f(x),$$

$\max(c_{k+1} - c_k) < \varepsilon$. Then

$$\begin{aligned}
\int f(x)\mu_n(dx) &\geq \sum_{k=0}^{N-1} c_k \mu_n(\{x : c_k < f(x) \leq c_{k+1}\}) \\
&= \sum_{k=0}^{N-1} c_k[\mu_n(\{x : f(x) \leq c_{k+1}\}) - \mu_n(\{x : f(x) \leq c_k\})] \\
&= c_{N-1}\mu_n(X) - \sum_{k=0}^{N-2} (c_{k+1} - c_k)\mu_n(\{x : f(x) \leq c_{k+1}\}).
\end{aligned}$$

Whence using (1.14) and the closure of the set $\{x : f(x) \leq c\}$ we obtain

$$\begin{aligned}
\varliminf_{n\to\infty} \int f(x)\mu_n(dx) &\geq c_{N-1}\mu(X) - \varlimsup_{n\to\infty} \sum_{k=0}^{N-2} (c_{k+1} - c_k)\mu_n(\{x : f(x) \\
&\leq c_{k+1}\}) \geq c_{N-1}\mu(X) - \sum_{k=0}^{N-2} (c_{k+1} - c_k)\mu(\{x : f(x) \leq c_{k+1}\}) \\
&= \sum_{k=0}^{N-1} c_k \mu(\{x : c_k < f(x) \leq c_{k+1}\}) \geq \int f(x)\mu(dx) - \varepsilon\mu(X).
\end{aligned}$$

This implies the validity of the lemma for a bounded function f. In the general case we utilize the fact that the function $[f \wedge N]$ is also lower semi-continuous. Thus in view of the proven above

$$\varliminf_{n\to\infty} \int f(x)\mu_n(dx) \geq \varliminf_{n\to\infty} \int [f(x) \wedge N]\mu_n(dx) \geq \int [f(x) \wedge N]\mu(dx).$$

The l.h.s. of this inequality does not depend on N. Approaching the limit as $N \to \infty$ we obtain the assertion of the lemma. □

We shall say that *a sequence of points* $\mathbf{x}^{(n)} \in X^N$ *converges to* $\mathbf{x}^{(0)}$ if for all k, $x_k^{(n)} \to x_k^{(0)}$, where $\mathbf{x}^{(n)} = \{x_0^{(n)}, x_1^{(n)}, \ldots\}$, $n = 0, 1, 2, \ldots$. A function $F(\mathbf{x}, \mathbf{u})$ defined on $X^N \times U^N$ is called *lower semi-continuous* if for any sequences $\mathbf{x}^{(n)}$ and $\mathbf{u}^{(n)}$ such that $\mathbf{x}^{(n)} \to \mathbf{x}^{(0)}$ and $\mathbf{u}^{(n)} \to \mathbf{u}^{(0)}$

$$\varliminf_{n\to\infty} F(\mathbf{x}^{(n)}, \mathbf{u}^{(n)}) \geq F(\mathbf{x}^{(0)}, \mathbf{u}^{(0)}).$$

Theorem 1.3. *Let U be a compact set, let X be a complete separable metric space and let for a controlled object $\mu(\cdot/\mathbf{u})$ the following condition be satisfied:*

(A) for all $g \in C_X$ the function

$$\int g(x) q_n(dx \mid x_0, \ldots, x_{n-1}; u_0, \ldots, u_{n-1})$$

is continuous jointly in the variables.

If the control cost $F(\mathbf{x}, \mathbf{u})$ is bounded from below and is lower semi-continuous then an optimal control exists in the class of all controls.

Proof (due to V. A. Polyvjanyĭ). Let $v^{(n)}(\cdot/\cdot)$ be a sequence of controls, and $(\xi^{(n)}, \eta^{(n)})$ be a controlled process with controlled object $\mu(\cdot/\cdot)$ and control $v^{(n)}(\cdot/\cdot)$. Assume furthermore that the marginal distributions of the process $\{(\xi_k^{(n)}, \eta_k^{(n)}), k = 0, 1, \ldots\}$ converge to the marginal distributions of a process $\{(\xi_k^0, \eta_k^0), k = 0, 1, \ldots\}$. Then if condition A is satisfied, the limiting process will also be a controlled process with the same controlled object $\mu(\cdot/\cdot)$. Indeed, for any k and continuous functions $g_k(x) \in C_X$ and

$$\psi(x_0, \ldots, x_{k-1}, u_0, \ldots, u_{k-1})$$

on $X^k \times U^k$ the following equalities are satisfied:

$$\lim_{n\to\infty} \mathbf{E} g_k(\xi_k^{(n)}) \psi(\xi_0^{(n)}, \ldots, \xi_{k-1}^{(n)}, \eta_0^{(n)}, \ldots, \eta_{k-1}^{(n)})$$
$$= \mathbf{E} g_k(\xi_k^{(0)}) \psi(\xi_0^{(0)}, \ldots, \xi_{k-1}^{(0)}, \eta_0^{(0)}, \ldots, \eta_{k-1}^{(0)}),$$

$$\lim_{n\to\infty} \mathbf{E} \int g_k(x) p_k(dx/\xi_0^{(n)}, \ldots, \xi_{k-1}^{(n)}; \eta_0^{(n)}, \ldots, \eta_{k-1}^{(n)})$$
$$\times \psi(\xi_0^{(n)}, \ldots, \xi_{k-1}^{(n)}, \eta_0^{(n)}, \ldots, \eta_{k-1}^{(n)})$$
$$= \mathbf{E} \int g_k(x) p_k(dx/\xi_0^{(0)}, \ldots, \xi_{k-1}^{(0)}; \eta_0^{(0)}, \ldots, \eta_{k-1}^{(0)})$$
$$\times \psi(\xi_0^{(0)}, \ldots, \xi_{k-1}^{(0)}; \eta_0^{(0)}, \ldots, \eta_{k-1}^{(0)}).$$

(Here we have utilized the convergence of the joint distributions of $\xi_i^{(n)}, \eta_i^{(n)}$; $i = 0, \ldots, k$, to the distributions of $\xi_i^{(0)}, \eta_i^{(0)}; i = 0, \ldots, k$ and the continuity of functions g_k, ψ and $\int g_k(x) p_k(dx/x_0, \ldots, x_{k-1}; u_0, \ldots, u_{k-1})$. Moreover for all $n > 0$

$$\mathbf{E} g_k(\xi_k^{(n)}) \psi(\xi_0^{(n)}, \ldots, \xi_{k-1}^{(n)}; \eta_0^{(n)}, \ldots, \eta_{k-1}^{(n)})$$
$$= \mathbf{E} \int g_k(x) p_k(dx/\xi_0^{(n)}, \ldots, \xi_{k-1}^{(n)}; \eta_0^{(n)}, \ldots, \eta_{k-1}^{(n)}) \tag{1.15}$$
$$\times \psi(\xi_0^{(n)}, \ldots, \xi_{k-1}^{(n)}; \eta_0^{(n)}, \ldots, \eta_{k-1}^{(n)}).$$

Consequently equality (1.15) is satisfied also for $n = 0$. This implies that

$$\mathbf{E}g_k(\xi_k^{(0)}/\xi_0^{(0)}, \ldots, \xi_{k-1}^{(0)}; \eta_0^{(0)}, \ldots, \eta_{k-1}^{(0)})$$
$$= \int g_k(x)p_k(dx/\xi_0^{(0)}, \ldots, \xi_{k-1}^{(0)}, \eta_0^{(0)}, \ldots, \eta_{k-1}^{(0)})$$

and

$$\mathbf{P}\{\xi_k^{(0)} \in E/\xi_0^{(0)}, \ldots, \xi_{k-1}^{(0)}, \eta_0^{(0)}, \ldots, \eta_{k-1}^{(0)}\}$$
$$= p_k(E/\xi_0^{(0)}, \ldots, \xi_{k-1}^{(0)}; \eta_0^{(0)}, \ldots, \eta_{k-1}^{(0)}).$$

We have thus demonstrated that $\{(\xi_k^{(0)}, \eta_k^{(0)}), k = 0, 1, \ldots\}$ is a controlled process with a controlled object $\mu(\cdot/\cdot)$. Now let the controls $\nu^{(n)}$ be such that

$$\mathbf{E}_{\nu^{(n)}}F(\xi, \eta) = \mathbf{E}F(\xi^{(n)}, \eta^{(n)}) \leq \inf_{\nu} \mathbf{E}_\nu F(\xi, \eta) + \varepsilon_n, \tag{1.16}$$

where $\varepsilon_n \downarrow 0$.

We show that the marginal distributions of the sequence $\{(\xi_k^{(n)}, \eta_k^{(n)}), k = 0, 1, \ldots\}$ are compact. It follows from condition A of the theorem that for any compact sets $K_0, K_1, \ldots, K_{k-1}$ in X, the family of distributions $\{p_k(\cdot/x_0, \ldots, x_{k-1}; u_0, \ldots, u_{k-1}); x_0 \in K_0, \ldots, x_{k-1} \in K_{k-1}; u_0, \ldots, u_{k-1} \in U\}$ is compact.

Indeed, given any sequence $(x_0^{(n)}, \ldots, x_{k-1}^{(n)}; u_0^{(n)}, \ldots, u_{k-1}^{(n)})$ by selecting a subsequence n_i such that the sequence $(x_0^{(n_i)}, \ldots, x_{k-1}^{(n_i)}; u_0^{(n_i)}, \ldots, u_{k-1}^{(n_i)})$ converges to $(x_0^{(0)}, \ldots, x_{k-1}^{(0)}; u_0^{(0)}, \ldots, u_{k-1}^{(0)})$ we obtain a weakly convergent subsequence of measures $p_k(\cdot/x_0^{(n_i)}, \ldots, x_{k-1}^{(n_i)}; u_0^{(n_i)}, \ldots, u_{k-1}^{(n_i)})$. Utilizing the condition of weak compactness of a family of measures (cf. [20], Vol 1, p. 362, theorem 1) we verify that for any sequence of compact sets $K_0, K_1, \ldots, K_{k-1}$ and $\varepsilon > 0$ there exists a compact set $K_k \subset X$ such that

$$p_k(K/x_0, \ldots, x_{k-1}; u_0, \ldots, u_{k-1}) \geq 1 - \varepsilon.$$

Choose an $\varepsilon > 0$ and construct compacts K_k in the following manner: K_0 is such that $p_0(K_0) \geq 1 - \varepsilon/2$; after $K_0, K_1, \ldots, K_{k-1}$ are selected we choose K_k so that

$$p_k(K_k/x_0, \ldots, x_{k-1}; u_0, \ldots, u_{k-1}) \geq 1 - \frac{\varepsilon}{2^{k+1}}$$

for $x_0 \in K_0, \ldots, x_{k-1} \in K_{k-1}$. Under such a choice of K_i, we have

$$\mathbf{P}\{\xi_0^{(n)} \in K_0, \ldots, \xi_k^{(n)} \in K_k\} \geq \left(1 - \frac{\varepsilon}{2^{k+1}}\right)\mathbf{P}\{\xi_0^{(n)} \in K_0, \ldots, \xi_{k-1}^{(n)} \in K_{k-1}\}$$
$$\geq 1 - \varepsilon\sum_{i=0}^{k} \frac{1}{2^{i+1}} \geq 1 - \varepsilon,$$

for any k and n. From here in view of the compactness of U (taking theorem 1 on p. 362 in Vol 1 of [20] into account) the compactness of marginal distributions of the sequence $\{(\xi_k^{(n)}, \eta_k^{(n)}), k = 0, 1, \ldots\}$ follows. Hence we can assume without loss of generality that the marginal distributions of

$$\{(\xi_k^{(n)}, \eta_k^{(n)}); k = 0, 1, \ldots\}$$

converge to the marginal distributions of a controlled process $\{(\xi_k^{(0)}, \eta_k^{(0)}); k = 0, 1, \ldots\}$ with the controlled object $\mu(\cdot/\cdot)$ (this was verified above) and a control $\bar{v}$. Then

$$\mathbf{E}_{\bar{v}} F(\xi, \eta) = \mathbf{E}F(\xi^{(0)}, \eta^{(0)}) \leq \lim_{n \to \infty} \mathbf{E}F(\xi^{(n)}, \eta^{(n)}) = \inf_{v} S(v)$$

in view of lemma 1.3 and inequality (1.16). □

Corollary. *If the conditions of theorems* 1.2 *and* 1.3 *are satisfied, then there exists an optimal non-randomized control.*

Remark 1.2. Under the conditions of theorem 1.3 the set of measures on $X^N \times U^N$ which corresponds to the controlled processes $(\xi^{(n)}, \eta^{(n)})$ with a given controlled object is compact.

Assume that U is locally compact and let U^0 be a compactification of U obtained by adding a single point, and the negative function $F(\mathbf{x}, \mathbf{u})$ is lower semi-continuous on $X^N \times U^N$ and vanishes on $X^N \times (U^0)^N \backslash X^N \times U^N$.

Let condition (A) of theorem 1.3 be satisfied for the controlled object $\mu(\cdot/\cdot)$ and for all $f \in C_X$ the function

$$\int f(x) p_n(dx/x_0, \ldots, x_{n-1}; u_0, \ldots, u_{n-1})$$

be continuously extendable onto $X^n \times (U^0)^n$. This means that the measure $p_n(\cdot/x_0, \ldots, x_{n-1}; u_0, \ldots, u_{n-1})$ is continuously extendable (in the sense of weak convergence on $X^n \times (U^0)^n$). This continuation extends the controlled object $\mu(\cdot/\cdot)$ onto the space $X^N \times (U^0)^N$. We shall denote this extension by the same symbol $\mu(\cdot/\cdot)$ and assume that condition (A) of theorem 1.3 is satisfied for the extended $\mu(\cdot/\cdot)$. Then in view of this theorem and the corollary following it there exists a non-randomized optimal control $u_k = \varphi_k(x_0, \ldots, x_k)$. Since

$$F(\mathbf{x}, \mathbf{u}) = 0 \quad \text{for } \mathbf{u} \in (U^0)^N \backslash U^N,$$

$$F(\mathbf{x}, \mathbf{u}) \leq 0 \quad \text{for } \mathbf{x} \in X^N, \mathbf{u} \in U^N,$$

this optimal control can be chosen in such a manner that

$$\varphi_k(x_0, x_1, \ldots, x_k) \in U$$

for all k with probability 1 (by assuming that $\varphi_k(x_0, \ldots, x_k) = u_k \in U$ for $\varphi_k \notin U$, we do not decrease the value of $\mathbf{E}_v F(\xi, \eta)$). The result obtained may be stated as

Theorem 1.4. *Let X be a complete metric separable space, U be locally compact and $\mu(\cdot/\cdot)$ satisfy the condition: for all $f \in C_X$ and $n \geq 0$ the function*

$$\int f(x)p_n(dx/x_0, \ldots, x_{n-1}; u_0, \ldots, u_{n-1})$$

is continuous jointly in the variables and there exists, uniformly in $x_0, \ldots, x_{n-1}$ a limit in $(u_0, \ldots, u_{n-1})$, provided this point in U^n approaches a point in $(U^0)^n \backslash U^n$.

If $F(\mathbf{x}, \mathbf{u})$ is non-positive, lower semi-continuous and $\lim F(\mathbf{x}, \mathbf{u}) = 0$ uniformly in $\mathbf{x}$, and if $\mathbf{u}$ approaches to a point $\mathbf{u}^0 \in (U^0)^N \backslash (U)^N$, then there exists an optimal non-randomized control.

3 Construction of Optimal and ε-Optimal Controls

Let X be a complete metric separable space and U be a compact set. Assume that condition (A) of theorem 1.3 is satisfied for the controlled object $\mu(\cdot/\cdot)$. We are interested in the question of how efficiently one can construct an optimal or ε-optimal control. In view of the above these controls can be sought among non-randomized controls.

First we shall assume that $F(\mathbf{x}, \mathbf{u}) = \Phi(x_0, \ldots, x_n; u_0, \ldots, u_n)$ where Φ is a function on $X^n \times U^n$ bounded from below and lower semi-continuous. To construct an optimal control in this case (whose existence follows from theorem 1.3) certain auxiliary assertions are required.

Lemma 1.4. *Let function $f(x, u)$ be bounded from below and lower semi-continuous for $x \in X$ and $u \in U$. Then the function $\bar{f}(x) = \inf_u f(x, u)$ is also lower semi-continuous and there exists a Borel function φ from X into U such that*

$$\bar{f}(x) = f(x, \varphi(x)). \tag{1.17}$$

PROOF. Since U is a compact set and the function $f(x, u)$ is lower semi-continuous in u, for each x there exists

$$\min_u f(x, u) = \inf_u f(x, u).$$

Let $x_n \to x_o$ and u_n be chosen so that

$$\inf_u f(x_n, u) \geq f(x_n, u_n) - \varepsilon.$$

Then

$$\begin{aligned} \underline{\lim}_{n\to\infty} \bar{f}(x_n) &= \underline{\lim}_{n\to\infty} \inf_u f(x_n, u) \geq \underline{\lim}_{n\to\infty} f(x_n, u_n) - \varepsilon \\ &= \lim_{k\to\infty} f(x_{n_k}, u_{n_k}) - \varepsilon \geq f(x_0, u_0) - \varepsilon \geq \bar{f}(x_0) - \varepsilon. \end{aligned}$$

Here n_k is a sequence, u_0 is the limit point of the sequence u_{n_k}. Since $\varepsilon > 0$ is arbitrary the first assertion of the theorem follows. We now prove the existence of a Borel function satisfying (1.17). Assume that $\bar{f} > 0$.

Let $B_1^{(n)}, \ldots, B_n^{(n)}$ be closed sets in U satisfying the conditions:

1. $\bigcup_{k=1}^{n} B_k^{(n)} = U$;

2. $\lim_{n\to\infty} \max_{1\le k\le n} \operatorname{diam}(B_k^{(n)}) = 0$,

where diam (B) is the diameter of the set B; (3) each one of the sets $B_i^{(n+1)}$ is totally contained in one and only one set $B_k^{(n)}$ and moreover if $B_i^{(n+1)} \subset B_k^{(n)}$, then $B_{i+1}^{(n+1)}$ is contained only in $B_k^{(n)} \cup B_{k+1}^{(n)}$, while $B_1^{(n+1)} \subset B_1^{(n)}$. Set

$$\Delta_{ki}^{(n;\, m)} = \left\{x : \exists u\left(-\chi_{B_k^{(n)}}(u) f(x, u) \ge \frac{i}{m}\right)\right\} \cap \left\{x : -\bar{f}(x) < \frac{i+1}{m}\right\}.$$

The function $\chi_{B_k^{(n)}}(u)$ being an indicator function of a closed set is upper semi-continuous, hence the set

$$\left\{x : \exists u\left(-\chi_{B_k^{(n)}}(u) f(x, u) \ge \frac{i}{n}\right)\right\}$$

is closed. The function $\bar{f}(x)$ is lower semi-continuous, therefore $\{x : (-\bar{f}(x) < i + 1/n)\}$ is open. The set $\Delta_{ki}^{(n,\, m)}$ is a Borel set. It is easy to see that

$$\bigcup_{k,\, i} \Delta_{ki}^{(n,\, m)} = X.$$

Set $\Delta_k^{(n,\, m)} = \bigcup_i \Delta_{ki}^{(n,\, m)}$, $\Delta_k^{(n)} = \bigcap_m \Delta_k^{(n,\, m)}$. All these sets are also Borel sets. If $x \in \Delta_k^{(n)}$, then there exists for each m, $u_m \in B_k^{(n)}$ such that

$$\bar{f}(x) - f(x, u_m) \le \frac{1}{m}.$$

Selecting from u_m a convergent sequence we verify that there exists $\bar{u} \in B_k^{(n)}$ such that $\bar{f}(x) = f(x, \bar{u})$. Conversely, if there exists $\bar{u} \in B_k^{(n)}$ such that $\bar{f}(x) = f(x, \bar{u})$ then $x \in \Delta_{ki}^{(n,\, m)}$ for $i = [m\bar{f}(x)]$ for any m; ($[\cdot]$ denotes the integral part). Hence

$$\Delta_k^{(n)} = \{x : \exists u \in B_k^{(n)}, \bar{f}(x) = f(x, u)\}.$$

Choose a point $u_k^{(n)}$ in the set $B_k^{(n)}$. Set

$$\varphi_n(x) = u_k^{(n)}, \qquad x \in \Delta_k^{(n)} \Big\backslash \bigcup_{j=1}^{k-1} \Delta_j^{(n)}.$$

Condition 3 implies that as we pass from n to $n + 1$ one of the sets $B_i^{(n)}$ is subdivided into two sets (possibly overlapping): $B_i^{(n)} = B_i^{(n+1)} \cup B_{i+1}^{(n+1)}$. Therefore if $\varphi_n(x) = u_k^{(n)}$, $\varphi_{n+1}(x) = u_i^{(n+1)}$, then $B_k^{(n)} \supset B_i^{(n+1)}$. We thus verify that for each n the values of $\varphi_n(x)$ and $\varphi_{n+1}(x)$ belong to the very same set $B_k^{(n)}$. Hence there exists

$$\varphi(x) = \lim_{n\to\infty} \varphi_n(x).$$

For a given x $\varphi_n(x) \in B_{k_n}^{(n)}$ and $B_{k_1}^{(1)} \supset B_{k_2}^{(2)} \supset \cdots$. If $u_n \in B_{k_n}^{(n)}$, $\bar{f}(x) = f(x, u_n)$, and $\bar{u} = \lim u_n$, then $\bar{f}(x) = f(x, \bar{u})$. Since $\varphi_n(x)$ and u_n belong to $B_{k_n}^{(n)}$ and diam $B_{k_n}^{(n)} \to 0$, it follows that $\bar{u} = \varphi(x)$. Consequently $\bar{f}(x) = f(x, \varphi(x))$. The functions $\varphi_n(x)$ are Borel functions, hence so is φ. □

We introduce successively the following functions:

$$\begin{aligned}
&\Phi_n(x_0, \ldots, x_n, u_0, \ldots, u_n) = \Phi(x_0, \ldots, x_n, u_0, \ldots, u_n);\\
&\hat{\Phi}_n(x_0, \ldots, x_n, u_0, \ldots, u_{n-1})\\
&\quad = \inf_{u_n} \Phi(x_0, \ldots, x_n, u_0, \ldots, u_n);\\
&\Phi_{n-1}(x_0, \ldots, x_{n-1}, u_0, \ldots, u_{n-1})\\
&\quad = \int \hat{\Phi}_n(x_0, \ldots, x_n, u_0, \ldots, u_{n-1})\\
&\qquad \times p_n(dx_n/x_0, \ldots, x_{n-1}, u_0, \ldots, u_{n-1});\\
&\hat{\Phi}_k(x_0, \ldots, x_k, u_0, \ldots, u_{k-1}) \qquad\qquad (1.18)\\
&\quad = \inf_{u_k} \Phi(x_0, \ldots, x_k, u_0, \ldots, u_k);\\
&\Phi_{k-1}(x_0, \ldots, x_{k-1}, u_0, \ldots, u_{k-1})\\
&\quad = \int \hat{\Phi}_k(x_0, \ldots, x_{k-1}, x_k, u_0, \ldots, u_{k-1})\\
&\qquad \times p_k(dx_k/x_0, \ldots, x_{k-1}, u_0, \ldots, u_{n-1});\\
&\hat{\Phi}_0(x_0) = \inf_{u_0} \Phi(x_0, u_0);\\
&\Phi = \int \hat{\Phi}_0(x_0) p_0(dx_0).
\end{aligned}$$

All the functions $\Phi_k(x_0, \ldots, x_k, u_0, \ldots, u_k)$ and $\hat{\Phi}_k(x_0, \ldots, x_k, u_0, \ldots, u_{k-1})$ are lower semi-continuous. Indeed, lower semi-continuity of $\hat{\Phi}_k(x_0, \ldots, x_k, u_0, \ldots, u_{k-1})$ follows from lemma 1 provided only Φ_k is lower semi-continuous.

To prove that Φ_k is lower semi-continuous given that $\hat{\Phi}_{k+1}$ is lower semi-continuous we use the following assertion:

Lemma 1.5. *Let $\Phi(x, x_1)$ be bounded from below and a lower semi-continuous function on $X \times X_1$, where X and X_1 are complete separable metric spaces* *and μ_n is a sequence of finite measures on X weakly convergent to a measure μ_0. If $x_1^{(n)} \to x_1^{(0)}$, then*

$$\varliminf_{n\to\infty} \int \Phi(x, x_1^{(n)})\mu_n(dx) \geq \int \Phi(x, x_1^{(0)})\mu_0(dx). \tag{1.19}$$

PROOF. Define measures $\bar{\mu}_n$ on $X \times X_1$ by means of the formula: for Borel sets $A \subset X$ and $A_1 \subset X_1$

$$\bar{\mu}_n(A \times A_1) = \mu_n(A)\chi_{A_1}(x_1^{(n)}), \qquad n = 0, 1, \ldots,$$

where χ_{A_1} is the indicator of the set A_1. It is easy to see that a sequence of measures $\bar{\mu}_n$ is weakly convergent to a measure $\bar{\mu}_0$. Therefore in view of lemma 1.3

$$\varliminf_{n\to\infty} \int \Phi(x, x_1)\bar{\mu}_n(dx \times dx_1) \geq \int \Phi(x, x_1)\bar{\mu}_0(dx \times dx_1).$$

It remains only to observe that

$$\int \Phi(x, x_1)\bar{\mu}_n(dx \times dx_1) = \int \Phi(x, x_1^{(n)})\mu_n(dx), \qquad n = 0, 1, \ldots \qquad \square$$

Remark 1.2. If under the conditions of the lemma the function $\Phi(x, x_1)$ is continuous and bounded, then

$$\lim_{n\to\infty} \int \Phi(x, x_1^{(n)})\mu_n(dx) = \int \Phi(x, x_1^{(0)})\mu_0(dx).$$

In such a case $-\Phi(x, x_1)$ is also lower semi-continuous, hence in addition to inequality (1.19) the reverse inequality is also valid.

Now let sequences $x_i^{(m)}, u_i^{(m)}, i = 0, \ldots, k$ converge to $x_i^{(0)}$ and $u_i^{(0)}$ correspondingly. Then the measures $p_{k+1}(dx_{k+1}/x_0^{(m)}, \ldots, x_k^{(m)}, u_0^{(m)}, \ldots, u_k^{(m)})$ are weakly convergent to the measure $p_{k+1}(dx_{k+1}/x_0^{(0)}, \ldots, x_k^{(0)}, u_0^{(0)}, \ldots, u_k^{(0)})$ in view of the conditions imposed on the controlled object. Therefore if $\hat{\Phi}_{k+1}(x_0, \ldots, x_{k+1}, u_0, \ldots, u_k)$ is lower semi-continuous we obtain in view of lemma 1.5

$$\begin{aligned}
&\varliminf_{n\to\infty} \Phi_k(x_0^{(m)}, \ldots, x_k^{(m)}, u_0^{(m)}, \ldots, u_k^{(m)}) \\
&\quad = \varliminf_{m\to\infty} \int \hat{\Phi}_{k+1}(x_0^{(m)}, \ldots, x_k^{(m)}, x_{k+1}, u_0^{(m)}, \ldots, u_k^{(m)}) \\
&\qquad \times p_{k+1}(dx_{k+1}/x_0^{(m)}, \ldots, x_k^{(m)}, u_0^{(m)}, \ldots, u_k^{(m)}) \\
&\quad \geq \int \hat{\Phi}_{k+1}(x_0^{(0)}, \ldots, x_k^{(0)}, x_{k+1}, u_0^{(0)}, \ldots, u_k^{(0)}) \\
&\qquad \times p_{k+1}(dx_{k+1}/x_0^{(0)}, \ldots, x_k^{(0)}, u_0^{(0)}, \ldots, u_k^{(0)}) \\
&\quad = \Phi_k(x_0^{(0)}, \ldots, u_k^{(0)}).
\end{aligned}$$

Hence in such a case $\Phi_k(x_0, \ldots, x_k, u_0, \ldots, u_k)$ is also lower semi-continuous. Since in the chain of functions (1.18) the first one is lower semi-continuous and the lower semi-continuity of a preceding one implies lower semi-continuity of the succeeding function, all the functions in this chain are lower semi-continuous. Lemma 1.4 implies the existence of Borel functions $\varphi_k(x_0, \ldots, x_k; u_0, \ldots, u_{k-1})$, $k = 0, \ldots, n$ such that

$$\begin{aligned}\hat{\Phi}_k&(x_0, \ldots, x_k, u_0, \ldots, u_{k-1})\\ &= \Phi_k(x_0, \ldots, x_k, u_0, \ldots, u_{k-1}), \varphi_k(x_0, \ldots, x_k, u_0, \ldots, u_{k-1}).\end{aligned} \tag{1.20}$$

Theorem 1.5. *Let functions Φ_k, $\hat{\Phi}_k$ and the number Φ be defined by equation* (1.18) *and functions φ_k by* (1.20). *Define successively the functions*

$$\begin{aligned}\bar{\varphi}_0(x_0) = \varphi_0(x_0), \bar{\varphi}_1(x_0, x_1) &= \varphi_1(x_0, x_1, \bar{\varphi}_0(x_0)), \ldots, \bar{\varphi}_k(x_0, \ldots, x_k)\\ &= \varphi_k(x_0, \ldots, x_k), \bar{\varphi}_0(x_0), \ldots, \bar{\varphi}_{k-1}(x_0, \ldots, x_{k-1}), \ldots .\end{aligned}$$

A non-randomized control $\bar{v}$ given by the sequence $\{u_k = \varphi_k(x_1, \ldots, x_k), k = 0, 1, \ldots\}$ is an optimal control and the quantity Φ is the optimal control cost.

PROOF. Let a non-randomized control v be defined by the functions $u_k = \psi_k(x_0, \ldots, x_k)$. Denote by $\{(\xi_k, \eta_k), k = 0, 1, \ldots, n\}$ the corresponding controlled sequence. Then

$$\begin{aligned}\mathbf{E}_v \Phi(\xi_0, \ldots, \xi_n, \eta_0, \ldots, \eta_n) &\geq \mathbf{E}_v \hat{\Phi}_n(\xi_0, \ldots, \xi_n, \eta_0, \ldots, \eta_{n-1})\\ &= \mathbf{E}_v(\mathbf{E}_v(\hat{\Phi}_n(\xi_0, \ldots, \xi_n, \eta_0, \ldots, \eta_{n-1})/\xi_0, \ldots, \xi_{n-1}))\\ &= \mathbf{E}_v \int \hat{\Phi}_n(\xi_0, \ldots, \xi_{n-1}, x_n, \eta_0, \ldots, \eta_{n-1})\\ &\quad \times p_n(dx_n/\xi_0, \ldots, \xi_{n-1}, \eta_0, \ldots, \eta_{n-1})\\ &= \mathbf{E}_v \Phi_{n-1}(\xi_0, \ldots, \xi_{n-1}, \eta_0, \ldots, \eta_{n-1}).\end{aligned}$$

Here we utilized the inequality $\Phi_n \geq \hat{\Phi}_n$ and the fact that $\eta_0, \ldots, \eta_{n-1}$ are functions of $\xi_0, \ldots, \xi_{n-1}$. Analogously we have for all $k > 0$

$$\begin{aligned}\mathbf{E}_v \Phi_k&(\xi_0, \ldots, \xi_k, \eta_0, \ldots, \eta_k)\\ &\geq \mathbf{E}_v \Phi_{k-1}(\xi_0, \ldots, \xi_{k-1}, \eta_0, \ldots, \eta_{k-1}).\end{aligned}$$

Thus

$$\begin{aligned}\mathbf{E}_v \Phi(\xi_0, \ldots, \xi_n, \eta_0, \ldots, \eta_n) &\geq \mathbf{E}_v \Phi_0(\xi_0, \eta_0) \geq \mathbf{E}_v \hat{\Phi}_0(\xi_0)\\ &= \int \hat{\Phi}_0(x_0) p_0(dx_0) = \Phi.\end{aligned} \tag{1.21}$$

Now let

$$\psi_k = \bar{\varphi}_k(x_0, \ldots, x_k).$$

Then

$$\eta_k = \varphi_k(\xi_0, \ldots, \xi_k, \eta_0, \ldots, \eta_{k-1}).$$

Therefore in view of (1.18) we have for all $k > 0$

$$\begin{aligned}
&\mathbf{E}_{\tilde{v}} \Phi_k(\xi_0, \ldots, \xi_k, \eta_0, \ldots, \eta_k) \\
&\quad = \mathbf{E}_{\tilde{v}} \Phi_k(\xi_0, \ldots, \xi_k, \eta_0, \ldots, \eta_{k-1}, \varphi_k(\xi_0, \ldots, \xi_{k-1}, \eta_0, \ldots, \eta_{k-1})) \\
&\quad = \mathbf{E}_{\tilde{v}} \hat{\Phi}_k(\xi_0, \ldots, \xi_k, \eta_0, \ldots, \eta_{k-1}) \\
&\quad = \mathbf{E}_{\tilde{v}} \mathbf{E}_{\tilde{v}}(\hat{\Phi}_k(\xi_0, \ldots, \xi_k, \eta_0, \ldots, \eta_{k-1})/\xi_0, \ldots, \xi_{k-1}) \\
&\quad = \mathbf{E}_{\tilde{v}} \int \hat{\Phi}_k(\xi_0, \ldots, \xi_{k-1}, x_k, \eta_0, \ldots, \eta_{k-1}) \\
&\qquad\qquad \times p_k(dx_k/\xi_0, \ldots, \xi_{k-1}, \eta_0, \ldots, \eta_{k-1}) \\
&\quad = \mathbf{E}_{\tilde{v}} \Phi_{k-1}(\xi_0, \ldots, \xi_{k-1}, \eta_0, \ldots, \eta_{k-1}).
\end{aligned}$$

Hence,

$$\begin{aligned}
\mathbf{E}_{\tilde{v}} \Phi(\xi_0, \ldots, \xi_n, \eta_0, \ldots, \eta_n) &= \mathbf{E}_{\tilde{v}} \Phi_0(\xi_0, \varphi_0(\xi_0)) \\
&= \mathbf{E}_{\tilde{v}} \hat{\Phi}_0(\xi_0) = \int \hat{\Phi}_0(x_0) p_0(dx_0) = \Phi.
\end{aligned} \tag{1.22}$$

Relations (1.21) and (1.22) yield the proof of the theorem. □

Remark 1.3. Functions Φ_k defined by equations (1.18) have the following meaning. Assume that at the times $0, \ldots, k$ controls $u_0, \ldots, u_k$ were chosen and the initial process takes on values $x_0, \ldots, x_k$. Then Φ_k becomes the optimal cost of continuation of control:

$$\begin{aligned}
\Phi_k(x_0, \ldots, x_k, u_0, \ldots, u_k) = \min_{v \in \mathfrak{N}_1} \mathbf{E}_v(\Phi(x_0, \ldots, x_k, \xi_{k+1}, \\
\ldots, \xi_n, u_0, \ldots, u_k, \eta_{k+1}, \ldots, \eta_n)/\xi_i = x_i) \qquad i = 0, \ldots, k,
\end{aligned}$$

where $\mathfrak{N}_1$ is the set of all controls such that $\eta_i = u_i$, $i = 0, \ldots, k$. This assertion is proved in the same manner as theorem 1.5.

Clearly the optimal control cost for functionals of control costs Φ_n and Φ_k is the same. This follows from the fact that an optimal control can be constructed by first using an optimal control on the first k steps and then continuing optimally this control. Formulas (1.18) show how one should continue optimally for one step (the transition from Φ_k to $\hat{\Phi}_k$) and how the optimal cost of controlling changes (the transition from Φ_k to Φ_{k-1}) in this case.

Remark 1.4. By altering slightly the bounded function Φ, using the procedure described in theorem 1.5, one can construct an optimal control in a class of controls with constraints.

Assume that there exists a sequence of closed sets $\Gamma_k \subset X^{k+1} \times U^{k+1}$ such that the following conditions are satisfied:

1. if $(x_0, \ldots, x_{k+1}, u_0, \ldots, u_{k+1}) \in \Gamma_{k+1}$, then $(x_0, \ldots, x_k, u_0, \ldots, u_k) \in \Gamma_k$;
2. for any k, point $(x_0, \ldots, x_k, u_0, \ldots, u_k) \in \Gamma_k$ and $x_{k+1}, \ldots, x_n \in X$ there exist Borel functions $g_{k,j}(x_0, \ldots, x_k, u_0, \ldots, u_k, x_{k+1}, \ldots, x_j)$ with values in $U (j > k)$ such that

$$(x_0, \ldots, x_j, u_0, \ldots, u_k), g_{k,k+1}(x_0, \ldots, x_k, u_0, \ldots, u_k, x_{k+1}),$$
$$\ldots, g_{k,j}(x_0, \ldots, x_k, u_0, \ldots, u_k, x_{k+1}, \ldots, x_j) \in \Gamma_j \quad \text{for} \quad j = k+1, \ldots, n.$$

Denote by $\mathfrak{N}$ the set of controls v satisfying $\mathbf{P}_v\{(\xi_0, \ldots, \xi_n; \eta_0, \ldots, \eta_n) \in \Gamma_n\} = 1$. There exists an optimal control in the class $\mathfrak{N}$. This control can be constructed as follows: Let

$$F^{(1)}(\mathbf{x}, \mathbf{u}) = \chi_{\Gamma_n}(x_0, \ldots, x_n, u_0, \ldots, u_n)$$
$$\times \Phi(x_0, \ldots, x_n, u_0, \ldots, u_n) + \lambda(1 - \chi_{\Gamma_n}),$$

where

$$\lambda > \sup \Phi(x_0, \ldots, x_n, u_0, \ldots, u_n).$$

An optimal control $\bar{v}$ for this functional can be chosen in such a manner that $\bar{v} \in \mathfrak{N}$.

Indeed, let this control be determined by the functions $u_k = \varphi_k(x_0, \ldots, x_k)$. For $(x_0, \ldots, x_n; \varphi_0(x_0), \ldots, \varphi_n(x_0, \ldots, x_n)) \notin \Gamma_n$ denote by k the smallest integer such that

$$(x_0, \ldots, x_k; \varphi_0(x_0), \ldots, \varphi_k(x_0, \ldots, x_k)) \notin \Gamma_k.$$

Set $\bar{\varphi}_j = \varphi_j$ for $j < k$, and for $j \geq k$ set

$$\bar{\varphi}_j(x_0, \ldots, x_j) = g_{k-1,j}(x_0, \ldots, x_{k-1}, \varphi_0(x_0),$$
$$\ldots, \varphi_{k-1}(x_0, \ldots, x_{k-1}), x_k, \ldots, x_j).$$

Then

$$F^{(1)}(x_0, \ldots, x_n, \varphi_0(x_0), \ldots, \varphi_n(x_0, \ldots, x_n)) = \lambda,$$
$$F^{(1)}(x_0, \ldots, x_n, \bar{\varphi}_0, \ldots, \bar{\varphi}_n(x_0, \ldots, x_n)) < \lambda.$$

Therefore by choosing the control $u_k = \bar{\varphi}_k$ we do not increase the control cost. However if the control $v \in \mathfrak{N}$, then

$$\mathbf{P}_v\{F^{(1)}(\xi, \eta) = F(\xi, \eta)\} = 1.$$

Consequently this control is optimal is the class $\mathfrak{N}$.

Consider the general case of a cost functional under the assumptions of theorem 1.3. To begin with we consider together with the initial controlled object a family of "shifted" controlled objects

$$\{\mu(\cdot/\cdot)_{\bar{x}_0, \ldots, \bar{x}_n, \bar{u}_0, \ldots, \bar{u}_n}, n = 0, 1, \ldots; \bar{x}_k \in X, \bar{u}_k \in U, k = 0, \ldots, n\},$$

defined by the conditional probabilities

$$p_k(dx_k/x_0, \ldots, x_{k-1}, u_0, \ldots, u_{k-1})_{\bar{x}_0, \ldots, \bar{x}_n, \bar{u}_0, \ldots, \bar{u}_n}$$
$$= p_{n+k}(dx_k/\bar{x}_0, \ldots, \bar{x}_n, x_0, \ldots, x_{k-1}, \bar{u}_0, \ldots, \bar{u}_n, u_0, \ldots, u_{k-1}).$$

Denote

$$F(\mathbf{x}, \mathbf{u})_{\bar{x}_0, \ldots, \bar{x}_n, \bar{u}_0, \ldots, \bar{u}_n} = F((\bar{x}_0, \ldots, \bar{x}_n, x_0, \ldots), (\bar{u}_0, \ldots, \bar{u}_n, u_0, \ldots)).$$

This is a family of shifted functionals. Let $\mathbf{E}_v^{\bar{x}_0, \ldots, \bar{x}_n, \bar{u}_0, \ldots, \bar{u}_n}$ be the mathematical expectation of a random sequence defined by the controlled object $\mu(\cdot/\cdot)_{\bar{x}_0, \ldots, \bar{x}_n, \bar{u}_0, \ldots, \bar{u}_n}$ and control v. Set

$$\left.\begin{aligned} &\Phi_n(\bar{x}_0, \ldots, \bar{x}_n, \bar{u}_0, \ldots, \bar{u}_n; v) \\ &\qquad = \mathbf{E}_v^{\bar{x}_0, \ldots, \bar{x}_n, \bar{u}_0, \ldots, \bar{u}_n} F(\xi, \eta)_{\bar{x}_0, \ldots, \bar{x}_n, \bar{u}_0, \ldots, \bar{u}_n}; \\ &\Phi_n(\bar{x}_0, \ldots, \bar{x}_n, \bar{u}_0, \ldots, \bar{u}_n) \\ &\qquad = \inf_v \Phi_n(\bar{x}_0, \ldots, \bar{x}_n, \bar{u}_0, \ldots, \bar{u}_n; v). \end{aligned}\right\} \tag{1.23}$$

The function $\Phi_n(\bar{x}_0, \ldots, \bar{x}_n, \bar{u}_0, \ldots, \bar{u}_n)$ is naturally called *the conditional optimal cost of controlling* provided on the first n steps the control $\bar{u}_0, \ldots, \bar{u}_n$ is chosen and the basic process takes on values $\bar{x}_0, \ldots, \bar{x}_n$.

Remark 1.5. Let $F(\mathbf{x}, \mathbf{u}) = \Phi_N(x_0, \ldots, x_N, u_0, \ldots, u_N)$ $(N > n)$. Then

$$F(\mathbf{x}, \mathbf{u})_{\bar{x}_0, \ldots, \bar{x}_n, \bar{u}_0, \ldots, \bar{u}_n}$$
$$= \Phi_N(\bar{x}_0, \ldots, \bar{x}_n, x_0, \ldots, x_{N-n-1}, \bar{u}_0, \ldots, \bar{u}_n, u_0, \ldots, u_{N-n-1}).$$

Theorem 1.5 implies that the optimal cost of controlling for the controlled object $\mu(\cdot/\cdot)_{\bar{x}_0, \ldots, \bar{x}_n, \bar{u}_0, \ldots, \bar{u}_n}$ with this functional of cost of controlling coincides with the function $\tilde{\Phi}_n(\bar{x}_0, \ldots, \bar{x}_n, \bar{u}_0, \ldots, \bar{u}_n)$ defined by means of the recurrence relations:

$$\tilde{\Phi}_N(\bar{x}_0, \ldots, \bar{x}_n, x_0, \ldots, x_{N-n-1}, \bar{u}_0, \ldots, \bar{u}_n, u_0, \ldots, u_{N-n-1})$$
$$= \Phi_N(\bar{x}_0, \ldots, \bar{x}_n, x_0, \ldots, x_{N-n-1}, \bar{u}_0, \ldots, \bar{u}_n, u_0, \ldots, u_{N-n-1});$$

for $n \le k < N$

$$\tilde{\Phi}_k(\bar{x}_0, \ldots, \bar{x}_n, x_0, \ldots, x_{k-n-1}, \bar{u}_0, \ldots, \bar{u}_n, u_0, \ldots, u_{k-n-1})$$
$$= \int p_{k+1}(dx_{k-n}/\bar{x}_0, \ldots, \bar{x}_n, x_0, \ldots, x_{k-n-1}, \bar{u}_0, \ldots, \bar{u}_n, u_0, \ldots, u_{k-n-1})$$
$$\inf_{u_{k-n}} \tilde{\Phi}_{k+1}(\bar{x}_0, \ldots, \bar{x}_n, x_0, \ldots, x_{k-n}, \bar{u}_0, \ldots, \bar{u}_n, u_0, \ldots, u_{k-n})$$

(for $k = n$ $\tilde{\Phi}$ and $p_{k+1}(dx_{k+1}/\cdot)$ do not depend on x_i and u_i). Clearly $\tilde{\Phi}_n(\bar{x}_0, \ldots, \bar{x}_n, \bar{u}_0, \ldots, \bar{u}_n)$ coincides with $\Phi_n(\bar{x}_0, \ldots, \bar{x}_n, \bar{u}_0, \ldots, \bar{u}_n)$ and $\tilde{\Phi}_{n+1}(\bar{x}_0, \ldots, \bar{x}_n, x_0, \bar{u}_0, \ldots, \bar{u}_n, u_0)$ with $\Phi_{n+1}(\bar{x}_0, \ldots, \bar{x}_n, x_0, \bar{u}_0, \ldots, \bar{u}_n,$

u_0). Therefore for functions $F(\mathbf{x}, \mathbf{u})$ of the form indicated above (which depend only on a finite number of coordinates) the relation

$$\begin{aligned}\Phi_n(x_0, \ldots, x_n, u_0, \ldots, u_n) \\ = \int p_{n+1}(dx_{n+1}/x_0, \ldots, x_n, u_0, \ldots, u_n) \\ \times \left[\inf_{u_{n+1}} \Phi_{n+1}(x_0, \ldots, x_{n+1}, u_0, \ldots, u_{n+1})\right]\end{aligned} \tag{1.24}$$

is satisfied.

We now proceed to cost functionals dependent on an infinite number of coordinates. Here some properties of semi-continuous and continuous functions will be required.

Lemma 1.6. *For any lower semi-continuous function $F(x)$ bounded from below on a separable complete metric space X one can find an increasing sequence of bounded continuous functions $F_n(x)$ such that $F_n(x) \uparrow F(x)$ for all $x \in X$.*

A proof of this lemma is given in [27] p. 237.

Lemma 1.7. *Let $F(\mathbf{x}, \mathbf{u})$ be a continuous function on $X^N \times U^N$ (X is a complete separable metric space and U is a compactum). For any sequence of compacta $K_i \subset X$ there exist continuous functions $\Psi_n(x_0, \ldots, x_n, u_0, \ldots, u_n)$ such that a sequence of functions $F_n(\mathbf{x}, \mathbf{u}) = \Psi_n(x_0, \ldots, x_n, u_0, \ldots, u_n)$ converges uniformly to $F(\mathbf{x}, \mathbf{u})$ on the set*

$$\tilde{K} = \{(\mathbf{x}, \mathbf{u}); x_i \in K_i, i = 1, \ldots\}.$$

PROOF. Consider the space $X^N \times U^N$ as a complete metric space with the metric

$$r((\mathbf{x}, \mathbf{u}), (\mathbf{x}', \mathbf{u}')) = \sum_{n=0}^{\infty} 2^{-n}(1 - \exp\{-r_X(x_n, x_n) - r_U(u_n, u_n)\}),$$

where r_X and r_U are the distances in X and U respectively. Choose a fixed point $\bar{x} \in X$, $\bar{u} \in U$ and set

$$\begin{aligned}\Psi_n(x_0, \ldots, x_n, u_0, \ldots, u_n) \\ = F((x_0, \ldots, x_n, \bar{x}, \ldots, \bar{x}, \ldots), (u_0, \ldots, u_n, \bar{u}, \ldots, \bar{u}, \ldots)).\end{aligned}$$

Clearly, in view of the continuity of $F(\mathbf{x}, \mathbf{u})$, $\lim_{n\to\infty} \Psi_n(x_0, \ldots, x_n, u_0, \ldots, u_n) = F(\mathbf{x}, \mathbf{u})$. Furthermore,

$$F_n(\mathbf{x}, \mathbf{u}) = F(\mathbf{x}^{(n)}, \mathbf{u}^{(n)}),$$

where

$$\mathbf{x}^{(n)} = (x_0, \ldots, x_n, \bar{x}, \ldots, \bar{x}, \ldots), \qquad \mathbf{u}^{(n)} = (u_0, \ldots, u_n, \bar{u}, \ldots, \bar{u}, \ldots)$$

and $r[(\mathbf{x}, \mathbf{u}), (\mathbf{x}^{(n)}, \mathbf{u}^{(n)})] \to 0$ as $n \to \infty$. Hence by virtue of the uniform continuity of the function $F(\mathbf{x}, \mathbf{u})$ on the compact set $\tilde{K}$ we have

$$\lim_{n\to\infty} \sup_{(\mathbf{x}, \mathbf{u}) \in K} |F(\mathbf{x}, \mathbf{u}) - F(\mathbf{x}^{(n)}, \mathbf{u}^{(n)})| = 0. \qquad \square$$

Corollary. *If $F(\mathbf{x}, \mathbf{u})$ is a lower semi-continuous function bounded from below, one can find an increasing sequence of continuous functions $F_n(\mathbf{x}, \mathbf{u}) = \Psi_n(x_0, \ldots, x_n, u_0, \ldots, u_n)$ such that*

$$F(\mathbf{x}, \mathbf{u}) = \lim_{n\to\infty} F_n(\mathbf{x}, \mathbf{u}), \qquad (\mathbf{x}, \mathbf{u}) \in \tilde{K}.$$

Lemma 1.8. *Let a sequence of continuous functionals of the cost of controlling $F^{(m)}(\mathbf{x}, \mathbf{u})$ be jointly bounded and monotonically increasing as m increases to a functional of the cost of controlling $F(\mathbf{x}, \mathbf{u})$. Let S_m and S be the optimal costs of controlling for these functionals respectively. Then $S_m \uparrow S$ as $m \to \infty$.*

PROOF. Let v be a control. We have

$$\mathbf{E}_v F^{(m)}(\xi, \eta) \le \mathbf{E}_v F(\xi, \eta).$$

Hence $S_m \le S$. However if the control v_n is chosen in such a manner that $\mathbf{E}_{v_n} F^{(n)}(\xi, \eta) \le S_n + \varepsilon$ then for all k

$$\lim_{n\to\infty} \mathbf{E}_{v_n} F^{(k)}(\xi, \eta) \le \lim_{n\to\infty} \mathbf{E}_{v_n} F^{(n)}(\xi, \eta) \le \lim_{n\to\infty} S_n + \varepsilon.$$

It was shown in theorem 1.3 that a sequence of measures corresponding to (ξ, η) on $X^N \times U^N$ under the choice of controls v_n is weakly compact, and it was also established that any limiting measure corresponds to a controlled sequence constructed in terms of the same control object and a strategy v. Hence

$$\mathbf{E}_{\bar{v}} F^{(k)}(\xi, \eta) \le \lim_{n\to\infty} S_n + \varepsilon,$$

where $\bar{v}$ is a control (which may depend on k). Choosing once again a limit point for the corresponding measures in $X^N \times U^N$ we verify that there exists a single control $\bar{v}$ (the same for all k) such that

$$\mathbf{E}_{\bar{v}} F^{(k)}(\xi, \eta) \le \lim_{n\to\infty} S_n + \varepsilon.$$

Utilizing Lebesgue's theorem we obtain $\mathbf{E}_{\bar{v}} F(\xi, \eta) \le \lim_{n\to\infty} S_n + \varepsilon$. It now remains only to observe that $S \le \mathbf{E}_{\bar{v}} F(\xi, \eta)$ and $\varepsilon > 0$ is arbitrary. $\square$

Below we shall use the fact that there exists an optimal control for a controlled object $\mu(\cdot/\cdot)_{\bar{x}_0, \ldots, \bar{u}_n}$ with the cost of controlling $F(\mathbf{x}, \mathbf{u})_{\bar{x}_0, \ldots, \bar{u}_n}$ measurably dependent on $\bar{x}_0, \ldots, \bar{u}_n$. This fact follows from

Lemma 1.9. *Let S be a complete separable space, the function $F(\mathbf{x}, \mathbf{u})_s$ be bounded and lower semi-continuous jointly in the variables $\mathbf{x}$, $\mathbf{u}$ and s and the family of controlled objects $\mu(\cdot/\mathbf{u})_s$ is such that for all n and $f \in C_X$*

$$\int f(x)P_{n+1}(dx \mid x_0, \ldots, x_n, u_0, \ldots, u_n)_s$$

is continuous in $x_0, \ldots, x_n, u_0, \ldots, u_n, s$. Then there exists a non-randomized optimal control v_s for the controlled object $\mu(\cdot/\cdot)_s$ with control cost $F(\cdot/\cdot)_s$ defined by a sequence of Borel functions $\varphi_k(x_0, \ldots, x_k, s)$.

Proof. Consider the set $\mathfrak{M}$ of all measures in $X^N \times U^N$, which correspond to a controlled sequence with controlled object $\mu(\cdot/\cdot)_s$ for $s \in S$ and an arbitrary control. As it follows from Remark 1.2 this set is weakly compact. It can be metrized in such a manner that the weak convergence of measures will correspond to convergence in the metric. In this case $\mathfrak{M}$ is compact.

Denote by $\mathfrak{M}_s$ the subset of $\mathfrak{M}$ consisting of measures which correspond to controlled sequences with the controlled object $\mu(\cdot/\cdot)_s$. In the course of the proof of theorem 1.3 it was shown that $\mathfrak{M}_s$ is a closed set. Consider the set of points in $S \times \mathfrak{M}$:

$$\mathfrak{M}^{(0)} = \{(s, \theta) : \theta \in \mathfrak{M}_s\}.$$

This set is closed. Let $s_n \to s_0$, $\theta_n \to \theta_0$ and $(s_n, \theta_n) \in \{(s, \theta) : \theta \in \mathfrak{M}_s\}$. Denote by $(\xi_0^{(n)}, \eta_0^{(n)}), \ldots, (\xi_k^{(n)}, \eta_k^{(n)})$ a random sequence with the distribution θ_n. For any continuous bounded function $\Psi(x_0, \ldots., x_k, u_0, \ldots, u_k)$ and $f(x)$ the relation

$$\mathbf{E}f(\xi_{k+1}^{(n)})\Psi(\xi_0^{(n)}, \ldots, \xi_k^{(n)}, \eta_0^{(n)}, \ldots, \eta_k^{(n)})$$

$$= \mathbf{E}\Psi(\xi_0^{(n)}, \ldots, \xi_k^{(n)}, \eta_0, \ldots, \eta_k^{(n)}) \int f(x_{k+1})$$

$$\times\, p_{k+1}(dx_{k+1}/\xi_0^{(n)}, \ldots, \eta_k^{(n)})_{s_n}$$

is valid. Approaching the limit as $n \to \infty$ in both sides of this equation we obtain

$$\mathbf{E}f(\xi_{k+1}^{(0)})\Psi(\xi_0^{(0)}, \ldots, \eta_k^{(0)}) = \mathbf{E}\Psi(\xi_0^{(0)}, \ldots, \eta_k^{(0)})$$

$$\times \int f(x_{k+1})p_{k+1}(dx_{k+1}/\xi_0^{(0)}, \ldots, \eta_k^{(0)})_{s_0}.$$

Since this relation is fulfilled for any bounded continuous functions f and Ψ, $\mathbf{P}\{\xi_{k+1}^{(0)} \in A/\xi_0^{(0)}, \ldots, \eta_k^{(0)}\} = p_{k+1}(A/\xi_0^{(0)}, \ldots, \eta_k^{(0)})_{s_0}$. Hence $\theta_0 \in \mathfrak{M}_{s_0}$. We have thus shown that $\mathfrak{M}^{(0)}$ is closed.

Let

$$c > \sup F(\mathbf{x}, \mathbf{u})_s,$$

$$\mathcal{T}(s, \theta) = c(1 - \chi_{\mathfrak{M}^{(0)}}(s, \theta)) + \chi_{\mathfrak{M}^{(0)}}(s, \theta)\mathbf{E}_\theta F(\xi, \eta)_s.$$

(Here $\mathbf{E}_\theta$ is the mathematical expectation for the sequence with distribution

θ). The function $\mathscr{T}(s, \theta)$ is lower semi-continuous. This follows from the fact that $\mathfrak{M}^{(0)}$ is closed and the lower semi-continuity of the function $\mathbf{E}_\theta F(\xi, \eta)_s$ (the latter follows from lemma 1.5). Clearly

$$\inf_{\theta \in \mathfrak{M}} \mathscr{T}(s, \theta) = c(s),$$

where $c(s)$ is the optimal cost of control for controlled object $\mu(\cdot/\cdot)_s$ and control cost $F(\cdot, \cdot)_s$. In view of lemma 1.4 there exists a Borel function θ_s such that

$$\inf_{\theta \in \mathfrak{M}} \mathscr{T}(s, \theta) = \inf_{\theta \in \mathfrak{M}_s} \mathbf{E}_\theta F(\xi, \eta)_s = \mathscr{T}(s, \theta_s) = c(s).$$

If

$$q_k(du_k/\xi_0, \ldots, \xi_k, \eta_0, \ldots, \eta_{n-1})_s = \mathbf{P}_{\theta_s}\{\eta_k \in du_k/\xi_0, \ldots, \xi_k, \eta_0, \ldots, \eta_{k-1}\},$$

where $\mathbf{P}_\theta$ is the probability distribution which corresponds to the measure θ, then the functions $q_k(du_k/\xi_0, \ldots, \xi_k, \eta_0, \ldots, \eta_{k-1})_s$ are Borel functions in s. Using this fact one can construct a non-randomized control (as it was done in theorem 1.2) which will be a Borel function in s. □

Theorem 1.6. *Let a controlled object satisfy the conditions of theorem* 1.3, *the functional of the cost of control* $F(\mathbf{x}, \mathbf{u})$ *be bounded and lower semi-continuous. Then:* I. *Functions* $\Phi_n(x_0, \ldots, x_n, u_0, \ldots, u_n)$ *satisfy the following conditions:*

1. *they are bounded and lower semi-continuous*
2. *for all* $n > 0$ *relation* (1.24) *is satisfied*
3. *for any* $\mathbf{x} \in X^N$ *and* $\mathbf{u} \in U^N$

$$\lim_{n\to\infty} \Phi_n(x_0, \ldots, x_n, u_0, \ldots, u_n) \geq F(\mathbf{x}, \mathbf{u});$$

if $\mathbf{x}, \mathbf{u}$ *is a continuity point of functional* F, *then*

$$\underline{\lim}_{n\to\infty} \Phi_n(x_0, \ldots, x_n, u_0, \ldots, u_n) = F(\mathbf{x}, \mathbf{u}).$$

II. *Let the Borel functions* $\varphi_n(x_0, \ldots, x_n, u_0, \ldots, u_{n-1})$ *satisfy the equality*

$$\Phi_n(x_0, \ldots, x_n, u_0, \ldots, u_{n-1}, \varphi_n(x_0, \ldots, x_n, u_0, \ldots, u_{n-1})) = \inf_{u_n} \Phi_n(x_0, \ldots, x_n, u_0, \ldots, u_n),$$

and

$$\bar{\varphi}_0(x_0) = \varphi(x_0),$$

$$\bar{\varphi}_1(x_0, x_1) = \varphi_1(x_0, x_1, \bar{\varphi}_0(x_0)),$$

$$\vdots$$

$$\bar{\varphi}_n(x_0, \ldots, x_n) = \varphi_n(x_0, \ldots, x_n, \bar{\varphi}_0(x_0), \ldots, \varphi_{n-1}(x_0, \ldots, x_{n-1}));$$

then a non-randomized control $\bar{v}: \{u_k = \bar{\varphi}_k(x_0, \ldots, x_k), k = 0, \ldots\}$ *will be optimal and the quantity*

$$S = \int \left[\inf_{u_0} \Phi_0(x_0, u_0)\right] p_0(dx_0)$$

will be the optimal cost of controlling.

PROOF. Given a control v we define on $X^N \times U^N$ measures by the relation

$$\mu^v_{\bar{x}_0, \ldots, \bar{x}_n, \bar{u}_0, \ldots, \bar{u}_n}(C) = \mathbf{E}_v^{\bar{x}_0, \ldots, \bar{x}_n, \bar{u}_0, \ldots, \bar{u}_n} \chi_C((\xi, \eta))$$

(C is a Borel set in $X^N \times U^N$, χ_C is the indicator of this set).

Let $\bar{x}_i^{(m)} \to \bar{x}_i^{(0)}$, $\bar{u}_i^{(m)} \to \bar{u}_i^{(0)}$, $i = 0, 1, \ldots, n$. Furthermore, let a sequence of controls v_l be chosen in such a manner that for all m

$$\Phi_n(\bar{x}_0^{(m)}, \ldots, \bar{x}_n^{(m)}, \bar{u}_0^{(m)}, \ldots, \bar{u}_n^{(m)}) = \varliminf_{l \to \infty} \mathbf{E}_{v_l}^{\bar{x}_0^{(m)}, \ldots, \bar{x}_n^{(m)}, u_0^{(m)}, \ldots, \bar{u}_n^{(m)}} F(\xi, \eta)_{\bar{x}_0^{(m)}, \ldots, \bar{x}_n^{(m)}, \bar{u}_0^{(m)}, \ldots, \bar{u}_n^{(m)}}. \tag{1.25}$$

As it follows from the proof of theorem 1.3 for any $\varepsilon > 0$ one can find compact sets $K_i \subset X$ such that for $x_i^{(m)} \in K_i$, $i = 0, \ldots, n$,

$$\mu^{v_l}_{\bar{x}_0^{(m)}, \ldots, \bar{u}_n^{(m)}}\left(\bigcap_{k=0}^{\infty} \{(\mathbf{x}, \mathbf{u}) : x_k \in K_{n+k+1}\}\right) \geq 1 - \varepsilon.$$

Therefore the family of measures

$$\{\mu^{v_l}_{\bar{x}_0^{(m)}, \ldots, \bar{x}_n^{(m)}}, m = 0, 1, \ldots\} \tag{1.26}$$

is compact. We select from this sequence a convergent subsequence; let this subsequence converge to a measure μ. Then in view of lemma 1.5

$$\varliminf_{l \to \infty} \int F(\mathbf{x}, \mathbf{u})_{\bar{x}_0^{(m_l)}, \ldots, \bar{u}_n^{(m_l)}} \mu^{v_{l_i}}_{\bar{x}_0^{(m_l)}, \ldots, \bar{u}_n^{(m_l)}}(d\mathbf{x} \times d\mathbf{u})$$

$$\geq \int F(\mathbf{x}, \mathbf{u})_{\bar{x}_0^{(0)}, \ldots, \bar{u}_n^{(0)}} \mu(d\mathbf{x} \times d\mathbf{u}).$$

If $g(x)$ and $\Psi(x_0, \ldots, x_{k-1}, u_0, \ldots, u_{k-1})$ are continuous bounded functions, then

$$\int g(x_k)\Psi(x_0, \ldots, x_{k-1}, u_0, \ldots, u_{k-1})\mu(d\mathbf{x} \times d\mathbf{u})$$

$$= \lim_{i \to \infty} \int g(x_k)\Psi(x_0, \ldots, x_{k-1}, u_0, \ldots, u_{k-1})\mu^{v_{l_i}}_{\bar{x}_0^{(m_i)}, \ldots, \bar{u}_n^{(m_i)}} \times (d\mathbf{x} \times d\mathbf{u})$$

$$= \lim_{i\to\infty} \int \Psi(x_0, \ldots, x_{k-1}, u_0, \ldots, u_{k-1})$$

$$\times \int g(x_k) p_{n+k+1}(dx_k / x_0^{(m_i)}, \ldots, x_n^{(m_i)}, x_0,$$

$$\ldots, x_{k-1}, u_0^{(m_i)}, \ldots, u_n^{(m_i)}, u_0, \ldots, u_{k-1}) \mu^{\nu_{l_i}}_{\bar{x}_0^{(m_i)}, \ldots, \bar{u}_n^{(m_i)}}$$

$$\times (d\mathbf{x} \times d\mathbf{u})$$

$$= \int \Psi(x_0, \ldots, x_{k-1}, u_0, \ldots, u_{k-1})$$

$$\times \int g(x_k) p_{n+k+1}(dx_k / \bar{x}_0^{(0)}, \ldots, \bar{x}_n^{(0)}, x_0, \ldots, x_{k-1}, \bar{u}_0^{(0)},$$

$$\ldots, \bar{u}_n^{(0)}, u_0, \ldots, u_{k-1}) \mu(d\mathbf{x} \times d\mathbf{u})$$

(we have utilized here Remark 1.2). Comparing the extreme terms of the last equality we verify that if $\{(\xi_k, \eta_k)\}$ is a random sequence with the corresponding measure μ in $X^N \times U^N$, then

$$\mathbf{P}\{\xi_k \in A / \xi_0, \ldots, \xi_{k-1}, \eta_0, \ldots, \eta_{k-1}\}$$
$$= p_{n+k+1}(A / \bar{x}_0^{(0)}, \ldots, \bar{x}_n^{(0)}, \xi_0, \ldots, \xi_{k-1}, \bar{u}_0^{(0)}, \ldots, \bar{u}_n^0, \eta_0, \ldots, \eta_{k-1}),$$

i.e. it is a controlled random sequence with controlled object $\mu(\cdot/\cdot)_{\bar{x}_0^{(0)}, \ldots, \bar{x}_n^{(0)}, \bar{u}_0^{(0)}, \ldots, \bar{u}_n^{(0)}}$ and hence

$$\int F(\mathbf{x}, \mathbf{u})_{\bar{x}_0^{(0)}, \ldots, \bar{u}_n^{(0)}} \mu(d\mathbf{x} \times d\mathbf{u}) \geq \Phi_n(\bar{x}_0^{(0)}, \ldots, \bar{x}_n^{(0)}, \bar{u}_0^{(0)}, \ldots, \bar{u}_n^{(0)}).$$

Thus for any weakly convergent sequence $\mu^{\nu_{l_i}}_{\bar{\mathbf{x}}^{(m_i)}, \ldots, \bar{\mathbf{u}}^{(m_i)}}$ the inequality

$$\varliminf_{i\to\infty} \int F(\mathbf{x}, \mathbf{u})_{\bar{x}_0^{(m_i)}, \ldots, \bar{u}_n^{(m_i)}} \mu^{\nu_{l_i}}_{\bar{x}_0^{(m_i)}, \ldots, \bar{u}_n^{(m_i)}}(d\mathbf{x} \times d\mathbf{u})$$
$$\geq \Phi_n(\bar{x}_0^{(0)}, \ldots, \bar{x}_n^{(0)}, \bar{u}_0^{(0)}, \ldots, \bar{u}_n^{(0)}).$$

is fulfilled. Hence in view of the compactness of the family of measures (1.26)

$$\varliminf_{m,l\to\infty} \int F(\mathbf{x}, \mathbf{u})_{\bar{x}_0^{(m)}, \ldots, \bar{u}_n^{(m)}} \mu^{\nu_{l}}_{\bar{x}_0^{(m_i)}, \ldots, \bar{u}_n^{(m)}}(d\mathbf{x} \times d\mathbf{u})$$
$$\geq \Phi_n(\bar{x}_0^{(0)}, \ldots, \bar{x}_n^{(0)}, \bar{u}_0^{(0)}, \ldots, \bar{u}_n^{(0)}).$$

However, in that case

$$\varliminf_{m\to\infty} \varliminf_{l\to\infty} \int F(\mathbf{x}, \mathbf{u})_{\bar{x}_0^{(m)}, \ldots, \bar{u}_n^{(m)}} \mu^{\nu_{l}}_{\bar{x}_0^{(m)}, \ldots, \bar{u}_n^{(m)}}(d\mathbf{x} \times d\mathbf{u})$$
$$\geq \Phi_n(\bar{x}_0^{(0)}, \ldots, \bar{x}_n^{(0)}, \bar{u}_0^{(0)}, \ldots, \bar{u}_n^{(0)}) \text{ also.}$$

Taking (1.25) into account we verify the validity of assertion I.1 (the boundedness of the functions Φ_n follows in an obvious manner from the boundedness of F).

To prove assertion I.2 we note that one can choose control v in such a manner that

$$\Phi_n(\bar{x}_0, \ldots, \bar{x}_n, \bar{u}_0, \ldots, \bar{u}_n) \geq \mathbf{E}_v^{\bar{x}_0, \ldots, \bar{x}_n, \bar{u}_0, \ldots, \bar{u}_n} F(\xi, \eta)_{\bar{x}_0, \ldots, \bar{u}_n} - \varepsilon.$$

Moreover,

$$\Phi_n(\bar{x}_0, \ldots, \bar{x}_n, \bar{u}_0, \ldots, \bar{u}_n) + \varepsilon \geq \mathbf{E}_v^{\bar{x}_0, \ldots, \bar{x}_n, \bar{u}_0, \ldots, \bar{u}_n}$$
$$\times \mathbf{E}_v^{\bar{x}_0, \ldots, \bar{x}_n, \bar{u}_0, \ldots, \bar{u}_n}(F(\xi, \eta)_{\bar{x}_0, \ldots, \bar{u}_n}/\xi_0; \eta_0)$$
$$= \mathbf{E}_v^{\bar{x}_0, \ldots, \bar{x}_n, \bar{u}_0, \ldots, \bar{u}_n} \mathbf{E}_{v'}^{\bar{x}_0, \ldots, \bar{x}_n, \xi_0, \bar{u}_0, \ldots, \bar{u}_n, \eta_0} F(\xi', \eta')_{\bar{x}_0, \ldots, \bar{x}_n, \xi_0, \bar{u}_0, \ldots, \bar{u}_n, \eta_0},$$

where the control v' is constructed from the control v as follows: for v', $\xi' = (\xi_1, \ldots)$, $\eta' = (\eta_1, \ldots)$

$$q'_k(du_k/x_0, \ldots, x_k, u_0, \ldots, u_{k-1})$$
$$= q_{k+1}(du_{k+1}/\xi_0, x_0, \ldots, x_k, \eta_0, u_0, \ldots, u_{k-1}).$$

Utilizing the inequalities

$$\mathbf{E}_{v'}^{\bar{x}_0, \ldots, \bar{x}_n, \xi_0, \bar{u}_0, \ldots, \bar{u}_n, \eta_0} F(\xi', \eta')_{\bar{x}_0, \ldots, \bar{x}_n, \xi_0, \bar{u}_0, \ldots, \bar{u}_n, \eta_0}$$
$$\geq \Phi_{n+1}(\bar{x}_0, \ldots, \bar{x}_n, \xi_0, \bar{u}_0, \ldots, \bar{u}_n, \eta_0)$$

and

$$\Phi_{n+1}(\bar{x}_0, \ldots, \bar{x}_n, \xi_0, \bar{u}_0, \ldots, \bar{u}_n, \eta_0)$$
$$\geq \inf_{u_{n+1}} \Phi_{n+1}(\bar{x}_0, \ldots, \bar{x}_n, \xi_0, \bar{u}_0, \ldots, \bar{u}_n, u_{n+1}),$$

we obtain

$$\Phi_n(\bar{x}_0, \ldots, \bar{x}_n, \bar{u}_0, \ldots, \bar{u}_n) + \varepsilon$$
$$\geq \mathbf{E}_v^{\bar{x}_0, \ldots, \bar{x}_n, \bar{u}_0, \ldots, \bar{u}_n} \inf_{u_{n+1}} \Phi_{n+1}(\bar{x}_0, \ldots, \bar{x}_n, \xi_0, \bar{u}_0, \ldots, \bar{u}_n, u_{n+1})$$
$$= \int p_{n+1}(dx_{n+1}/\bar{x}_0, \ldots, \bar{x}_n, \bar{u}_0, \ldots, \bar{u}_n)$$
$$\times \left[\inf_{u_{n+1}} \Phi_{n+1}(\bar{x}_0, \ldots, \bar{x}_n, x_{n+1}, \bar{u}_0, \ldots, \bar{u}_n, u_{n+1})\right].$$

Since $\varepsilon > 0$ is arbitrary we have thus shown that

$$\Phi_n(x_0, \ldots, x_n, u_0, \ldots, u_n)$$
$$\geq \int p_{n+1}(dx_{n+1}/x_0, \ldots, x_n, u_0, \ldots, u_n)$$
$$\times \left[\inf_{u_{n+1}} \Phi_{n+1}(x_0, \ldots, x_n, x_{n+1}, u_0, \ldots, u_n, u_{n+1})\right]. \tag{1.27}$$

Let $K_i \subset X$ be an increasing sequence of compact sets such that for any control v

$$\mu^{v}_{x_0, \ldots, x_n, u_0, \ldots, u_n}\left(\bigcap_{k=0}^{\infty} \{(\mathbf{x}, \mathbf{u}) : x_k \in K_{n+k+1}\} \right) \geq 1 - \varepsilon,$$

provided only $x_i \in K_i$ for $i = 0, \ldots, n$ (the existence of such a sequence of compact sets was discussed in the proof of assertion I.1).

Denote by $F^{(\varepsilon)}(\mathbf{x}, \mathbf{u})$ the functional

$$F^{(\varepsilon)}(\mathbf{x}, \mathbf{u}) = c(1 - \chi_{\tilde{K}}(\mathbf{x}, \mathbf{u})) + \chi_{\tilde{K}}(\mathbf{x}, \mathbf{u})F(\mathbf{x}, \mathbf{u}), \quad c > \sup F(\mathbf{x}, \mathbf{u}),$$

where $\tilde{K}$ is a compact set in $X^N \times U^N$:

$$\tilde{K} = \{(\mathbf{x}, \mathbf{u}) : x_i \in K_i, \quad i = 0, 1, \ldots\}.$$

It is easy to verify that for any n and control v

$$0 \leq \mathbf{E}_v^{\bar{x}_0, \ldots, \bar{u}_n} F^{(\varepsilon)}(\xi, \eta)_{\bar{x}_0, \ldots, \bar{u}_n} - \mathbf{E}_v^{\bar{x}_0, \ldots, \bar{u}_n} F(\xi, \eta)_{\bar{x}_0, \ldots, \bar{u}_n} \leq |c|\varepsilon,$$

if $\bar{x}_i \in K_i$. Therefore if $\Phi_n^{(\varepsilon)}(x_0, \ldots, x_n, u_0, \ldots, u_n)$ are defined by relations (1.23) (F should be replaced by $F^{(\varepsilon)}$) then

$$\begin{aligned} 0 &\leq \Phi_n^{(\varepsilon)}(x_0, \ldots, x_n, u_0, \ldots, u_n) - \Phi_n(x_0, \ldots, x_n, u_0, \ldots, u_n) \\ &\leq |c|\varepsilon, \end{aligned} \tag{1.28}$$

if $x_i \in K_i$. It follows from the corollary to Lemma 1.7 that there exist continuous functionals $F^{(\varepsilon, m)}(\mathbf{x}, \mathbf{u})$ which depend on $x_0, \ldots, x_m; u_0, \ldots, u_m$ only and such that

$$F^{(\varepsilon, m)}(\mathbf{x}, \mathbf{u}) \uparrow F^{(\varepsilon)}(\mathbf{x}, \mathbf{u}).$$

Define $\Phi_n^{(\varepsilon, m)}(x_0, \ldots, x_n, u_0, \ldots, u_n)$ by means of relationship (1.23) where F is replaced by $F^{(\varepsilon, m)}$. It then follows from lemma 1.8 that

$$\Phi_n^{(\varepsilon, m)}(x_0, \ldots, x_n, u_0, \ldots, u_n) \uparrow \Phi_n^{(\varepsilon)}(x_0, \ldots, x_n, u_0, \ldots, u_n).$$

Since $F^{(\varepsilon, m)}$ depends only on a finite number of coordinates in view of Remark 1.6 relation (1.24) is valid for $\Phi_n^{(\varepsilon, m)}$. Hence

$$\begin{aligned} &\Phi_n^{(\varepsilon, m)}(x_0, \ldots, x_n, u_0, \ldots, u_n) \\ &\quad = \int p_{n+1}(dx_{n+1}/x_0, \ldots, x_n, u_0, \ldots, u_n) \\ &\qquad \times \left[\inf_{u_{n+1}} \Phi_{n+1}^{(\varepsilon, m)}(x_0, \ldots, x_{n+1}, u_0, \ldots, u_{n+1}) \right]. \end{aligned}$$

Approaching the limit as $m \to \infty$ in this equation, we obtain by virtue of the Lebesgue theorem that

$$\Phi_n^{(\varepsilon)}(x_0, \ldots, x_n, u_0, \ldots, u_n)$$
$$= \int p_{n+1}(dx_{n+1}/x_0, \ldots, x_n, u_0, \ldots, u_n)$$
$$\times \lim_{m\to\infty} \left[\inf_{u_{n+1}} \Phi_{n+1}^{(\varepsilon, m)}(x_0, \ldots, x_{n+1}, u_0, \ldots, u_{n+1})\right]$$
$$\leq \int p_{n+1}(dx_{n+1}/x_0, \ldots, x_n, u_0, \ldots, u_n)$$
$$\times \left[\inf_{u_{n+1}} \Phi_{n+1}^{(\varepsilon)}(x_0, \ldots, x_{n+1}, u_0, \ldots, u_{n+1})\right],$$

since for any increasing sequence of functions $\Psi_n(u)$ inequality $\lim_{n\to\infty} \inf \Psi_n(u) \leq \inf \lim_{n\to\infty} \Psi_n(u)$ is clearly valid. Utilizing (1.28) we obtain

$$\Phi_n(x_0, \ldots, x_n, u_0, \ldots, u_n)$$
$$\leq \int p_{n+1}(dx_{n+1}/x_0, \ldots, x_n, u_0, \ldots, u_n)$$
$$\times \left[\inf_{u_{n+1}} \Phi_{n+1}^{(\varepsilon)}(x_0, \ldots, x_{n+1}, u_0, \ldots, u_{n+1})\right]$$
$$\leq \int p_{n+1}(dx_{n+1}/x_0, \ldots, x_n, u_0, \ldots, u_n)$$
$$\times \left[\inf_{u_{n+1}} \Phi_{n+1}(x_0, \ldots, x_{n+1}, u_0, \ldots, u_{n+1}) + \varepsilon\right].$$

Since $\varepsilon > 0$ is arbitrary this inequality in conjunction with (1.27) yields expression (1.24). The latter is proved so far under the condition that $x_i \in K_i$ for $i = 0, \ldots, n$. Clearly compacta K_i can always be extended so that this condition will be fulfilled. Assertion I.2 is thus proved.

To prove assertion I.3 we introduce in $X^N \times U^N$ controlled sequences $(\xi_0^{(n)}, \eta_0^{(n)}), (\xi_1^{(n)}, \eta_1^{(n)}), \ldots$ whose distribution is obtained if we choose optimal control for the controlled object $\mu(\cdot/\cdot)_{x_0, \ldots, x_n, u_0, \ldots, u_n}$ with the cost of controlling $F(\cdot, \cdot)_{x_0, \ldots, u_n}$. In this case

$$\Phi_n(x_0, \ldots, x_n, u_0, \ldots, u_n) = \mathbf{E}_{\mathbf{v}_n}^{x_0, \ldots, u_n} F(\xi, \eta)_{x_0, \ldots, u_n}$$
$$= \mathbf{E}F((x_0, \ldots, x_n, \xi_0^{(n)}, \xi_1^{(n)}, \ldots), (u_0, \ldots, u_n, \eta_0^{(n)}, \eta_1^{(n)}, \ldots)).$$

Since $((x_0, \ldots, x_n, \xi_0^{(n)}, \xi_1^{(n)}, \ldots), (u_0, \ldots, u_n, \eta_0^{(n)}, \eta_1^{(n)}, \ldots)) \to (\mathbf{x}, \mathbf{u})$ in $X^N \times U^N$ with probability 1 as $n \to \infty$, it follows that

$$\underline{\lim}_{n\to\infty} F((x_0, \ldots, x_n, \xi_0^{(n)}, \xi_1^{(n)}, \ldots), (u_0, \ldots, u_n, \eta_0^{(n)}, \eta_1^{(n)}, \ldots))$$
$$\geq F(\mathbf{x}, \mathbf{u})$$

and at the points of continuity of F

$$\lim_{n\to\infty} F((x_0, \ldots, x_n, \xi_0^{(n)}, \xi_1^{(n)}, \ldots), (u_0, \ldots, u_n, \eta_0^{(n)}, \eta_1^{(n)}, \ldots)) = F(\mathbf{x}, \mathbf{u}).$$

Applying mathematical expectation in the last relationship and utilizing Fatou's and Lebesgue's theorems we conclude the proof of assertion I.3.

With the aid of lemma 1.9 one can construct an optimal control defined by the Borel functions

$$\{u_k = g_k^{(n)}(\bar{x}_0, \ldots, \bar{x}_n, \bar{u}_0, \ldots, \bar{u}_n, x_0, \ldots, x_k), k = 0, 1, \ldots\}$$

for the controlled object $\mu(\cdot/\cdot)_{\bar{x}_0, \ldots, \bar{u}_n}$ with the cost of controlling $F(\cdot, \cdot)_{\bar{x}_0, \ldots, \bar{u}_n}$.

Denote this control by $v_{\bar{x}_0, \ldots, \bar{x}_n, \bar{u}_0, \ldots, \bar{u}_n}$. We have

$$\Phi_n(\bar{x}_0, \ldots, \bar{x}_n, \bar{u}_0, \ldots, \bar{u}_n) = \mathbf{E}^{\bar{x}_0, \ldots, \bar{u}_n}_{v_{\bar{x}_0, \ldots, \bar{x}_n, \bar{u}_0, \ldots, \bar{u}_n}} F(\xi, \eta)_{\bar{x}_0, \ldots, \bar{u}_n}.$$

Now choosing the control v_n:

$$u_i = \bar{\varphi}_i(x_0, \ldots, x_i), \qquad i \le n,$$

$$u_i = g^n_{i-n}(x_0, \ldots, x_n, \bar{\varphi}_0(x_0), \ldots, \bar{\varphi}_n(x_0, \ldots, x_n), x_{n+1}, \ldots, x_{i-n})$$

we verify that

$$\mathbf{E}_{v_n} F(\xi, \eta) = \mathbf{E}_{v_n}(F(\xi, \eta)/\xi_0, \ldots, \xi_n, \eta_0, \ldots, \eta_n)$$
$$= \mathbf{E}_{v_n} \Phi_n(\xi_0, \ldots, \xi_n, \eta_0, \ldots, \eta_n).$$

In exactly the same manner as in the proof of theorem 1.5 one can show that for the control v_n the equality

$$\mathbf{E}_{v_n} \Phi_n(\xi_0, \ldots, \zeta_n, \eta_0, \ldots, \eta_n) = \Phi$$

is valid (here Φ is the optimal control cost). Consequently, we have for all n

$$\Phi = \mathbf{E}_{v_n} F(\xi, \eta).$$

Since the sequence of measures μ_{v_n}, which corresponds in $X^N \times U^N$ to the sequence $(\xi_0, \eta_0), (\xi_1, \eta_1), \ldots$ with controlled object $\mu(\cdot/\cdot)$ and control v_n is weakly convergent to the measure $\mu_{\bar{v}}$ corresponding in $X^N \times U^N$ to the sequence $(\xi_0, \eta_0), (\xi_1, \eta_1), \ldots$ with the same controlled object and control v, by virtue of lemma 1.3 one can write

$$\varliminf_{n\to\infty} \mathbf{E}_{v_n} F(\xi, \eta) \ge \mathbf{E}_{\bar{v}} F(\xi, \eta).$$

Therefore $\Phi \ge \mathbf{E}_{\bar{v}}\, F(\xi, \eta)$. However, $\Phi \le \mathbf{E}_{\bar{v}}\, F(\xi, \eta)$. Thus $\Phi = \mathbf{E}_{\bar{v}} F(\xi, \eta)$ and assertion II is verified. □

Remark 1.6. In the course of the proof of the theorem we have described the method for determining the functions $\Phi_n(x_0, \ldots, x_n, u_0, \ldots, u_n)$ if X is locally compact. For this purpose it is necessary to construct a sequence of continuous bounded functions $F^{(m)}$ which depend only on a finite number of coordinates of points $\mathbf{x}$ and $\mathbf{u}$ and such that $F^{(m)}(\mathbf{x}, \mathbf{u}) \uparrow F(\mathbf{x}, \mathbf{u})$. If $\Phi_n^{(m)}(x_0, \ldots, x_n, u_0, \ldots, u_n)$ are functions defined for $F^{(m)}$ by relations (1.23) then

$$\Phi_n^{(m)}(x_0, \ldots, x_n, u_0, \ldots, u_n) \uparrow \Phi_n(x_0, \ldots, x_n, u_0, \ldots, u_n).$$

Functions $\Phi_n^{(m)}$ are effectively determined in the manner described in theorem 1.5. These can be utilized also for the construction of ε-optimal solutions: for m sufficiently large an optimal solution for functional $F^{(m)}$ will be ε-optimal for functional F.

Remark 1.7. The condition of semi-continuity of F as well as the condition of theorem 1.3 imposed on the controlled object were utilized only for the proof of semi-continuity of Φ_n and the existence of functions φ_k. If the functions φ_k which satisfy condition II of the theorem do exist, the sequence $u_k = \bar{\varphi}_k(x_0, \ldots, x_k)$ will be an optimal control all the same. Such is the case when the set U is finite.

We shall say that a functional of the cost of controlling is *regular* if for all $\mathbf{x}$ and $\mathbf{u}$

$$\lim_{n \to \infty} \Phi_n(x_0, \ldots, x_n, u_0, \ldots, u_n) = F(\mathbf{x}, \mathbf{u}). \tag{1.29}$$

Remark 1.8. If a functional of the cost of controlling is regular then conditions I.2 and (1.29) uniquely determine the sequence of functions Φ_n.

Indeed let $\bar{\Phi}_n$ be another sequence of functionals satisfying I.2 and (1.29). Let the inequality

$$|\Phi_n(x_0, \ldots, x_n, u_0, \ldots, u_n) - \bar{\Phi}_n(x_0, \ldots, x_n, u_0, \ldots, u_n)| > \delta$$

be valid for $x_0, \ldots, x_n, u_0, \ldots, u_n$. Then there exist x_{n+1} and u_{n+1} such that

$$|\Phi_n(x_0, \ldots, x_n, x_{n+1}, u_0, \ldots, u_n, u_{n+1}) \\ - \bar{\Phi}_n(x_0, \ldots, x_n, u_0, \ldots, u_n)| > \delta,$$

since

$$|\Phi_n(x_0, \ldots, x_n, u_0, \ldots, u_n) - \bar{\Phi}_n(x_0, \ldots, x_n, u_0, \ldots, u_n)|$$

$$\leq \int \sup_{u_{n+1}} |\Phi_n(x_0, \ldots, x_n, x_{n+1}, u_0, \ldots, u_{n+1}) \\ - \bar{\Phi}_n(x_0, \ldots, x_n, x_{n+1}, u_0, \ldots, u_{n+1})| p_{n+1}(dx_{n+1}/x_0, \ldots, u_n)$$

$$\leq \sup_{u_{n+1}, x_{n+1}} |\Phi_n(x_0, \ldots, x_{n+1}, u_0, \ldots, u_{n+1}) \\ - \bar{\Phi}_n(x_0, \ldots, x_{n+1}, u_0, \ldots, u_{n+1})|.$$

Therefore one can find sequences $x_0, \ldots, x_N, \ldots, u_0, \ldots, u_N, \ldots,$ such that for all $N > n$

$$|\Phi_N(x_0, \ldots, x_N, u_0, \ldots, u_N) - \bar{\Phi}_N(x_0, \ldots, x_N, u_0, \ldots, u_N)| > \delta.$$

Approaching the limit as $N \to \infty$ and taking (1.29) into account we arrive at a contradiction.

It turns out that it is possible to replace a functional of the cost of controlling by a regular one without changing the optimal control and the optimal cost of controlling. In this case the values of functions Φ_n and that of the functional at the continuity points is not altered, however the latter value may increase at the discontinuity points. Such a functional is called a *regularization* of a given functional. We now show how to construct such a functional.

Let

$$F^*(\mathbf{x}, \mathbf{u}) = \lim_{n\to\infty} \Phi_n(x_0, \ldots, x_n, u_0, \ldots, u_n) \tag{1.30}$$

be satisfied at all the points of the existence of the limit and let

$$F^*(\mathbf{x}_1, \mathbf{u}_1) = \underline{\lim}_{\mathbf{x}\to\mathbf{x}_1,\, \mathbf{u}\to\mathbf{u}_1} F^*(\mathbf{x}, \mathbf{u}) \tag{1.31}$$

at the remaining points $(\mathbf{x}_1, \mathbf{u}_1)$ (the limit on the r.h.s. is taken over these points $\mathbf{x}$ and $\mathbf{u}$ which satisfy (1.30)).

Let $D \subset X^N \times U^N$ be the set of $\mathbf{x}$ and $\mathbf{u}$ for which the limit (1.30) exists. We have $\mu_v(D) = 1$ for an arbitrary control v. Indeed, let $(\xi_0, \eta_0), (\xi_1, \eta_1), \ldots$ be a controlled sequence obtained for the control v. Then

$$\begin{aligned}
&\mathbf{E}[\Phi_{n+1}(\xi_0, \xi_1, \ldots, \xi_{n+1}, \eta_0, \ldots, \eta_{n+1})/\xi_0, \ldots, \xi_n, \eta_0, \ldots, \eta_n] \\
&\quad \geq \mathbf{E}\Big[\inf_{u_{n+1}} \Phi_{n+1}(\xi_0, \ldots, \xi_{n+1}, \eta_0, \ldots, \eta_n, u_{n+1})/\xi_0, \\
&\qquad \ldots, \xi_n, \eta_0, \ldots, \eta_n \Big] \\
&\quad = \int p_{n+1}(dx_{n+1}/\xi_0, \ldots, \xi_n, \eta_0, \ldots, \eta_n) \\
&\qquad \times \inf_{u_{n+1}} \Phi_{n+1}(\xi_0, \ldots, \xi_n, x_{n+1}, \eta_0, \ldots, \eta_n, u_{n+1}) \\
&\quad = \Phi_n(\xi_0, \ldots, \xi_n, \eta_0, \ldots, \eta_n).
\end{aligned}$$

Therefore the sequence $\Phi_0(\xi_0, \eta_0), \Phi_1(\xi_0, \xi_1, \eta_0, \eta_1), \ldots, \Phi_n(\xi_0, \ldots, \xi_n, \eta_0, \ldots, \eta_n)$ is a bounded submartingale with respect to measure μ_v. Hence the limit

$$\lim_{n\to\infty} \Phi_n(\xi_0, \ldots, \xi_n, \eta_0, \ldots, \eta_n)$$

exists with probability 1, i.e. $\mu_v(D) = 1$.

In view of assertion I.3 of theorem 1.6, at all points of continuity the equality $F^*(\mathbf{x}, \mathbf{u}) = F(\mathbf{x}, \mathbf{u})$ is valid. At all the other points of D in view of the same assertion $F^*(\mathbf{x}, \mathbf{u}) \geq F(\mathbf{x}, \mathbf{u})$. Utilizing the lower semi-continuity of function $F(\mathbf{x}, \mathbf{u})$ and relation (1.31) we verify that the inequality $F^*(\mathbf{x}, \mathbf{u}) \geq F(\mathbf{x}, \mathbf{u})$ is fulfilled on the closure of D. However the closure of D coincides with $X^N \times U^N$, since D contains all the points of continuity of $F(\mathbf{x}, \mathbf{u})$ and in view of the lower semi-continuity of $F(\mathbf{x}, \mathbf{u})$ the set of discontinuities of F has no interior points (see, e.g. [27] p. 237).

We now show that functions Φ_n^* constructed for the cost function F^* by means of formulas (1.23) coincide with Φ_n.

Let the control v for $k > n$ be determined by the functions $u_k = \bar{\varphi}_k(x_0, \ldots, x_k)$ (cf. assertion II of theorem 1.6). Then for $k > n$

$$\begin{aligned} &\mathbf{E}_v(\Phi_k(\xi_0, \ldots, \xi_k, \eta_0, \ldots, \eta_k)/\xi_0, \ldots, \xi_{k-1}, \eta_0, \ldots, \eta_{k-1}) \\ &\quad = \Phi_{k-1}(\xi_0, \ldots, \xi_{k-1}, \eta_0, \ldots, \eta_{k-1}) \end{aligned}$$

(this was verified in the course of the proof of theorem 1.6). Therefore

$$\{\Phi_n(\xi_0, \ldots, \xi_n, \eta_0, \ldots, \eta_n), \Phi_{n+1}(\xi_0, \ldots, \xi_{n+1}, \eta_0, \ldots, \eta_{n+1}), \ldots\}$$

will be a bounded martingale relative to the probability P_v. Hence

$$\begin{aligned} \Phi_n(\xi_0, \ldots, \xi_n, \eta_0, \ldots, \eta_n) &= \mathbf{E}_v(F^*(\xi, \eta)/\xi_0, \ldots, \xi_n, \eta_0, \\ &\ldots, \eta_n) \geq \Phi_n^*(\xi_0, \ldots, \xi_n, \eta_0, \ldots, \eta_n), \end{aligned}$$

however since $F^* \geq F$ it follows that $\Phi_n^* \geq \Phi_n$. The equality $\Phi_n^* = \Phi_n$ is thus verified.

4 Control of Processes with Incomplete Observations

In a number of problems of controlled random processes when constructing an optimal control one can utilize only a part of the information concerning the states of the basic process at the previous times. This means that the conditional probabilities

$$q_n(du_n/x_0, \ldots, x_n, u_0, \ldots, u_{n-1})$$

are measurable with respect to $\mathfrak{A}'_{n+1} \times \mathfrak{B}^n$ (where $\mathfrak{A}'_{n+1} \subset \mathfrak{A}^{n+1}$ is a subalgebra of the algebra $\mathfrak{A}^{n+1}$) rather than with respect to $\mathfrak{A}^{n+1} \times \mathfrak{B}^n$. It follows from theorem 1.2 that in this case optimal and ε-optimal controls can be found among non-randomized controls. However for arbitrary $\mathfrak{A}'_{n+1}$ it is difficult to settle the question of the existence of an optimal control and the methods for its determination. Below we describe a rather general class of controlled random processes with incomplete observations. A study of these processes reduces to a study of a controlled process whose characteristics are constructed from the characteristics of the initial process. This process is of the form which was discussed in Section 3.

Assume that the phase space $(X, \mathfrak{A})$ of the basic process is representable as a product of two spaces $(Y \times Z, \mathfrak{A}_Y \times \mathfrak{A}_Z)$, i.e. a point $x \in X$ is determined by a pair of components $x = (y; z)$, $y \in Y$, $z \in Z$. The component y of point x is the unobservable component and component z is the observable one. The control may depend only on the observable components of the basic process.

A controlled object in this case is conveniently defined by means of a sequence of pairs of conditional distributions:

$$\left.\begin{array}{l} p_k^{(1)}(dy_k/y_0, \ldots, y_{k-1}; z_0, \ldots, z_k; u_0, \ldots, u_{k-1}), \\ p_k^{(2)}(dz_k/y_0, \ldots, y_{k-1}; z_0, \ldots, z_{k-1}; u_0, \ldots, u_{k-1}). \end{array}\right\} \tag{1.32}$$

Distributions $p_k(\cdot/\cdot)$ which define the controlled object are expressed in terms $p_k^{(1)}$ and $p_k^{(2)}$ via the formula

$$\begin{aligned} & p_k(A \times B/(y_0, z_0), \ldots, (y_{k-1}, z_{k-1}), u_0, \ldots, u_{k-1}) \\ &\quad = \int_B p_k^{(1)}(A/y_0, \ldots, y_{k-1}; z_0, \ldots, z_k; u_0, \ldots, u_{k-1}) \\ &\quad\quad \times p_k^{(2)}(dz_k/y_0, \ldots, y_{k-1}; z_0, \ldots, z_{k-1}; u_0, \ldots, u_{k-1}). \end{aligned} \tag{1.33}$$

Conversely if the conditional probabilities $p_k(\cdot/\cdot)$ which define the controlled object are given then

$$\begin{aligned} & p_k^{(2)}(B/y_0, \ldots, y_{k-1}; z_0, \ldots, z_{k-1}; u_0, \ldots, u_{k-1}) \\ &\quad = p_k(Y \times B/(y_0, z_0), \ldots, (y_{k-1}, z_{k-1}), u_0, \ldots, u_{k-1}); \end{aligned}$$

$p_k^{(1)}$ is determined from relation (1.33) as the density of a measure with respect to B which appears in the r.h.s. of (1.33) with respect to $p_k^{(2)}(B/\cdot)$.

Admissible controls v are determined by the conditional probabilities

$$q_k(du_k/z_0, \ldots, z_k, u_0, \ldots, u_{k-1}),$$

and non-randomized admissible controls are given by the sequence of functions $u_k = \varphi_k(z_0, \ldots, z_k)$.

We construct a new controlled object with phase space of the basic process $(Z, \mathfrak{A}_Z)$ for which for any non-randomized control v the distribution of the controlled sequence $(\xi_0^{(2)}, \eta_0), \ldots, (\xi_n^{(2)}, \eta_n), \ldots$ is the same as that of $(\bar{\xi}_0^{(2)}, \bar{\eta}_0), \ldots, (\bar{\xi}_n^{(2)}, \bar{\eta}_n)$ where $((\bar{\xi}_0^{(1)}, \bar{\xi}_0^{(2)}), \bar{\eta}_0), \ldots, ((\bar{\xi}_n^{(1)}, \bar{\xi}_n^{(2)}), \bar{\eta}_n), \ldots$ is a controlled sequence with a controlled object defined by the equalities (1.32) and the same control v.

Set for $E_0, \ldots, E_k \in \mathfrak{A}_Z$

$$\begin{aligned} & R_k(u_0, \ldots, u_{k-1}, E_0, \ldots, E_k) \\ &\quad = \int_{E_0} p_0^{(2)}(dz_0) \int_Y p_0^{(1)}(dy_0/z_0) \int_{E_1} p_1^{(2)}(dz_1/y_0, z_0, u_0) \\ &\quad\quad \times \int_Y p_1^{(1)}(dy_1/y_0, z_0, z_1, u_0) \\ &\quad \cdots \int_{E_k} p_k^{(2)}(dz_k/y_0, \ldots, y_{k-1}, z_0, \ldots, z_{k-1}, u_0, \ldots, u_{k-1}). \end{aligned} \tag{1.34}$$

Consider the measure r_k on $\mathfrak{A}_Z^k$: if $C = E_0 \times E_1 \times \cdots \times E_{k-1}$, then

$$r_k(E_k/C, u_0, \ldots, u_{k-1}) = R_k(u_0, \ldots, u_{k-1}, E_0, \ldots, E_k).$$

This measure depends on $E_k \in \mathfrak{A}_Z$ and $u_0, \ldots, u_{k-1}$ as on parameters. Let

$$r_k^*(C; u_0, \ldots, u_{k-2}) = R_{k-1}(u_0, \ldots, u_{k-2}, E_0, \ldots, E_{k-1}),$$

if $C \in \mathfrak{A}_Z^k$, is of the form given above. Since

$$\begin{aligned} r_k(E_k/C, u_0, \ldots, u_{k-1}) &\le r_k(Z/C, u_0, \ldots, u_{k-1}) \\ &= r_k^*(C; u_0, \ldots, u_{k-1}), \end{aligned}$$

the measure r_k is absolutely continuous with respect to r_k^*. Denote by

$$\bar{p}_k(E_k/z_0, \ldots, z_{k-1}, u_0, \ldots, u_{k-1})$$

the density of r_k with respect to r_k^*. This is a $\mathfrak{A}_Z^k \times \mathfrak{B}^k$-measurable function. Assume furthermore that $\bar{p}_k(E_k/\cdot)$ as a function E_k is a measure on $\mathfrak{A}_Z$; this is valid for example if Z is a separable complete metric space and $\mathfrak{A}_Z$ is a σ-algebra of Borel sets (cf. [20], Vol I, p. 36, theorem 3).

Consider a controlled object $\bar{\mu}(\cdot/\cdot)$ with the phase space of the basic process $(Z, \mathfrak{A}_Z)$ and the phase space of controls $(U, \mathfrak{B})$ defined by a sequence of probabilities $\bar{p}_k(\cdot/\cdot)$. Now let v defined by the sequence $\{u_k = \varphi_k(z_0, \ldots, z_k), k = 0, 1, \ldots\}$ be an arbitrary non-randomized control. Consider the joint distribution of the variables $\xi_0^{(2)}, \ldots, \xi_n^{(2)}$ under this control and the controlled object defined by the conditional probabilities (1.32) ($\xi_k = (\xi_k^{(1)}, \xi_k^{(2)})$ is the state of the basic process at time k, $\xi_k^{(2)}$ is the observable component at that time). Clearly for any bounded $\mathfrak{A}_Y^{n+1} \times \mathfrak{A}_Z^{n+1} \times \mathfrak{B}^{n+1}$-measurable function $f(y_0, \ldots, y_n, z_0, \ldots, z_n, u_0, \ldots, u_n)$ the relation

$$\begin{aligned} &\mathbf{E}_v f(\xi_0^{(1)}, \ldots, \xi_n^{(1)}, \xi_0^{(2)}, \ldots, \xi_n^{(2)}, \eta_0, \ldots, \eta_n) \\ &\quad = \iint f(y_0, \ldots, y_n, z_0, \ldots, z_n, \varphi_0(z_0), \ldots, \varphi_n(z_0, \ldots, z_n)) \\ &\qquad \times p_0^{(2)}(dz_0)p_0^{(1)}(dy_0/z_0), \ldots, p_0^{(2)}(dz_n/y_0, \ldots, y_{n-1}, z_0, \\ &\qquad \ldots, z_{n-1}, \varphi_0(z_0), \ldots, \varphi_n(z_0, \ldots, z_n))p_0^{(1)}(dy_n/y_0, \ldots, y_{n-1}, z_0, \\ &\qquad \ldots, z_n, \varphi_0(z_0), \ldots, \varphi_n(z_0, \ldots, z_n)) \end{aligned}$$

is fulfilled. Therefore choosing in place of function f the function

$$\prod_{k=0}^{n} \chi_{E_k}(z_k),$$

we obtain

$$\mathbf{P}_\nu\{\xi_0^{(2)} \in E_0, \ldots, \xi_n^{(2)} \in E_n\} = \int_{E_0} p_0^{(2)}(dz_0) \int_Y p_0^{(1)}(dy_0/z_0)$$

$$\times \int_{E_1} p_1^{(2)}(dz_1/y_0, z_0, \varphi_0(z_0)) \int_Y p_1^{(1)}(dy_1/y_0, z_0, z_1, \varphi_0(z_0)) \times$$

$$\cdots \times \int_{E_n} p_n^{(2)}(dz_n/y_0, \ldots, y_{n-1}, z_0, \ldots, z_{n-1}, \varphi_0(z_0), \ldots,$$

$$\varphi_{n-1}(z_0, \ldots, z_{n-1})). \tag{1.35}$$

Hence

$$\mathbf{P}_\nu\{\xi_n^{(2)} \in E_n/\xi_0^{(2)}, \ldots, \xi_{n-1}^{(2)}\}_{\xi_k^{(2)}=z_k, k=0, \ldots, n-1}$$
$$= \bar{p}_n(E_n/z_0, \ldots, z_{n-1}, \varphi_0(z_0), \ldots, \varphi_{n-1}(z_0, \ldots, z_{n-1})).$$

Now let $(\bar{\xi}_0^{(2)}, \bar{\eta}_0)$, $(\bar{\xi}_1^{(2)}, \bar{\eta}_1)$ be a controlled sequence with controlled object $\bar{\mu}(\cdot/\cdot)$ and the same control. Then

$$\mathbf{P}_\nu\{\bar{\xi}_n^{(2)} \in E_n/\bar{\xi}_0^{(2)}, \ldots, \bar{\xi}_{n-1}^{(2)}\}_{\bar{\xi}_k^{(2)}=z_k, k=0, \ldots, n-1}$$
$$= \bar{p}_n(E_n/z_0, \ldots, z_{n-1}, \varphi_0(z_0), \ldots, \varphi_{n-1}(z_0, \ldots, z_{n-1})).$$

We have thus proved that

$$\mathbf{P}_\nu\{\xi_0^{(2)} \in E_0, \ldots, \xi_n^{(2)} \in E_n\} = \mathbf{P}_\nu\{\bar{\xi}_0^{(2)} \in E_0, \ldots, \bar{\xi}_n^{(2)} \in E_n\}. \tag{1.36}$$

Now construct a linear mapping Γ of the set of functions $F(\mathbf{y}, \mathbf{z}, \mathbf{u})$ defined on $Y^N \times Z^N \times U^N$ and measurable with respect to $\mathfrak{A}_Y^N \times \mathfrak{A}_Z^N \times \mathfrak{B}^N$ into the set of functions defined on $Z^N \times U^N$ and measurable with respect to $\mathfrak{A}_Z^N \times \mathfrak{B}^N$ such that for any nonrandomized control ν the relation

$$\mathbf{E}_\nu F(\xi^{(1)}, \xi^{(2)}, \eta) = \mathbf{E}_\nu \Gamma F(\xi^{(2)}, \eta) \tag{1.37}$$

is satisfied.

Formula (1.37) in a certain sense uniquely determines the operator Γ. First assume that the function F is of the form

$$F(\mathbf{y}, \mathbf{z}, \mathbf{u}) = f_k(y_0, \ldots, y_k, z_0, \ldots, z_k, u_0, \ldots, u_k),$$

and the control ν is defined by a sequence of functions $\{u_n = \varphi_n(z_0, \ldots, z_k), n = 0, \ldots\}$. Then

$$\mathbf{E}_\nu F(\xi^{(1)}, \xi^{(2)}, \eta) = \mathbf{E}_\nu \mathbf{E}(F(\xi^{(1)}, \xi^{(2)}, \eta)/\xi^{(2)}, \eta).$$

Thus in order that formula (1.37) be valid it is necessary that

$$\Gamma F(\xi^{(2)}, \eta) = \mathbf{E}(F(\xi^{(1)}, \xi^{(2)}, \eta)/\xi^{(2)}, \eta). \tag{1.38}$$

The r.h.s. in general depends on the choice of control v. The l.h.s. should not depend on it. If one can actually choose a conditional mathematical expectation in (1.38) such that it would not depend on v, then determining Γ by means of relation (1.38) we shall achieve that (1.37) will be fulfilled. If (1.38) is fulfilled for any control then it is also fulfilled for a fixed sequence of controls $u_0, \ldots, u_k$. Hence for F of the form indicated above the relation

$$\Gamma F(\mathbf{z}, \mathbf{u}) = \frac{p_0^{(2)}(dz_0) \int p_0^{(1)}(dy_0/z_0) \cdots \int f_k(y_0, \ldots, y_k, z_0, \ldots, z_k, u_0, \ldots, u_k) \times p_k^{(1)}(dy_k/y_0, \ldots, y_{k-1}, z_0, \ldots, z_k, u_0, \ldots, u_k)}{p_0^{(2)}(dz_0) \int p_0^{(1)}(dy_0/z_0) \cdots p_k^{(2)}(dz_k/y_0, \ldots, y_{k-1}, z_0, \ldots, z_{k-1}, u_0, \ldots, u_{k-1})} \tag{1.39}$$

is valid (here a density of measures in $dz_0 \times \cdots \times dz_k$ is written). Observe now that the quantity $\Gamma F(z_0, \ldots, z_k, \varphi_0(z_0), \ldots, \varphi_k(z_0, \ldots, z_k))$ is a density of a measure defined on the set $E_0 \times \cdots \times E_k$,

$$\int_{E_0} p_0^{(2)}(dz_0) \int p_0^{(1)}(dy_0/z_0) \cdots \int f_k(y_0, \ldots, y_k, z_0, \ldots, z_k, \varphi_0(z_0), \ldots, \varphi_k(z_0, \ldots, z_k)) \times p_k^{(1)}(dy_k/y_0, \ldots, y_{k-1}, z_0, \ldots, z_{k-1}, \varphi_0(z_0), \ldots, \varphi_k(z_0, \ldots, z_{k-1}))$$

with respect to the same measure but with $f_k = 1$ (this follows from formula (1.39) if we substitute there $u_i = \varphi_i(z_0, \ldots, z_i)$. Thus for functions F which depend only on a finite number of coordinates the l.h.s. of (1.39) may be chosen to be independent of the choice of the control v given that ΓF in this equality is determined by means of relation (1.39). Note that (1.39) yields a definition of ΓF which does not depend on the representation of function F as a function in the coordinates.

Indeed, let $F = f_k(y_0, \ldots, y_k, z_0, \ldots, z_k, u_0, \ldots, u_k)$ and let f_k be actually independent of the coordinates y_k, z_k and u_k : $f_k(y_0, \ldots, y_k, z_0, \ldots, z_k, u_0, \ldots, u_k) = f_{k-1}(y_0, \ldots, y_{k-1}, z_0, \ldots, z_{k-1}, u_0, \ldots, u_{k-1})$. Then

$$\Gamma F(\mathbf{z}, \mathbf{u}) = \frac{p_0^{(2)}(dz_0) \int p_0^{(1)}(dy_0/z_0) \cdots \int f_{k-1}(y_0, \ldots, y_{k-1}, z_0, \ldots, z_{k-1}, u_0, \ldots, u_{k-1}) p_{k-1}^{(1)}(dy_{k-1}/y_0, \ldots, y_{k-2}, z_0, \ldots, z_{k-1}, u_0, \ldots, u_{k-1})}{p_0^{(2)}(dz_0) \int p_0^{(1)}(dy_0/z_0) \cdots p_{k-1}^{(2)}(dz_{k-1}/y_0, \ldots, y_{k-2}, z_0, \ldots, z_{k-2}, u_0, \ldots, u_{k-2})},$$

since integrating in the numerator of (1.39) with respect to dy_k we obtain $p_k^{(1)}(Y/y_0, \ldots, y_{k-1}, z_0, \ldots, z_k, u_0, \ldots, u_k) = 1$ and the terms $p_k^{(2)}(dz_k/y_0, \ldots, y_{k-1}, z_0, \ldots, z_{k-1}, u_0, \ldots, u_{k-1})$ appearing in both numerator and denominator of (1.39) may be cancelled.

ΓF is defined on the linear space of all functions F which depend only on a finite number of coordinates. Utilizing the fact that $\Gamma F \geq 0$ for F one can extend—using monotonicity—the operator Γ to all $\mathfrak{A}_Y^N \times \mathfrak{A}_Z^N \times \mathfrak{B}^N$-measurable functions bounded from below.

Thus to each controlled object $\mu(\cdot/\cdot)$ with incomplete data and cost function $F(\mathbf{y}, \mathbf{z}, \mathbf{u})$ there corresponds a controlled object $\bar{\mu}(\cdot/\cdot)$ whose phase space is the phase space of the observable component Z and a cost function $\Gamma F(\mathbf{z}, \mathbf{u})$ defined on $Z^N \times U^N$ such that for any non-randomized control v relation (1.37) is satisfied. Therefore an optimal (ε-optimal) control for object $\bar{\mu}(\cdot/\cdot)$ with functional of the cost of controlling $\Gamma F(\mathbf{z}, \mathbf{u})$ will be an optimal (ε-optimal) admissible control for the initial controlled object with incomplete data $\mu(\cdot/\cdot)$ and the functional of the cost of controlling $F(\mathbf{y}, \mathbf{z}, \mathbf{u})$.

If the transformed controlled object and the functional of the cost of controlling possess regularity conditions stipulated in theorem 1.6, then this theorem may be used for the determination of an optimal admissible control with incomplete data as well.

We present a sufficient condition for such a regularity.

Lemma 1.10. *Let $(Y, \mathfrak{A}_Y)$, $(Z, \mathfrak{A}_Z)$ be separable complete metric spaces with σ-algebras of Borel sets, $(U, \mathfrak{B})$ be a compact space with a σ-algebra of Borel sets. Let the following conditions be satisfied:*

(1) there exists in $(Z, \mathfrak{A}_Z)$ a countably-finite measure $m(dz)$ such that for all $k, y_0, \ldots, y_{k-1}, z_0, \ldots, z_{k-1}, u_0, \ldots, u_{k-1}$ the measures $p_k^{(2)}(\cdot/y_0, \ldots, y_{k-1}, z_0, \ldots, z_{k-1}, u_0, \ldots, u_{k-1})$ are absolutely continuous with respect to measure m and the density $p_k^{(2)}(\cdot/\cdot)$ with respect to m

$$p_k(z_k/y_0, \ldots, y_{k-1}, z_0, \ldots, z_{k-1}, u_0, \ldots, u_{k-1})$$

is bounded and continuous jointly in the variables; (2) for all k and any bounded continuous function $g(y_0, \ldots, y_k, z_0, \ldots, z_{k-1}; u_0, \ldots, u_{k-1})$ the function

$$\int g(y_0, \ldots, y_k, z_0, \ldots, z_{k-1}, u_0, \ldots, u_{k-1}) \times p_k^{(1)}(dy_k/y_0, \ldots, y_{k-1}, z_0, \ldots, z_k, u_0, \ldots, u_{k-1}).$$

is continuous jointly in the variables. Then:

A. *for any continuous bounded function $g(z)$ the function*

$$\int g(z)\bar{p}_k(dz/z_0, \ldots, z_{k-1}, u_0, \ldots, u_{k-1})$$

is bounded and continuous.

B. *for any lower semicontinuous function $F(\mathbf{y}, \mathbf{z}, \mathbf{u})$ bounded from below the function $\Gamma F(\mathbf{z}, \mathbf{u})$ is bounded from below and semicontinuous.*

PROOF. In view of proposition 1

$$\int g(z)\bar{p}_k(dz/z_0, \ldots, z_{k-1}, u_0, \ldots, u_{k-1})$$
$$= \int p_0^{(1)}(dy_0/z_0)p_1(z_1/z_0, y_0, u_0) \cdots$$
$$\cdots \int p_{k-1}^{(1)}(dy_{k-1}/y_0, \ldots, y_{k-2}, z_0, \ldots, z_{k-1}, u_0, \ldots, u_{k-2})$$
$$= \frac{\times \int g(z)p_k^{(2)}(dz/y_0, \ldots, y_{k-1}, z_0, \ldots, z_{k-1}, u_0, \ldots, u_{k-1})}{\int p_0^{(1)}(dy_0/z_0)p_1(z_1/z_0, y_0, u_0) \cdots \cdots p_{k-1}^{(2)}(z_{k-1}/y_0, \ldots, y_{k-2}, z_0, \ldots, z_{k-2}, u_0, \ldots, u_{k-2})}.$$

The boundedness of this function follows from the fact that $\bar{p}_k(Z/\cdots) = 1$ and its continuity from the conditions of the lemma. Assertion A is thus proved.

Furthermore under condition 1 the function $\Gamma F(\mathbf{z}, \mathbf{u})$ can be explicitly written as:

$$\Gamma F(\mathbf{z}, \mathbf{u}) = \frac{\begin{aligned}&\int p_0^{(1)}(dy_0/z_0)p_1(z_1/z_0, y_0, u_0) \cdots p_k(z_k/y_0, \ldots, y_{k-1}, z_0, \\ &\ldots, z_{k-1}, u_0, \ldots, u_{k-1}) \int f_k(y_0, \ldots, y_k, z_0, \ldots, z_k, u_0, \ldots, u_k) \\ &\times p_k^{(1)}(dy_k/y_0, \ldots, y_{k-1}, z_0, \ldots, z_{k-1}, u_0, \ldots, u_{k-1})\end{aligned}}{\begin{aligned}&\int p_0^{(1)}(dy_0/z_0)p_1(z_1/z_0, y_0, u_0) \cdots \\ &\cdots p_k(z_k/y_0, \ldots, y_{k-1}, z_0, \ldots, z_{k-1}, u_0, \ldots, u_{k-1})\end{aligned}}.$$

provided

$$F(\mathbf{y}, \mathbf{z}, \mathbf{u}) = f_k(y_0, \ldots, y_k, z_0, \ldots, z_k, u_0, \ldots, u_k).$$

Therefore $\Gamma F(\mathbf{z}, \mathbf{u})$ is continuous and bounded ($\|\Gamma\| < 1$) for any continuous and bounded function $F(\mathbf{y}, \mathbf{z}, \mathbf{u})$. To complete the proof of assertion B it is sufficient to utilize the corollary of lemma 1.7. □

5 Optimal Stopping Problems

One of the most important problems of optimal control which admits a solution in the general case is the problem of choosing an optimal value by a random sequence.

Let a random sequence $\xi_1, \ldots, \xi_n, \ldots$ be observed. The process of observing the sequence can be stopped at any instant of time but it is desirable that at the stopping time the value of the observed random variable be as large as possible. How should we select the stopping time? A more precise statement of the problem is as follows.

Let $\{\Omega, \mathfrak{S}, \mathbf{P}\}$ be a probability space, $\{\xi_k = \xi_k(\omega), k = 1, 2, \ldots\}$ be a sequence of random variables on this space. Denote by $\mathfrak{F}_n (n = 1, 2, \ldots)$ the

σ-algebra generated by random variables $\xi_1, \xi_2, \ldots, \xi_n$. The random variables τ taking on values $n = 1, 2, \ldots$ such that $\{\tau = n\} \in \mathfrak{F}_n$ are called *stopping times* (*stopping rules*). It is required to find $\sup_\tau \xi_\tau$ where the supremum taken over all the stopping times, and those τ at which this supremum is attained.

It would seem natural that a solution to this problem could be based on the following consideration: if the sequence $\{\xi_n\}$ is not stopped before time n and

$$\xi_n < \sup_{\tau \in T_n} \mathbf{E}\{\xi_\tau/\mathfrak{F}_n\}, \tag{1.40}$$

where T_n is the class of all stopping times τ such that $n \le \tau < \infty$, then it is not desirable to stop at time n. If, however

$$\xi_n \ge \sup_{\tau \in T_n} \mathbf{E}\{\xi_\tau/\mathfrak{F}_n\},$$

then it is advisable to stop at time n.

We now show that these intuitive considerations indeed lead us to a solution of the problem.

Let a sequence of random variables $\xi_0, \xi_1, \ldots, \xi_N$ with finite expectations be given. Set $\eta_N = \xi_N$ and

$$\eta_k = \max[\xi_k, \mathbf{E}\{\eta_{k+1}/\mathfrak{F}_k\}], \qquad k = N-1, \ldots, 0.$$

Since $\mathbf{E}\{\eta_{k+1}/\mathfrak{F}_k\} \le \eta_k$, then η_k is a supermartingale relative to the current $\{\mathfrak{F}_k\}$, $\eta_k \ge \xi_k$. If η'_k is a supermartingale such that $\eta'_k \ge \xi_k$ then for $k = N$, $\eta'_k \ge \eta_k$; if, however, for some k, $\eta'_k \ge \eta_k$ then

$$\eta'_{k-1} \ge \mathbf{E}(\eta'_k/\mathfrak{F}_{k-1}) \ge \mathbf{E}(\eta_k/\mathfrak{F}_{k-1}),$$

$$\eta'_{k-1} \ge \xi_{k-1} \quad \text{and hence} \quad \eta'_{k-1} \ge \eta_{k-1}.$$

Thus the sequence $\{\eta_k, k = 0, \ldots, N\}$ is a minimal supermartingale dominating the sequence ξ_k.

We now define the stopping time $\tau_0 = \min(n: \xi_n = \eta_n, 0 \le n \le N)$. Note that $\eta_{\tau_0 \wedge n}$ is a martingale relative to the current $\mathfrak{F}_n$:

$$\begin{aligned} \mathbf{E}(\eta_{\tau_0 \wedge n+1}/\mathfrak{F}_n) &= \mathbf{E}(\eta_{\tau_0 \wedge n+1}\chi_{\{\tau_0 > n\}}/\mathfrak{F}_n) \\ &\quad + \mathbf{E}(\eta_{\tau_0 \wedge n+1}\chi_{\{\tau_0 \le n\}}/\mathfrak{F}_n) \\ &= \mathbf{E}(\eta_{n+1}/\mathfrak{F}_n)\chi_{\{\tau_0 > n\}} + \eta_{\tau_0 \wedge n}\chi_{\{\tau_0 \le n\}} = \eta_{\tau_0 \wedge n} \end{aligned}$$

(if $\tau_0 > n$, then $\xi_n < \eta_n$ and $\eta_n = \mathbf{E}(\eta_{n+1}/\mathfrak{F}_n)$.) Since $\eta_{\tau_0 \wedge n}$ is a martingale it follows that

$$\eta_0 = \mathbf{E}\{\eta_{\tau_0}/\mathfrak{F}_0\} = \mathbf{E}\{\xi_{\tau_0}/\mathfrak{F}_0\}.$$

However for any stopping time τ

$$\eta_0 \ge \mathbf{E}\{\eta_\tau/\mathfrak{F}_0\} \ge \mathbf{E}\{\xi_\tau/\mathfrak{F}_0\}.$$

Thus τ_0 is an optimal stopping time and

$$\eta_0 \geq \mathbf{E}\{\xi_\tau/\mathfrak{F}_0\} \qquad (\text{mod } \mathbf{P}), \qquad \forall\tau,$$

$$\eta_0 = \mathbf{E}\{\xi_{\tau_0}/\mathfrak{F}_0\}.$$

We have obtained a solution of the problem in the case of a finite sequence. The case of an infinite sequence involves several difficulties.

To begin with it is necessary to clarify the meaning of the variable appearing in the r.h.s. of (1.40). In the first place the random variables $\mathbf{E}\{\xi_\tau/\mathfrak{F}_n\}$ are defined only (mod **P**) and therefore their upper bound is not defined; moreover the class T_n is in general uncountable and therefore $\sup_{\tau \in T_n} f_\tau(\omega)$ where $f_\tau(\omega)$ is a $\mathfrak{S}$-measurable function, may turn out to be non-measurable. To overcome these difficulties we shall consider below the *essential supremum*

$$\operatorname*{ess\,sup}_{\tau \in T_n} \mathbf{E}\{\xi_\tau/\mathfrak{F}_n\},$$

and we shall define this notion in the following manner.

Let a set B of $\mathfrak{S}$-measurable functions $f = f(\omega)$, $\omega \in \Omega$, be given. A $\mathfrak{S}$-measurable function $g(\omega)$ possessing the properties:

1. $g(\omega) \geq f(\omega) \qquad (\text{mod } \mathbf{P}), \qquad \forall f \in B;$
2. if the $\mathfrak{S}$-measurable function $h(\omega)$ is such that

$$h(\omega) \geq f(\omega) \qquad (\text{mod } \mathbf{P}) \quad \forall f \in B,$$

then

$$h(\omega) \geq g(\omega) \qquad (\text{mod } \mathbf{P})$$

is called the *essential supremum* of the set B and is denoted by

$$g(\omega) = \text{ess sup } B.$$

We show that for an arbitrary family B of real-valued functions $f(\omega)$ the quantity ess sup B exists (and may take the value $+\infty$) and is unique (mod **P**).

The uniqueness is evident. If there were two functions $g_1(\omega)$ and $g_2(\omega)$ possessing properties 1 and 2 then setting $h = g_1$, $g = g_2$ we obtain $g_2(\omega) \geq g_1(\omega)$ (mod **P**). Interchanging indices 1 and 2 we arrive at $g_1(\omega) \geq g_2(\omega)$ (mod **P**) whence $g_1(\omega) = g_2(\omega)$ (mod **P**). To prove the existence of the upper bound we first observe that it may be assumed without loss of generality that $\|f(\omega)\| \leq 1$ otherwise we could have considered the family of functions $\{2/\pi \text{ arctg } f, f \in B\}$.

For any countable sequence $G = \{f_k(\omega), k = 1, \ldots, n, \ldots\}$ we set $f_G(\omega) = \sup_k f_k(\omega)$ and let $\alpha = \sup_{G \subset B} \mathbf{E} f_G$. There exists a sequence G_n such that $\alpha = \lim \mathbf{E} f_{G_n}$ where $G_n \subset B$. Let $G^* = \bigcup_n G_n$. We show that $\text{ess sup}_{f \in B} f = f_{G^*}$.

Choose an arbitrary function $f \in B$. If $f(\omega) > f_{G^*}(\omega)$ on a set of positive measure then $\mathbf{E}(f \vee f_{G^*}) > \alpha$. Since $f \vee f_{G^*} = f_{G_1}$, where $G_1 = \{f\} \cup G^*$ is a countable set of functions, it follows that $\mathbf{E}f_{G_1} > \alpha$ which contradicts the definition of α. Thus $f_{G^*} \geq f \pmod{\mathbf{P}}$, $\forall f \in B$. If however $h(\omega) \geq f(\omega)\ \forall f \in B$ for some $h(\omega)$, it then follows from the definition of f_{G^*} that $h(\omega) \geq f(\omega)$ (mod **P**). Therefore the function f_{G^*} satisfies the definition of the essential supremum of a set of functions B. We have thus proved the following theorem

Theorem 1.7. *An arbitrary set of real-valued measurable functions possesses a unique measurable essential supremum.*

Remark 1.9. If the set B possesses the property: for arbitrary two functions f_1 and f_2 belonging to B there exists a function $f_3 \in B$ such that $f_3 \geq f_1 \vee f_2$, then clearly there exists a monotone non-decreasing sequence $f_n \in B$ such that

$$\text{ess sup } B = \sup\{f_n\} \qquad (\text{mod } \mathbf{P}). \tag{1.41}$$

The following theorem presents a solution for the optimal stopping problem.

Theorem 1.8. *Let* $\{\xi_n,\ n = 1, 2, \ldots\}$ *be a sequence of random variables with* $\mathbf{E}\sup|\xi_n| < \infty$. *Then the random variables*

$$\eta_n = \operatorname*{ess\ sup}_{\tau \in T_n} (\mathbf{E}\{\xi_\tau/\mathfrak{F}_n\}) \tag{1.42}$$

are integrable and satisfy the relation

$$\eta_n = \max(\xi_n, \mathbf{E}\{\eta_{n+1}/\mathfrak{F}_n\}) \qquad (\text{mod } \mathbf{P}), \qquad n = 1, 2, \ldots \tag{1.43}$$

Proof. Since

$$\operatorname*{ess\ sup}_{\tau \in T_n} \mathbf{E}\{\xi_\tau/\mathfrak{F}_n\} \leq \mathbf{E}\{\operatorname*{ess\ sup}_{\tau \in T_n} \xi_\tau/\mathfrak{F}_n\} < \mathbf{E}\{\operatorname*{ess\ sup}_{k \geq n} \xi_k/\mathfrak{F}_n\},$$

it follows that $\mathbf{E}$ ess sup $\mathbf{E}\{\xi_\tau/\mathfrak{F}_n\} \leq \mathbf{E}(\sup \xi_n) < \infty$ by the condition of the theorem. Thus the variables η_n are integrable. For any $\tau \in T_n$ we set

$$\xi_\tau = \xi_n \chi_{(\tau = n)} + \xi_{\tau_1} \chi_{(\tau > n)},$$

where $\tau_1 = \tau \vee (n + 1) \in T_{n+1}$. Therefore

$$\mathbf{E}\{\xi_\tau/\mathfrak{F}_n\} = \xi_n \chi_{(\tau = n)} + \mathbf{E}\{\xi_{\tau_1}/\mathfrak{F}_n\}\chi_{(\tau > n)} = \xi_n \chi_{(\tau = n)}$$
$$+ \mathbf{E}\{\mathbf{E}\{\xi_{\tau_1}/\mathfrak{F}_{n+1}\}/\mathfrak{F}_n\}\chi_{(\tau > n)} \leq \xi_n \chi_{(\tau = n)} + \mathbf{E}\{\eta_{n+1}/\mathfrak{F}_n\}\chi_{(\tau > n)}.$$

This shows that for any τ

$$\mathbf{E}\{\xi_\tau/\mathfrak{F}_n\} \leq \max[\xi_n, \mathbf{E}\{\eta_{n+1}/\mathfrak{F}_n\}] \qquad (\text{mod } \mathbf{P}).$$

Therefore,

$$\eta_n = \text{ess sup } \mathbf{E}\{\xi_\tau/\mathfrak{F}_n\} \leq \max[\xi_n, \mathbf{E}\{\eta_{n+1}/\mathfrak{F}_n\}].$$

To verify the inequality

$$\eta_n \geq \max[\xi_n, \mathbf{E}\{\eta_{n+1}/\mathfrak{F}_n\}],$$

we first observe that the set of random variables $(\mathbf{E}\{\xi_\tau/\mathfrak{F}_n\}, \tau \in T_n\}$ is closed relative to the operation of taking the supremum. Indeed let

$$A = \{\omega : \mathbf{E}(\xi_{\tau_1}/\mathfrak{F}_n) < \mathbf{E}(\xi_{\tau_2}/\mathfrak{F}_n)\}, \qquad \tau_i \in T_n, i = 1, 2.$$

The set A is $\mathfrak{F}_n$-measurable. Set $\tau_3 = \tau_1$ on $\bar{A} = \Omega - A$ and $\tau_3 = \tau_2$ on A. The variable τ_3 is a stopping time:

$$\{\tau_3 \le k\} = (\{\tau_1 \le k\} \cap \bar{A}) \cup (\{\tau_2 \le k\} \cap A) \in \mathfrak{F}_k, \qquad \forall k \ge n.$$

However

$$\mathbf{E}\{\xi_{\tau_3}/\mathfrak{F}_n\} = \mathbf{E}\{\xi_{\tau_1}/\mathfrak{F}_n\}\chi_{(\bar{A})} + \mathbf{E}\{\xi_{\tau_2}/\mathfrak{F}_n\}\chi_{(A)} = \max_{i=1,2} \mathbf{E}\{\xi_{\tau_i}/\mathfrak{F}_n\}.$$

By virtue of the Remark 1.9 there exists a sequence $\tau_k \in T_{n+1}$, $k = 1, 2, \ldots,$ such that

$$\text{ess sup } \mathbf{E}\{\xi_\tau/\mathfrak{F}_{n+1}\} = \lim \mathbf{E}\{\xi_{\tau_k}/\mathfrak{F}_{n+1}\}. \tag{1.44}$$

For any $B \in \mathfrak{F}_n$ we set $\tau_k' = n\chi_{(B)} + \tau_k \chi_{(\bar{B})}$. Then

$$\eta_n = \operatorname*{ess\,sup}_{\tau \in T_n} \mathbf{E}\{\xi_\tau/\mathfrak{F}_n\} \ge \sup_k \{\xi_n \chi_{(B)} + \mathbf{E}\{\xi_{\tau_k'}/\mathfrak{F}_n\}\chi_{(\bar{B})}\}$$

$$\ge \xi_n \chi_{(B)} + \mathbf{E}\{\eta_{n+1}/\mathfrak{F}_n\}\chi_{(\bar{B})}.$$

Since the set B is an arbitrary set in $\mathfrak{F}_n$, we obtain

$$\eta_n \ge \max[\xi_n, \mathbf{E}\{\eta_{n+1}/\mathfrak{F}_n\}].$$ □

Corollary. *If ξ_n are non-negative, $n = 1, 2, \ldots$, then $\{\eta_n\}$ is a minimal supermartingale which dominates the sequence $\{\xi_n, n = 1, 2, \ldots\}$.*

Indeed, if ξ_n are non-negative so are, by definition, η_n, and the relation (1.43) shows that $\{\eta_n\}$ is a supermartingale dominating $\{\xi_n\}$. If $\{\eta_n'\}$ is another supermartingale dominating sequence $\{\xi_n\}$ then for any $\tau \in T_n$

$$\eta_n' \ge \mathbf{E}\{\eta_\tau'/\mathfrak{F}_n\} \ge \mathbf{E}\{\xi_\tau/\mathfrak{F}_n\},$$

whence $\eta_n' \ge \operatorname*{ess\,sup}_{\tau \in T_n} \mathbf{E}\{\xi_\tau/\mathfrak{F}_n\} = \eta_n$ which proves the assertion.

We introduce the stopping time τ_0:

$$\tau_0 = \begin{cases} \inf\{n : \eta_n = \xi_n\}, \\ \infty \qquad \text{if } \eta_n > \xi_n, \quad \forall n. \end{cases} \tag{1.45}$$

Recall that as it follows from the definition, $\eta_n \ge \xi_n$ (mod $\mathbf{P}$) $\forall n$. In view of the above one would expect that τ_0 is an optimal stopping time. We shall investigate the necessary conditions for this to be true.

Let τ be an arbitrary stopping time satisfying the condition $\tau \le \tau_0$. Consider the sequence $\{\eta_{\tau \wedge n}, n = 1, 2, \ldots\}$. We show that this sequence is a martingale.

Indeed, $\{\tau_0 > n\}$ implies that $\xi_n < \eta_n$ and $\eta_n = \mathbf{E}\{\eta_{n+1}/\mathfrak{F}_n\}$. Furthermore, we have

$$\begin{aligned}\mathbf{E}\{\eta_{\tau\wedge(n+1)}/\mathfrak{F}_n\} &= \mathbf{E}\{\eta_{\tau\wedge(n+1)}\chi_{(\tau\le n)} + \eta_{\tau\wedge(n+1)}\chi_{(\tau>n)}/\mathfrak{F}_n\}\\ &= \eta_\tau\chi_{(\tau\le n)} + \chi_{(\tau>n)}\mathbf{E}\{\eta_{n+1}/\mathfrak{F}_n\} = \eta_\tau\chi_{(\tau\le n)} + \eta_n\chi_{(\tau>n)} = \eta_{\tau\wedge n},\end{aligned}$$

since $\{\tau > n\}$ implies $\{\tau_0 > n\}$. Thus $\eta_{\tau\wedge n}$ is a martingale. Consequently, $\mathbf{E}\eta_{\tau\wedge n} = \mathbf{E}\eta_0$. Since $|\eta_n| \le \mathbf{E}(\sup_n|\xi_n|/\mathfrak{F}_n)$ the sequence $\{\eta_n\}$ is uniformly integrable and hence so is the sequence $\{\eta_{\tau\wedge n}, n = 1, 2, \ldots\}$. To verify this observe that

$$\int_{\{\eta_{\tau\wedge n}>N\}} \eta_{\tau\wedge n}\,d\mathbf{P} = \sum_{k=1}^{n-1}\int_{\{\eta_{\tau\wedge n}>N\}\cap\{\tau=k\}} \eta_k\,d\mathbf{P} + \int_{\{\eta_{\tau\wedge n}>N\}\cap\{\tau\ge n\}} \eta_n\,d\mathbf{P}.$$

Since the event

$$\{\eta_{\tau\wedge n} > N\} \cap \{\tau = k\} = \{\eta_k > N\} \cap \{\tau = k\}$$

is $\mathfrak{F}_k$-measurable $(k = 1, 2, \ldots, n)$ it follows that

$$\begin{aligned}\int_{\{\eta_{\tau\wedge n}>N\}} \eta_{\tau\wedge n}\,d\mathbf{P} &= \sum_{k=1}^{n=1}\int_{\{\eta_{\tau\wedge n}>N\}\cap\{\tau=k\}} \eta_k\,d\mathbf{P} + \int_{\{\eta_{\tau\wedge n}>N\}\cap\{\tau\ge n\}} \eta_n\,d\mathbf{P}\\ &= \int_{\{\eta_{\tau\wedge n}>N\}} \eta_n\,d\mathbf{P} \le \int_{\{\eta_{\tau\wedge n}>N\}} |\eta_n|\,d\mathbf{P}.\end{aligned}$$

An analogous inequality is obtained when we consider integration over the set $\{\eta_{\tau\wedge n} < -N\}$. Thus,

$$\int_{\{|\eta_{\tau\wedge n}|>N\}} |\eta_{\tau\wedge n}|\,d\mathbf{P} \le \int_{\{|\eta_{\tau\wedge n}|>N\}} |\eta_n|\,d\mathbf{P}.$$

Since

$$\mathbf{P}\{|\eta_{\tau\wedge n}| > N\} \le \frac{\mathbf{E}|\eta_{\tau\wedge n}|}{N} \le \frac{\mathbf{E}(\sup|\xi_n|)}{N} \to 0$$

as $N \to \infty$ uniformly in n, the uniform integrability of $|\eta_n|$ implies that

$$\int_{\{|\eta_{\tau\wedge n}|>N\}} |\eta_{\tau\wedge n}|\,d\mathbf{P} \to 0 \quad \text{as } N \to \infty,$$

which proves the uniform integrability of the martingale $\eta_{\tau\wedge n}$. Therefore as $n \to \infty$ the equality $\mathbf{E}\{\eta_{\tau\wedge n}/\mathfrak{F}_0\} = \eta_0$ yields

$$\mathbf{E}\{\eta_\tau/\mathfrak{F}_0\} = \eta_0.$$

Assume that τ_0 is finite with probability 1. Then

$$\mathbf{E}\{\xi_{\tau_0}/\mathfrak{F}_0\} = \mathbf{E}\{\eta_{\tau_0}/\mathfrak{F}_0\} = \eta_0 = \sup_{\tau \in T_0} \mathbf{E}\{\xi_\tau/\mathfrak{F}_0\} \qquad (\text{mod } \mathbf{P}),$$

where T_0 is the class of all stopping times. Thus τ_0 is an optimal stopping time:

$$\mathbf{E}\{\xi_{\tau_0}/\mathfrak{F}_0\} = \operatorname*{ess\,sup}_{\tau \in T_0} \mathbf{E}\{\xi_\tau/\mathfrak{F}_0\} \qquad (\text{mod } \mathbf{P}). \tag{1.46}$$

Theorem 1.9. *In order that there exist an optimal stopping time it is necessary and sufficient that the stopping time τ_0 defined by equation* (1.45) *be finite with probability* 1. *In this case τ_0 is the optimal stopping time.*

Proof. The sufficiency of the stated condition for the existence of an optimal stopping rule was proven above. We now show that if an optimal stopping rule τ' exists (τ' is finite with probability 1) then τ_0 is also finite with probability 1.

For this purpose we shall verify that for any finite (mod $\mathbf{P}$) stopping time ν_1 the relation

$$\eta_{\nu_1} = \operatorname*{ess\,sup}_{\nu \in T_{\nu_1}} \mathbf{E}\{\xi_\nu/\mathfrak{F}_{\nu 1}\} \tag{1.47}$$

is valid where T_{ν_1} is the class of all finite stopping times such that $\nu_1 \le \nu$.

Set

$$\zeta = \operatorname*{ess\,sup}_{\nu \in T_{\nu_1}} \mathbf{E}\{\xi_\nu/\mathfrak{F}_{\nu_1}\}.$$

Let $\nu \in T_{\nu_1}$. Observe that $\nu \vee n \in T_n$ and that $\nu_1 = n$ implies $\nu \ge n$. Therefore

$$\mathbf{E}\{\xi_\nu/\mathfrak{F}_{\nu_1}\} = \mathbf{E}\{\xi_\nu/\mathfrak{F}_n\}_{n=\nu_1} = \mathbf{E}\{\xi_{\nu \vee n}/\mathfrak{F}_n\}_{n=\nu_1} \le \eta_n|_{n=\nu_1} = \eta_{\nu_1}.$$

Whence $\zeta \le \eta_{\nu_1}$, if however $\nu' \in T_n$ and $\{\nu_1 = n\}$ then $\nu' \vee \nu_1 \in T_{\nu_1}$ and $\nu' = \nu' \vee \nu_1$ respectively. Therefore for

$$\nu_1 = n$$

$$\mathbf{E}\{\xi_{\nu_1}/\mathfrak{F}_n\} = \mathbf{E}\{\xi_{\nu' \vee \nu_1}/\mathfrak{F}_n\},$$

$$\begin{aligned}\zeta \ge \mathbf{E}\{\xi_{\nu' \vee \nu_1}/\mathfrak{F}_{\nu_1}\} &= \mathbf{E}\{\xi_{\nu' \vee \nu_1}/\mathfrak{F}_n\}_{n=\nu_1} = \mathbf{E}\{(\chi_{\{\nu_1 = n\}} \\ &+ \chi_{\{\nu_1 \ne n\}})\xi_{\nu' \vee \nu_1}/\mathfrak{F}_n\}_{n=\nu_1} = \chi_{\{\nu_1 = n\}}\mathbf{E}\{\xi_{\nu' \vee n}/\mathfrak{F}_n\}_{n=\nu_1} \\ &= \mathbf{E}\{\xi_{\nu'}/\mathfrak{F}_n\}_{n=\nu_1} = \mathbf{E}\{\xi_{\nu'}/\mathfrak{F}_{\nu_1}\}.\end{aligned}$$

Thus on the set $\nu_1 = n$

$$\zeta \ge \operatorname*{ess\,sup}_{\nu' \in T_n} \mathbf{E}\{\xi_{\nu'}/\mathfrak{F}_n\} = \eta_n,$$

i.e. $\zeta \ge \eta_{\nu_1}$(mod $\mathbf{P}$). Equality (1.47) is thus proved.

Observe that the equality

$$\mathbf{E}\{\eta_{\nu_1}/\mathfrak{F}_0\} = \sup_{\nu \in T_{\nu_1}} \mathbf{E}\{\xi_\nu/\mathfrak{F}_0\}$$

is proved in the same manner.

Let τ' be an optimal stopping rule. Then

$$\mathbf{E}\{\eta_{\tau'}/\mathfrak{F}_0\} = \sup_{\nu \in T_{\nu'}} \mathbf{E}\{\xi_\nu/\mathfrak{F}_0\} \leq \sup_{\nu \in T_0} \mathbf{E}\{\xi_\nu/\mathfrak{F}_0\} = \mathbf{E}\{\xi_{\tau'}/\mathfrak{F}_0\}.$$

However $\eta_{\tau'} \geq \xi_{\tau'}$. Therefore $\eta_{\tau'} = \xi_{\tau'}$ (mod $\mathbf{P}$). Now from the definition of τ_0, $\tau_0 \leq \tau'$ and the stopping time τ_0 is almost everywhere finite. □

Theorem 1.10. *Let the assumptions of theorem* 1.9 *be satisfied. For any* $\varepsilon > 0$ *the variable* τ_ε, *defined by relation*

$$\tau_\varepsilon = \inf\{n : \eta_n < \xi_n + \varepsilon\},$$

defines an ε*-optimal stopping time, i.e.*

$$\sup_\nu \mathbf{E}\{\xi_\nu/\mathfrak{F}_0\} \leq \mathbf{E}\{\xi_{\tau_\varepsilon}\} + \varepsilon.$$

PROOF. First we show that

$$\overline{\lim_{n\to\infty}}\, \eta_n = \overline{\lim_{n\to\infty}}\, \xi_n. \tag{1.48}$$

Indeed, if $n \geq m$, then

$$\eta_n = \operatorname{ess\,sup}_{\nu \in T_n} \mathbf{E}\{\xi_\nu/\mathfrak{F}_n\} \leq \mathbf{E}\left\{\sup_{p \geq m} \xi_p/\mathfrak{F}_n\right\},$$

$$\overline{\lim}\, \eta_n \leq \mathbf{E}\left\{\sup_{p > m} \xi_p/\mathfrak{F}_n\right\} \leq \mathbf{E}(\overline{\lim}\, \xi_m/\mathfrak{F}_n).$$

Approaching n to infinity in the r.h.s. of the inequality above we obtain, in view of a well-known theorem on conditional mathematical expectations

$$\overline{\lim}\, \eta_n \leq \lim_n \mathbf{E}\{\overline{\lim}\, \xi_m/\mathfrak{F}_n\} = \overline{\lim}\, \xi_m.$$

Since $\eta_n \geq \xi_n$ the last equality immediately yields the validity of equation (1.48).

We now proceed to discuss the stopping rule τ_ε introduced above. We show that τ_ε is a finite (mod $\mathbf{P}$) random variable. Since $\tau_\varepsilon \leq \tau_0$, $\eta_{\tau_\varepsilon \wedge n}$ is an integrable martingale as it was shown above and moreover $\sup \mathbf{E}\,|\eta_{\tau_\varepsilon \wedge n}| < \infty$. Therefore there exists with probability 1 a finite limit $\lim_{n\to\infty} \eta_{\tau_\varepsilon \wedge n} = \eta_0$.

Consider the set $A : \{\tau_\varepsilon = \infty\}$. On this set $\eta_n \geq \xi_n + \varepsilon$, $\forall n$, and therefore $\overline{\lim}\, \eta_n \geq \overline{\lim}\, \xi_n + \varepsilon$, since on the set A $\lim \eta_n = \lim \eta_{\tau_\varepsilon \wedge n}$ exists (mod $\mathbf{P}$). In view of (1.48) the inequality obtained is possible if and only if $\mathbf{P}(A) = 0$. Thus $\tau_\varepsilon \to \infty$ (mod $\mathbf{P}$).

Utilizing now theorem 1.8 and the inequality $\eta_{\tau_\varepsilon} < \xi_{\tau_\varepsilon} + \varepsilon$ we obtain

$$\sup_{\nu \in T_0} \mathbf{E}\xi_\nu = \mathbf{E}\eta_0 = \mathbf{E}\eta_{\tau_\varepsilon} \leq \mathbf{E}\xi_{\tau_\varepsilon} + \varepsilon. \qquad \square$$

6 Controlled Markov Chains

Assume that conditional probabilities which define the probabilistic object are of the form

$$p_k(A_k/x_0, \ldots, x_{k-1}, u_0, \ldots, u_{k-1}) = P_k(x_{k-1}, A_k; u_{k-1}), \qquad k = 1, 2, \ldots. \tag{1.49}$$

Such a controlled object is *Markovian* (Markov) and controlled sequences with a Markovian controlled object are called *controlled Markov chains*. When considering Markov controlled objects the distribution $p_0(dx_0)$ is not fixed. Therefore a Markov controlled object is actually a family of controlled objects $\mu_{x_0}(\cdot/\cdot)$ which depends on x_0 as on a parameter. Measures $\mu_{x_0}(\cdot/\mathbf{u})$ on $\mathfrak{A}^N$ are defined by the equality

$$\mu_{x_0}(C/\mathbf{u}) = \int_{A_1} P_1(x_0, dx_1; u_0) \int_{A_2} P_2(x_1, dx_2; u_1) \cdots \int_{A_k} P_k(x_{k-1}, dx; u_{k-1}), \tag{1.50}$$

where $C = \{\mathbf{x} : x_1 \in A_1, \ldots, x_k \in A_k\}$. Formula (1.50) implies that for any $\mathbf{u}$ a measure $\mu_{x_0}(\cdot/\mathbf{u})$ corresponds to a Markov chain with one-step transition probability $P_k(x_{k-1}, A_k; u_{k-1})$ (on the k-th step). One can also define for this chain transition probabilities for several steps:

$$P(n, x, m, A; \mathbf{u}) = \int P_{n+1}(x, dx_{n+1}; u_n) \cdots P_m(x_{m-1}, A; u_{m-1}). \tag{1.51}$$

If $\mathfrak{B}_{n,k}$ denotes the σ-algebra of cylinders in $\mathfrak{B}^N$ with bases over $[n, k]$ then as it follows from expression (1.51), $P(n, x, m, A : \mathbf{u})$ is a $\mathfrak{B}_{n,m-1}$-measurable function of $\mathbf{u}$. We note that this fact will be utilized for the definition of continuous-time controlled Markov processes in Chapter 2. (In the continuous case one-step transition probability is not available, therefore it is required to utilize all the transition probabilities). If a family of Markov chains which depend on $\mathbf{u}$ as on a parameter is given and the transition probability $P(n, x, m, A; \mathbf{u})$ is $\mathfrak{B}_{n,m-1}$-measurable in $\mathbf{u}$, then setting

$$P_k(x_{k-1}, A_k; u_k) = P(k-1, x_{k-1}, k, A_k; \mathbf{u})$$

(by assumption the r.h.s. of this equality is $\mathfrak{B}_{k-1,k-1}$-measurable, i.e. it is a function of u_{k-1}) we obtain a sequence of functions (1.49) which define a Markov controlled object.

If the functional of the cost of controlling is arbitrary then the optimization problem for a controlled Markov chain is as complicated as for general controlled objects. However, for functionals of an evaluationary type the optimization problem is substantially simplified.

We say that a functional $F(\mathbf{x}, \mathbf{u})$ is of an *evolutionary type* if it is of the form

$$F(\mathbf{x}, \mathbf{u}) = \sum_{k=0}^{\infty} \left(\prod_{j=0}^{k-1} g_j(x_j, u_j) \right) f_k(x_k, u_k) \qquad \left(\prod_{j=0}^{-1} = 1 \right), \tag{1.52}$$

where $\{g_n(x, u)\}$ and $\{f_n(x, u)\}$ are two sequences of $\mathfrak{A} \times \mathfrak{B}$-measurable functions, $g_n \geq 0$.

With a functional (1.52) one can associate a sequence of functionals $F_m(\mathbf{x}, \mathbf{u})$:

$$F_m(\mathbf{x}, \mathbf{u}) = \sum_{k=m}^{\infty} \left(\prod_{j=m}^{k-1} g_j(x_j, u_j) \right) f_k(x_k, u_k).$$

Here $F = F_0$ and F_m are related by the following recurrence relation:

$$F_m(\mathbf{x}, \mathbf{u}) = f_m(x_m, u_m) + g_m(x_m, u_m) F_{m+1}(\mathbf{x}, \mathbf{u}).$$

Denote by

$$\Psi_{m-1}(x_{m-1}, u_{m-1}) = \inf \mathbf{E}_v[F_m(\xi, \eta)/\xi_{m-1} = x_{m-1}, \eta_{m-1} = u_{m-1}], \tag{1.53}$$

where the inf is taken over all controls v for which the controls $u_0, \ldots, u_{m-1}$ are fixed. Since $F_m(\xi, \eta)$ depends only on $\xi_m, \xi_{m+1}, \ldots, \eta_m, \eta_{m+1}, \ldots$ and in view of the Markovian property of the controlled object, the joint distribution of these variables for given $x_0, \ldots, x_{m-1}, u_0, \ldots, u_{m-1}$ depends only on x_{m-1} and u_{m-1}, it follows that the r.h.s. of (1.53) depends only on x_{m-1} and u_{m-1}. The functions $\Phi_{m-1}(x_0, \ldots, x_{m-1}, u_0, \ldots, u_{m-1})$ which were defined in Section 3 by relations (1.23) are expressed in terms of functions Ψ_{m-1} in a rather simple manner. Indeed,

$$F(\mathbf{x}, \mathbf{u})_{\bar{x}_0, \ldots, \bar{x}_{m-1}, \bar{u}_0, \ldots, \bar{u}_{m-1}} = \sum_{k=0}^{m-1} \left(\prod_{j=0}^{k-1} g_j(x_j, y_j) \right) f_k(x_k, u_k) + \prod_{k=0}^{m-1} g_k(x_k, u_k) F_m(\mathbf{x}, \mathbf{u}). \tag{1.54}$$

Therefore since $\prod_{k=0}^{m-1} g_k(x_k, u_k)$ is non-negative and the sum $\sum_{k=0}^{m-1}$ and the product $\prod_{k=0}^{m-1}$ appearing on the r.h.s. of (1.54) are constant for fixed $x_0, \ldots, x_{m-1}, u_0, \ldots, u_{m-1}$ we obtain

$$\Phi_{m-1}(x_0, \ldots, x_{m-1}, u_0, \ldots, u_{m-1}) = \sum_{k=0}^{m-1} \left(\prod_{j=0}^{k-1} g_j(x_j, u_j) \right) f_k(x_k, u_k) + \prod_{k=0}^{m-1} g_k(x_k, u_k) \Psi_{m-1}(x_{m-1}, u_{m-1}). \tag{1.55}$$

Utilizing equality (1.24) in Section 3 we write the following relation which connects Ψ_{m-1} and Ψ_m:

$$\sum_{k=0}^{m-1}\left(\prod_{j=0}^{k-1} g_j(x_j, u_j) f_k(x_k, u_k)\right) + \left(\prod_{k=0}^{m-1} g_k(x_k, u_k)\right)\Psi_{m-1}(x_{m-1}, u_{m-1})$$

$$= \int P_m(x_{m-1}, dx_m; u_{m-1}) \inf_{u_m}\left[\sum_{k=0}^{m}\left(\prod_{j=0}^{k-1} g_j(x_j, u_j)\right) f_k(x_k, u_k) + \prod_{k=0}^{m} g_k(x_k, u_k)\Psi_m(x_m, u_m)\right]$$

$$= \sum_{k=0}^{m-1}\left(\prod_{j=0}^{k-1} g_j(x_j, u_j)\right) f_k(x_k, u_k) + \int P_m(x_{m-1}, dx_m; u_{m-1})$$

$$\times \inf_{u_m}[f_m(x_m, u_m) + g_m(x_m, u_m)\Psi_m(x_m, u_m)] \prod_{k=0}^{m-1} g_k(x_k, u_k).$$

Whence

$$\Psi_{m-1}(x_{m-1}, u_{m-1})$$
$$= \int \inf_{u_m}[f_m(x_m, u_m) + g_m(x_m, u_m)\Psi_m(x_m, u_m)]P_m(x_{m-1}, dx_m; u_{m-1}). \tag{1.56}$$

Evidently if the inf in (1.56) is attained at a certain point $\bar{u}_m$, then the $\inf_{u_m} \Phi_m(\cdot, u_m)$ is also attained at that point, hence

$$\Phi_{m-1}(x_0, \ldots, x_{m-1}, u_0, \ldots, u_{m-1})$$
$$= \int P_m(x_{m-1}, dx_m; u_{m-1})\Phi_m(x_0, \ldots, x_m, u_0, \ldots, \bar{u}_m).$$

Let there exist measurable functions φ_m such that

$$\inf_{u_m}[f_m(x_m, u_m) + g_m(x_m, u_m)\Psi_m(x_m, u_m)]$$
$$= f_m(x_m, \varphi_m(x_m)) + g_m(x_m, \varphi_m(x_m))\Psi_m(x_m, \varphi_m(x_m)).$$

Then these functions (under suitable regularity conditions, on the functional of the cost of controlling) will be determined in view of theorem 1.6 the optimal control. We note that this optimal control is defined by the functions of the form $\{u_m = \varphi_m(x_m), m = 0, \ldots\}$, i.e. the control at time m depends on the value of the basic process at the time and is independent of the behavior of the process at the preceding times. Such a control is called (*non-randomized*) *Markovian*. A general Markovian control is determined by a sequence of conditional measures $q_k(du_k/x_k)$ which depend only on the state of the basic process at the present time.

Assume that the following conditions are fulfilled:

1. X is a separable complete metric space, $\mathfrak{A}$ is the σ-algebra of Borel sets and U is a compact set with a σ-algebra of Borel sets $\mathfrak{B}$;
2. transition probability $P_m(x, A; u)$ is such that for all $f \in C_X$

$$\int f(y)P_m(x, dy; u) \in C_{X \times U};$$

3. functions f_k and g_k which define functional $F(\mathbf{x}, \mathbf{u})$ by means of formula (1.51) are non-negative and lower semicontinuous, $g_k > 0$ and the product

$$\prod_{k=0}^{\infty} g_k(x_k, u_k)$$

converges on $X^N \times U^N$.

Condition 3 assures lower semicontinuity of the functional F.

We shall now investigate the conditions under which the functional $F(\mathbf{x}, \mathbf{u})$ is regular in the sense of the definition given in Section 3. Utilizing formula (1.55) we verify that for $\mathbf{u}$ and $\mathbf{x}$, such that the limit $\lim_{m\to\infty} \Phi_m(x_0, \ldots, x_m, u_0, \ldots, u_m)$ exists

$$\lim_{m\to\infty} \Phi_m(x_0, \ldots, x_m, u_0, \ldots, u_m)$$
$$= F(\mathbf{x}, \mathbf{u}) + \prod_{k=0}^{\infty} g_k(x_k, u_k) \lim_{m\to\infty} \Psi_m(x_m, u_m)$$

is valid. Whence the limit on the right-hand-side exists provided only the limit $\lim_{m\to\infty} \Psi_m(x_m, u_m)$ exists. Moreover in order that $F(\mathbf{x}, \mathbf{u})$ be regular it is necessary and sufficient that for all $\mathbf{x}$ and $\mathbf{u}$ the limit $\lim_{m\to\infty} \Psi_m(x_m, u_m) = 0$. Observe that if this condition is satisfied then also

$$\lim_{m\to\infty} \sup_{x,u} \Psi_m(x, u) = 0.$$

Indeed, let x_m^0 and u_m^0 be chosen in such a manner that

$$\Psi_m(x_m^0, u_m^0) > \delta_m,$$

where δ_m is an arbitrary sequence satisfying the inequality $\delta_m < \sup_m \Psi_m(x, u)$. Then setting $\mathbf{x}^0 = (x_1^0, x_2^0, \ldots)$, $\mathbf{u}^0 = (u_1^0, u_2^0, \ldots)$ we obtain $\lim_{m\to\infty} \Psi_m(x_m^0, u_m^0) = 0$; hence $\lim_{m\to\infty} \delta_m = 0$.

If however $\overline{\lim}_{m\to\infty} \sup_{x,u} \Psi_m(x, u) > \delta$, one could have chosen a subsequence δ_m such that $\overline{\lim}\, \delta_m > \delta$ also.

We now show that the sequence of non-negative functions $\Psi_m(x, u)$ satisfying relation (1.56) and condition $\lim_{m\to\infty} \sup_{x,u} \Psi_m(x, u) = 0$ is uniquely determined.

Let $\overline{\Psi}_m$ be another sequence of functions satisfying (1.49) such that $\lim_{m\to\infty} \sup_{x,u} \overline{\Psi}_m(x, u) = 0$. Then

$$
\begin{aligned}
&|\overline{\Psi}_{m-1}(x, u) - \Psi_{m-1}(x, u)| \\
&\le \int \Big| \inf_{u_m} [f_m(x_m, u_m) + g_m(x_m, u_m)\overline{\Psi}_m(x_m, u_m)] \\
&- \inf_{u_m} [f_m(x_m, u_m) + g_m(x_m, u_m)\Psi_m(x_m, u_m)] \Big| P_m(x, dx_m; u) \\
&\le \int \sup_{u_m} [g_m(x_m, u_m)|\overline{\Psi}_m(x_m, u_m) - \Psi_m(x_m, u_m)|] P_m(x, dx_m; u);
\end{aligned}
$$

$$
\begin{aligned}
&|\overline{\Psi}_{m-1}(x, u) - \Psi_{m-1}(x, u)| \\
&\le \sup_{\bar{x}, \bar{u}} |\overline{\Psi}_m(\bar{x}, \bar{u}) - \Psi_m(\bar{x}, \bar{u})| \int \sup_{u_m} g_m(x', u_m) P_m(x, dx'; u).
\end{aligned}
$$

Let $\bar{x}_m$ and $\bar{u}_m$ be such that

$$\sup_x \sup_u g_m(x, u) \le (1 + 2^{-m}) g_m(\bar{x}_m, \bar{u}_m).$$

Then

$$
\begin{aligned}
\sup_{x,u} |\overline{\Psi}_{m-1}(x, u) - \Psi_{m-1}(x, u)| &\le (1 + 2^{-m}) g_m(\bar{x}_m, \bar{u}_m) \\
&\times \sup_{x,u} |\overline{\Psi}_m(x, u) - \Psi_m(x, u)|.
\end{aligned}
$$

Hence,

$$
\begin{aligned}
\sup_{x,u} |\overline{\Psi}_{m-1}(x, u) - \Psi_{m-1}(x, u)| &\le \prod_{k=m}^{N} (1 + 2^{-k}) \prod_{k=m}^{N} g_k(\bar{x}_k, \bar{u}_k) \\
&\times \sup_{x,u} |\overline{\Psi}_N(x, u) - \Psi_N(x, u)|.
\end{aligned}
$$

Utilizing the convergence of the products

$$\prod_{k=m}^{\infty} (1 + 2^{-k}), \qquad \prod_{k=m}^{\infty} g_k(x_k, u_k)$$

(the second product converges in view of condition 3) and the fact that $\lim_{N\to\infty} \sup_{x,u} \overline{\Psi}_N(x, u) = \lim_{N\to\infty} \sup_{x,u} \Psi_N(x, u) = 0$ we obtain

$$\overline{\Psi}_{m-1}(x, u) = \Psi_{m-1}(x, u)$$

for all $m \ge 1$.

We now show that if condition 3 is fulfilled then

$$\lim_{m\to\infty} \sup_{\mathbf{x}, \mathbf{u}} F_m(\mathbf{x}, \mathbf{u}) = 0. \tag{1.57}$$

Let x_k^+, u_k^+ and x_k^-, u_k^- be chosen so that for some r

$$\left[\frac{1}{r} \vee \inf g_k(x, u)\right] \geq (1 - 2^{-k}) g_k(x_k^-, u_k^-),$$

$$[r \wedge \sup g_k(x, u)] \leq (1 + 2^{-k}) g_k(x_k^+, u_k^+).$$

Convergence of the products

$$\prod_{k=0}^{\infty} g_k(x_k^-, u_k^-) \quad \text{and} \quad \prod_{k=1}^{\infty} g_k(x_k^+, u_k^+)$$

implies that for N sufficiently large $\inf g_k(x, u) > 0$ and $\sup g_k(x, u) < \infty$ for $k \geq N$ and moreover

$$\prod_{k=N}^{\infty} \sup g_k(x, u) \quad \text{and} \quad \prod_{k=N}^{\infty} \inf g_k(x, u)$$

converge. Consequently one can find an N and constants c_1 and c_2 such that

$$c_1 \leq \prod_{k=m}^{l} g_k(x, u) \leq c_2 \quad \text{for } N \leq m \leq l.$$

Now let $m \geq N$ and x_k and u_k be chosen by condition

$$1 \wedge \sup_{x, u} f_k(x, u) \leq (1 + 2^{-k}) f_k(x_k, u_k).$$

Then

$$F_m(\mathbf{x}', \mathbf{u}') = \sum_{k=m}^{\infty} \prod_{j=m}^{k-1} g_k(x_k', u_k') f_k(x_k', u_k')$$

$$\geq c_1 \sum_{k=m}^{\infty} 1 \wedge \sup_{x, u} f_k(x, u).$$

Hence for k sufficiently large the supremum $\sup_{x, u} f_k(x, u) \leq 1$ and for m sufficiently large

$$\sum_{k=m}^{\infty} \sup_{x, u} f_k(x, u) < \infty.$$

Therefore for m sufficiently large

$$\sup_{x, u} F_m(x, u) \leq c_2 \sum_{k=m}^{\infty} \sup_{x, u} f_k(x, u).$$

Clearly the r.h.s. of the last inequality tends to 0 as $m \to \infty$. Relation (1.57) is verified.

Thus theorem 1.6 and the preceding deliberations yield the following theorem.

Theorem 1.11. *Let a controlled Markov chain satisfy conditions* 1 *and* 2 *and the cost functional* $F(\mathbf{x}, \mathbf{u})$ *be of the form* (1.52) *and satisfy condition* 3. *Then there exists*:

I. *A sequence of functions* $\Psi_m(x, u)$ *satisfying the conditions*:

(a) $\Psi_m(x, u)$ *is lower semicontinuous*;
(b) $\lim_{m\to\infty} \sup_{x,u} \Psi_m(x, u) = 0$;
(c) *for all* $m \geq 1$ *the relation* (1.56) *is fulfilled.*

Conditions (a)–(c) *determine uniquely the sequence of functions* Ψ_m.

II. *A sequence of Borel functions* $\varphi_m(x)$ *satisfying the relation*

$$\inf_u [f_m(x, u) + g_m(x, u)\Psi_m(x, u)] = f_m(x, \varphi_m(x)) + g_m(x, \varphi_m(x))\Psi_m(x, \varphi_m(x)). \tag{1.58}$$

For any sequence $\varphi_m(x)$ *satisfying relation* (1.58) *a non-randomized Markovian control* v, *defined by the equalities* $\{u_m = \varphi_m(x_m),\ m = 0, \ldots\}$ *will be optimal.*

III. *An optimal cost of controlling* $S(x)$ *given that* $\xi(0) = x$ *is defined by the equality*

$$S(x) = \inf_u [f_0(x, u) + g(x, y)\Psi_0(x, u)].$$

We shall describe a method for constructing functions $\Psi_m(x, u)$. We assume that functions $g_k(x, u)$ satisfy for all k and for some $\delta > 0$ the relation $1/r \leq g_k(x, u) \leq r$. It then follows from the preceding argument that for some c_1 and c_2

$$c_1 \leq \prod_{k=l}^{m} g_k(x_k, u_k) \leq c_2, \qquad 0 \leq l \leq m.$$

Denote by $\Psi_m^N(x, u)$, $m \leq N$, a sequence of functions determined by the recurrence relations

$$\Psi_N^N(x, u) = 0$$

$$\Psi_m^N(x, u) = \int \inf_{u'} [f_{m+1}(y, u') + g_{m+1}(y, u')\Psi_{m+1}^N(y, u')] \times P_{m+1}(x, dy; u). \tag{1.59}$$

In the same manner as the bound on the difference between Ψ_m and $\overline{\Psi}_m$ was obtained, we derive the bound

$$\sup_{x,u} |\Psi_m^N(x, u) - \Psi_m(x, u)| \leq c_2 \sup_{x,u} |\Psi_N^N(x, u) - \Psi_N(x, u)| = c_2 \sup_{x,u} |\Psi_N(x, u)|.$$

Approaching the limit as $N \to \infty$ and taking I.b into account (Ψ_N satisfies this condition) we have

$$\Psi_m(x, u) = \lim_{N \to \infty} \Psi_m^N(x, u).$$

Remark 1.10. Utilizing Remark 1.7 we verify that if Borel functions $\varphi_k(x)$ satisfying (1.58) exist, then a Markov control defined by the sequence $\{u_k = \varphi_k(x), k = 0, \ldots\}$ will be optimal even if conditions (a) and (b) are not fulfilled and the functions $f_k(x)$ which define F are only nonnegative and measurable: for example in the case when the set U is finite.

Remark 1.11. Let the cost functional $F(\mathbf{x}, \mathbf{u})$ be defined by the equality

$$F(\mathbf{x}, \mathbf{u}) = \sum_{k=0}^{\infty} \prod_{i=0}^{k-1} g_i(x_i, x_{i+1}, \ldots, x_{i+l}, u_i, \ldots, u_{i+l}) \times f_k(x_k, \ldots, x_{k+l}, u_k, \ldots, u_{k+l}),$$

where g_i, f_k are nonnegative lower semi-continuous functions on $X^{l+1} \times U^{l+1}$. All previous discussions are applicable to such a functional.

One can establish the existence of a sequence of lower semi-continuous functions

$$\Psi_m(x_1, \ldots, x_l, u_1, \ldots, u_l), \qquad m = 0, 1, \ldots,$$

which satisfy the relations

$$\limsup_{m \to \infty} \Psi_m(x_1, \ldots, x_l, u_1, \ldots, u_l) = 0$$

and

$$\begin{aligned}
&\Psi_{m-1}(x_1, \ldots, x_l, u_1, \ldots, u_l) \\
&\quad = \int \inf_{u_{l+1}} [f_m(x_1, \ldots, x_{l+1}, u_1, \ldots, u_{l+1}) \\
&\quad\quad + g_m(x_1, \ldots, x_{l+1}, u_1, \ldots, u_{l+1}) \\
&\quad\quad \times \Psi_m(x_2, \ldots, x_{l+1}, u_1, \ldots, u_{l+1})] P(x_l, dx_{l+1}; u_{l+1}).
\end{aligned} \tag{1.60}$$

If the functions $\varphi_k(x_1, \ldots, x_{l+1}, u_1, \ldots, u_{l+1})$ are such that

$$\begin{aligned}
&\inf_{u_{l+1}} [f_m(x_1, \ldots, x_{l+1}, u_1, \ldots, u_{l+1}) \\
&\quad + g_m(x_1, \ldots, x_{l+1}, u_1, \ldots, u_{l+1}) \\
&\quad \times \Psi_m(x_2, \ldots, x_{l+1}, u_1, \ldots, u_{l+1})] \\
&\quad = f_m(x_1, \ldots, x_{l+1}, u_1, \ldots, \varphi_{m+l}(\cdot)) \\
&\quad + g_m(x_1, \ldots, x_{l+1}, u_1, \ldots, \varphi_{m+l}(\cdot)) \\
&\quad \times \Psi_m(x_2, \ldots, x_{l+1}, u_1, \ldots, \varphi_{m+l}(\cdot)),
\end{aligned}$$

then $\{u_{m+1} = \varphi(\cdot)\}_{m+l}$ will be a conditional optimal control under the condition that $x_0, \ldots, x_l, u_0, \ldots, u_{l-1}$ are given.

Consider the important particular case when

$$F(\mathbf{x}, \mathbf{u}) = \sum_{k=0}^{\infty} \left(\prod_{i=0}^{k-1} g_i(x_i, x_{i+1}, u_i) f_k(x_k, x_{k+1}, u_k) \right). \tag{1.61}$$

Set

$$F_m(\mathbf{x}, \mathbf{u}) = \sum_{k=m}^{\infty} \left(\prod_{i=m}^{k-1} g_i(x_i, x_{i+1}, u_i) f_k(x_k, x_{k+1}, u_k) \right),$$

$$\rho_m(x) = \inf_{\nu} \mathbf{E}_\nu(F_m(\xi, \eta)/\xi_m = x).$$

Since

$$F_{m-1}(\mathbf{x}, \mathbf{u}) = f_{m-1}(x_{m-1}, x_m, u_{m-1}) + g_{m-1}(x_{m-1}, x_m, u_{m-1}) F_m(\mathbf{x}, \mathbf{u}),$$

it follows that

$$\begin{aligned} \mathbf{E}_\nu(F_{m-1}(\xi, \eta)/\xi_{m-1} = x) &= \mathbf{E}(\mathbf{E}_\nu(f_{m-1}(\xi_{m-1}, \xi_m, \eta_{m-1}) \\ &\quad + g_{m-1}(\xi_{m-1}, \xi_m, \eta_{m-1}) F_m(\xi, \eta)/\xi_{m-1}, \xi_m, \eta_{m-1})) \\ &= \mathbf{E}_\nu(f_{m-1}(\xi_{m-1}, \xi_m, \eta_{m-1}) + g_{m-1}(\xi_{m-1}, \xi_m, \eta_{m-1}) \\ &\quad \times \mathbf{E}_\nu(F_m(\xi, \eta)/\xi_m)). \end{aligned}$$

This relation implies the following equation for $\rho_m(x)$:

$$\rho_{m-1}(x) = \inf_u \int (f_{m-1}(x, y, u) + g_{m-1}(x, y, u)\rho_m(y)) P_m(x, dy; u) \tag{1.62}$$

Equation (1.62) may also be derived from (1.60) with $l = 1$ utilizing the chain of relations:

$$\begin{aligned} &\Psi_{m-1}(x_{m-1}, x_m, u_{m-1}, u_m) \\ &\quad = \inf_\nu \mathbf{E}_\nu(F_m(\xi, \eta)/\xi_{m-1} = x_{m-1}, \xi_m = x_m, \eta_{m-1} = u_{m-1}, \eta_m = u_m) \\ &\quad = \inf_\nu \mathbf{E}_\nu(F_m(\xi, \eta)/\xi_m = x_m, \eta_m = u_m). \end{aligned}$$

$\Psi_{m-1}(x_{m-1}, x_m, u_{m-1}, u_m)$ depend on x_m and u_m only; therefore

$$\rho_m(x) = \inf_u \Psi_{m-1}(u, x).$$

The functions $\rho_m(x)$ are used in place of functions Ψ_m since they depend only on one variable.

One can obtain the following assertion from theorem 1.6 in a manner analogous to theorem 1.11: There exists a unique sequence of functions which are non-negative and lower semicontinuous $\rho_m(x)$ satisfying (1.62) and the condition $\lim_{m\to\infty} \rho_m(x) = 0$. If Borel functions $\varphi_m(x)$ are determined from the equality

$$\rho_m(x) = \int [f_m(x, y, \varphi_m(x)) + g_m(x, y, \varphi_m(x))\rho_{m+1}(y)] \times P_{m+1}(x, dy; \varphi_m(x)),$$

then the control $v: \{u_m = \varphi_m(x_m), m = 0, \ldots\}$ is an optimal strategy and the quantity $\rho_0(x)$ is the optimal cost of controlling given $\xi_0 = x$.

Consider now one particular problem, the problem of an optimal "reaching" of a set. Let G be a closed set; $g(x) = 1$ for $x \notin G$ and $g(x) = 0$ for $x \in G$. We minimize the functional

$$F(\mathbf{x}, \mathbf{u}) = \sum_{k=0}^{\infty} \prod_{j=1}^{k-1} g(x_j) f_k(x_k, x_{k+1}, u_k) = \sum_{k=0}^{\tau-1} f_k(x_k, x_{k+1}, u_k), \quad (1.63)$$

where τ denotes the instant at which the sequence $x_0, \ldots, x_k \ldots$ reaches the set G for the first time. Functions f_k are assumed to be non-negative and lower semi-continuous. Let

$$F_m(\mathbf{x}, \mathbf{u}) = \sum_{k=m}^{\infty} \prod_{j=m}^{k-1} g(x_j) f_k(x_k, x_{k+1}, u_k),$$
$$\rho_m(x) = \inf_{v} \mathbf{E}_v(F_m(\xi, \eta)/\xi_m = x). \quad (1.64)$$

Functions $\rho_m(x)$ are non-negative lower semi-continuous and satisfy relation

$$\rho_{m-1}(x) = \inf_{u} \int [f_{m-1}(x, y, u) + g(y)\rho_m(y)]\, P_m(x, dy; u). \quad (1.65)$$

An optimal control is determined from relation

$$\rho_{m-1}(x) = \int [f_{m-1}(x, y, \varphi_{m-1}(x)) + g(y)\rho_m(y)] P_m(x, dy; \varphi_{m-1}(x)). \quad (1.66)$$

These assertions are valid provided the controlled Markov chain satisfies conditions 1 and 2. However, the functional F does not satisfy condition 3 (function g may vanish). Therefore the problem of uniqueness of the solution of equation (1.65) requires an additional investigation. Simultaneously we obtain sufficient conditions for $\rho_m(x)$ to be bounded.

Theorem 1.12. *Let the controlled Markov chain satisfy conditions* 1 *and* 2 *and functions* $f_k(x, y, u)$ *appearing in the definition of the functional* $F(\mathbf{x}, \mathbf{u})$ *be lower semi-continuous, jointly bounded and* $f_k \geq \delta > 0$, *and let an increasing sequence of integers* $n_0 < n_1 < \cdots < \cdots$ *exist for a given set* G *such that*

$$\sup_{l} \sum_{k=l}^{\infty} (n_{k+1} - n_l) \exp\left\{-\sum_{i=l}^{k-1} \inf_{x \notin G} \sup_{\mathbf{u}} P(n_i, x, n_{i+1}, G); \mathbf{u})\right\} < \infty \quad (1.67)$$

(here $P(n, x, l, A; \mathbf{u})$ are several-step transition probabilities for the controlled Markov chain defined by formulas (1.51)). *Then the functions $\rho_m(x)$ defined by* (1.64) *are bounded and equation* (1.65) *possesses a unique bounded solution.*

PROOF. Choose control v in such a manner that

$$\sup \sum_{k=l}^{\infty} (n_{k+1} - n_l)\exp\left\{-\sum_{i=l}^{k-1} \inf_{x \in X\setminus G} P^v(n_i, x, n_{i+1}, G)\right\} < \infty \tag{1.68}$$

(P^v is the transition probability after the control v was chosen). Such a choice is possible since $P(n_i, x, n_{i+1}; X\setminus G; \mathbf{u})$ is lower semi-continuous, (the function $\chi_{X\setminus G}(x)$ being the indicator of an open set is lower semi-continuous and

$$\int \chi_{X\setminus G}(y)P(n_i, x, n_{i+1}\, dy; \mathbf{u}) = P(n_i, x, n_{i+1}, X\setminus G; \mathbf{u});$$

(the lower semi-continuity of the function on the left-hand-side follows from lemma 1.5). In view of lemma 1.4 there exist functions $\varphi_{n_i}(x), \varphi_{n_i+1}(x), \ldots, \varphi_{n_{i+1}-1}(x)$ such that

$$\begin{aligned}\sup_{\mathbf{u}} & P(n_i, x, n_{i+1}, G; \mathbf{u}) \\ &= 1 - \inf_{u} P(n_i, x, n_{i+1}, X\setminus G; u_{n_i}, \ldots, u_{n_{i+1}-1}) \\ &= P(n_i, x, n_{i+1}, X\setminus G; \varphi_{n_i}(x), \ldots, \varphi_{n_{i+1}-1}(x)).\end{aligned}$$

If the control v is such that $u_k = \varphi_k(x_{n_i})$ for $n_i \le k < n_{i+1}$ the formula (1.68) is valid.

Let τ be the instant at which the sequence $\xi_0, \xi_1, \ldots$ reaches the set G for the first time. Then for $r < n_i$

$$\begin{aligned}\mathbf{P}_v\{\tau > n_i/\xi_r = x\} &\le \mathbf{P}_v\{\xi_{n_0} \in X\setminus G, \ldots, \xi_{n_i} \in X\setminus G/\xi_r = x\} \\ &= \int_{X\setminus G} \mathbf{P}_v\{\xi_0 \in X\setminus G, \ldots, \xi_{n_i-1} \in dy/\xi_r = y\}P^v(n_{i-1}, y, n_i, X\setminus G) \\ &\le \left[1 - \inf_{x \in X\setminus G} P^v(n_{i-1}, x, n_i, G)\right]\mathbf{P}_v\{\xi_{n_0} \in X\setminus G, \ldots, \xi_{n_i} \in X\setminus G/\xi_r = y\} \\ &\le \exp\left\{-\sum_{j=1}^{i} \inf_{x \in X\setminus G} P^v(n_{j-1}, x, n_j, G)\right\}.\end{aligned}$$

Therefore

$$\mathbf{E}_v(\tau/\xi_r = x) \le n_0 + \sum_{i=0}^{\infty} n_{i+1} \exp\left\{-\sum_{j=1}^{i} \inf_{x \in X\setminus G} P^v(n_{j-1}, x, n_j, G)\right\} < \infty.$$

Since $\rho_r(x) \le \mathbf{E}_v(\tau/\xi_r = x)c$, where v is an arbitrary control and $c \ge \sup_{k,x,u,y} f_k(x, y, u)$, $\rho_0(x)$ is bounded. Clearly for a shifted controlled object defined by transition probabilities $P_{n_l+k-1}(x, A; u)$ the conditions of the theorem are also satisfied; therefore functions $\rho_k(x)$ for $n_l \le k < n_{l+1}$ are also bounded by the quantity appearing on the l.h.s. of inequality (1.68).

We now show that jointly bounded functions $\rho_m(x)$ satisfying relation (1.65) are uniquely determined. For this purpose we shall prove that these functions necessarily satisfy relation (1.64).

Let v be an arbitrary control for which τ is finite with probability 1. Then if $\mathfrak{F}_{m-1}$ is a σ-algebra generated by the variables $\xi_0, \ldots, \xi_{m-1}, \eta_0, \ldots, \eta_{m-1}$ we obtain from (1.65)

$$\begin{aligned}\rho_{m-1}(\xi_{m-1}) &\le \mathbf{E}_v[f_{m-1}(\xi_{m-1}, \xi_m, \eta_{m-1}) + g_m(\xi_m)\rho_m(\xi_m)/\mathfrak{F}_{m-1}] \\ &= \mathbf{E}_v[f_{m-1}(\xi_{m-1}, \xi_m, \eta_{m-1}) + \chi_{\{\tau \ge m\}}\rho_m(\xi_m)/\mathfrak{F}_{m-1}].\end{aligned} \tag{1.69}$$

Multiplying this relation by $\chi_{\{\tau > m-1\}}$ and taking the conditional mathematical expectation with respect to $\mathfrak{F}_k$ $(k \le m-1)$ we arrive at

$$\begin{aligned}\mathbf{E}_v(\rho_{m-1}(\xi_{m-1})\chi_{\{\tau > m-1\}}/\mathfrak{F}_k) \le \mathbf{E}_v(f_{m-1}(\xi_{m-1}, \xi_m, \eta_{m-1})\chi_{\{\tau > m-1\}}/\mathfrak{F}_k) \\ + \mathbf{E}_v(\rho_m(\xi_m)\chi_{\{\tau > m\}}/\mathfrak{F}_k).\end{aligned}$$

Summing up these inequalities over m from $k-1$ up to n, we obtain

$$\begin{aligned}\rho_k(\xi_k)\chi_{\{\tau > k\}} \le \mathbf{E}_v\left(\sum_{m=k+1}^{n} f_{m-1}(\xi_{m-1}, \xi_m, \eta_{m-1})\chi_{\{\tau > m-1\}}/\mathfrak{F}_k\right) \\ + \mathbf{E}_v(\rho_n(\xi_n)\chi_{\{\tau > n\}}/\mathfrak{F}_k).\end{aligned}$$

Since τ is finite and ρ_n are bounded

$$\begin{aligned}\rho_k(\xi_k)\chi_{\{\tau > k\}} &\le \mathbf{E}_v\left(\sum_{m=k+1}^{\infty} f_{m-1}(\xi_{m-1}, \xi_m, \eta_{m-1})\chi_{\{\tau > m-1\}}/\mathfrak{F}_k\right) \\ &= \mathbf{E}_v\left(\sum_{m=k}^{\tau-1} f_m(\xi_m, \xi_{m+1}, \eta_m)/\mathfrak{F}_k\right).\end{aligned} \tag{1.70}$$

However if we choose control v defined by equality $\eta_{m-1} = \varphi_{m-1}(\xi_{m-1})$ where functions φ_{m-1} satisfy relation (1.66) then inequality (1.69) and hence also (1.70) become an equality. Thus for $x \notin G$

$$\rho_k(x) = \inf_v \mathbf{E}_v\left(\sum_{m=k}^{\tau-1} f_m(\xi_m, \xi_{m+1}, \eta_m)/\xi_k = x, \quad \tau > k\right)$$

(τ is finite for this control). □

7 Homogeneous Controlled Markov Chains

Assume that transition probabilities $P_n(x, E; u) = P(x; E; u)$, $n = 1, 2, \ldots$ defining controlled Markov objects do not depend on n. Then a controlled Markov chain is called *homogeneous*. We shall consider several questions related to optimal controlling by means of such sequences.

Let the cost function be of the form

$$F(\mathbf{x}, \mathbf{u}) = \sum_{n=0}^{\infty} \alpha^n f(x_n, x_{n+1}, u_n), \tag{1.71}$$

where $\alpha \in (0, 1)$ and $f(x, y, u)$ is a lower semi-continuous function. $F(\mathbf{x}, \mathbf{u})$ is called a *cost function with discounting.*

We utilize the notation and the results obtained in the preceding Section. The cost function is of the form (1.61). Functions $F_m(\mathbf{x}, \mathbf{u})$ are defined by the equalities.

$$F_m(\mathbf{x}, \mathbf{u}) = \sum_{k=m}^{\infty} \alpha^{k-m} f(x_k, x_{k+1}, u_k).$$

Let a non-randomized control v be given by a sequence of functions $u_k = \varphi_k(x_m, \ldots, x_k)$, $k \geq m$, and v' be given by a sequence of functions

$$u_k = \varphi_k(x_0, \ldots, x_k) = \varphi_{k+m}(x_0, \ldots, x_k).$$

Then taking the homogeneity of the controlled chain into account we may write

$$\mathbf{E}_v(F_m(\xi, \eta)/\xi_m = x) = \mathbf{E}_{v'}(F(\xi, \eta)/\xi_0 = x).$$

Clearly as v runs through all the non-randomized controls for the sequence $\xi_m, \xi_{m+1}, \ldots$, v' runs through all the non-randomized controls for the sequence $\xi_0, \xi_1, \ldots$. Therefore $\rho_m(x) = \rho_0(x)$. Equation (1.62) (with $g = \alpha$) will then be rewritten as

$$\rho_0(x) = \inf_u \int [f(x, y, u) + \alpha\rho_0(y)]P(x, dy; u). \tag{1.72}$$

Assume that the series on the r.h.s. of (1.71) converges for all $\mathbf{x}$ and $\mathbf{u}$. This is possible only if $f(x, y, u)$ is a bounded function. Indeed, otherwise there exist points x_{2n}, x_{2n+1}, u_{2n} such that

$$\alpha^{2n} | f(x_{2n}, x_{2n+1}, u_{2n}) | > 1,$$

and the series in (1.71) diverges. If, however, the function $f(x, y, u)$ is bounded so is $F(\mathbf{x}, \mathbf{u})$ and hence also $\rho_0(x)$. However, equation (1.72) possesses a unique bounded solution. Indeed if $\bar{\rho}(x)$ is also a solution of (1.72) then

$$|\rho_0(x) - \bar{\rho}(x)| \leq \sup_u \int \alpha |\rho_0(y) - \bar{\rho}(y)| P(x, dy; u),$$

$$\sup_x |\rho_0(x) - \bar{\rho}(x)| \leq \alpha \sup_y |\rho_0(y) - \bar{\rho}(y)|.$$

Consequently $\sup_x |\rho_0(x) - \bar{\rho}(x)| = 0$ since $0 < \alpha < 1$.

As far as the optimal control is concerned, the functions which define this control satisfy equation

$$\sup_u \int [\alpha\rho_0(y) + f(x, y, u)]P(x, dy; u)$$
$$= \int [\alpha\rho_0(y) + f(x, y, \varphi_k)]P(x, dy; \varphi_k). \tag{1.73}$$

Clearly a solution of the system (1.73) may be chosen to be independent of k. This is an important conclusion.

A strategy (control) determined by a sequence of functions $u_k = g(x_k)$ where g is a Borel function is called a *stationary Markov strategy*.

Theorem 1.13. *Let X be a complete separable metric space, U be a compact set and the transition probability $P(x, E; u)$ for a controlled Markov chain satisfy the condition*

$$\int f(y)P(x, dy, u) \in C_{X \times U}, \qquad \forall f \in C_X.$$

$F(\mathbf{x}, \mathbf{u})$ is of the form (1.71), *where f is a bounded and lower semicontinuous function. Then there exists an optimal Markov stationary control $\eta_k = g(\xi_k)$ where $u = g(x)$ is a solution of equation*

$$\int [\alpha\rho_0(y) + f(x, y, g)]P(x, dy, g) = \rho_0(x), \tag{1.74}$$

and $\rho_0(x)$ is the unique solution of equation (1.72).

We see that in order to define efficiently an optimal control, first of all effective methods for determination of the function $\rho_0(x)$ are required. We shall now discuss this problem.

Equation (1.72) can be written in the form

$$\rho_0(x) = \min_u \left[F(x, u) + \alpha \int \rho_0(y)P(x, dy, u)\right], \tag{1.75}$$

where

$$F(x, u) = \int f(x, y, u)P(x, dy, u) \tag{1.76}$$

is a bounded and continuous function. We shall show that a solution of equation (1.67) can be obtained by means of the method of successive approximations. Let $v_0(x)$ be an arbitrary bounded Borel function. Construct by induction the sequence $\{v_n(x), n = 1, 2, \ldots\}$ by setting

$$v_{n+1}(x) = \min_u \left[F(x, u) + \alpha \int v_n(y)P(x, dy, u)\right]. \tag{1.77}$$

Applying the inequality

$$\left|\inf_u g_1(u) - \inf_u g_2(u)\right| \leq \sup_u |g_1(u) - g_2(u)|$$

to the functions

$$g_1(u) = F(x, u) + \alpha \int v_n(y)P(x, dy, u),$$

$$g_2(u) = F(x, u) + \alpha \int \rho_0(y)P(x, dy, u),$$

we obtain

$$|v_{n+1}(x) - \rho_0(x)| \leq \alpha \int |v_n(y) - \rho_0(y)| P(x, dy, u).$$

Setting $\delta_n = \max |v_n(x) - \rho_0(x)|$ we have in view of the preceding inequality

$$\delta_{n+1} \leq \alpha\, \delta_n, \qquad \delta_n \leq \alpha^n\, \delta_0 \to 0 \quad \text{as } n \to \infty.$$

We shall describe a method for obtaining a sequence of "improving" approximate solutions of equation (1.75). It is called the *procedure for improving (correcting) strategies*. Choose an arbitrary stationary Markov strategy $\eta_k = h(\xi_k)$ and set

$$r(x, h) = \mathbf{E}\{F(\xi, h(\xi)) \,|\, \xi_0 = x\}.$$

We have

$$\begin{aligned} r(x, h) &= \sum_{n=0}^{\infty} \alpha^n \mathbf{E}\{f(\xi_n, \xi_{n+1}, h(\xi_n))/\xi_0 = x\} \\ &= \tilde{f}_h(x) + \alpha \mathbf{E}\left\{\sum_{n=1}^{\infty} \alpha^{n-1} \mathbf{E}[f(\xi_n, \xi_{n+1}, h(\xi_n))/\xi_1]/\xi_0 = x\right\} \\ &= \tilde{f}_h(x) + \alpha \mathbf{E}\{r(\xi_1, h)/\xi_0 = x\}, \end{aligned}$$

where

$$\tilde{f}_h(x) = \int f(x, y, h(x))P(x, dy; h(x)).$$

Thus the function $r(x, h)$ satisfies the equation

$$r(x, h) = \tilde{f}_h(x) + \alpha \int r(y, h)P(x, dy; h(x)).$$

Define for each x the set E_x:

$$E_x = \left\{u : \tilde{f}_h(x) + \alpha \int r(y, h)P(x, dy; h(x)) > \tilde{f}_u(x) + \alpha \int r(y, h)P(x, dy; u)\right\}.$$

Here $\tilde{f}_u(x)$ denotes the function $\tilde{f}_h(x)$ for $h(x) \equiv u$. To improve the strategy we introduce a new stationary Markov strategy $h_1(x)$ where $h_1(x)$ is constructed as follows: $h_1(x) = h(x)$ if $E_x = \varnothing$ and $h_1(x) = u_x$ ($u_x \in E_x$ for $E_x \neq \varnothing$). We show that strategy h_1 is indeed better than strategy h, and moreover if $E_x \neq \varnothing$, then $r(x, h_1) < r(x, h)$. Noting the expression for the function $\tilde{f}_h(x)$ and the definition of strategy h_1 we can write inequality

$$r(x, h) \geq \tilde{f}_{h_1}(x) + \alpha \int r(y, h)P(x, dy; h_1(x)). \tag{1.78}$$

Multiplying this inequality by α^n and integrating with respect to measure $P^{(n)}(z, dx; h_1)$, where $P^{(n)}(z, A; h_1)$ is the n-step transition probability in a Markov chain with one step transition probability $P(z, A; h_1(z))$, we obtain

$$\alpha^n \int r(x, h)P^{(n)}(z, dx; h_1) \geq \alpha^n \int \tilde{f}_{h_1}(x)P^{(n)}(z, dx; h_1)$$
$$+ \alpha^{n+1} \int r(y, h)P^{(n+1)}(z, dx; h_1).$$

For $n = 0$ this inequality coincides with (1.78).

Summing up the inequalities obtained over all $n \geq 0$, we arrive after some obvious cancellations at

$$r(z, h) \geq \sum_{n=0}^{\infty} \alpha^n \iint f(x, y, h_1(x))P(x, dy; h_1(x))P^{(n)}(z, dx; h_1)$$
$$= r(z, h_1).$$

A strict inequality sign is valid for all z such that $E_z \neq \varnothing$ since for these z strict inequality appears in relation (1.78).

Remark 1.12. If a Markov chain is finite and the set U is finite then the procedure of improving strategies leads after a finite number of steps to an optimal strategy.

Indeed, in this case the number of different stationary Markov strategies is also finite and the correcting procedure always leads to a new strategy.

Consider now the problem of minimizing the average cost of controlling for one step by means of a Markov chain. We shall assume that the control expenditures for one step are determined by a function $f(x, y, u)$ where x is the initial state, y is the state after one step and u is the selected control. The average cost of controlling for N steps assuming that the basic process takes on successively values $x_0, x_1, \ldots, x_N$ and the control takes on values $u_0, \ldots, u_{N-1}$ is determined by the quantity

$$\frac{1}{N} \sum_{k=0}^{N-1} f(x_k, x_{k+1}, u_k).$$

Denote

$$r_N(x) = \inf_v \mathbf{E}_v\left(\frac{1}{N}\sum_{k=0}^{N-1} f(\xi_k, \xi_{k+1}, \eta_k)/\xi_0 = x\right).$$

Then

$$\begin{aligned} r_{N+1}(x) &= \inf_v \mathbf{E}_v\left(\frac{1}{N+1} f(\xi_0, \xi_1, \eta_0) + \frac{N}{N+1}\left(\frac{1}{N}\sum_{k=1}^{N} f(\xi_k, \xi_{k+1}, \eta_k)\right)/\xi_0 = x\right) \\ &= \inf_v \mathbf{E}_v\left[\frac{1}{N+1} f(\xi_0, \xi_1, \eta_0) + \frac{N}{N+1}\mathbf{E}_v\left[\frac{1}{N}\sum_{k=1}^{N} f(\xi_k, \xi_{k+1}, \eta_k)/\xi_0, \xi_1, \eta_0\right]/\xi_0 = x\right] \\ &= \inf_v \mathbf{E}_v\left[\frac{1}{N+1} f(\xi_0, \xi_1, \eta_0) + \frac{N}{N+1} r_N(\xi_1)/\xi_0 = x\right] \\ &= \inf_v \int \left[\frac{1}{N+1} f(x, y, \eta_0) + \frac{N}{N+1} r_N(y)\right] P(x, dy; \eta_0) \\ &= \inf_u \int \left[\frac{1}{N+1} f(x, y, u) + \frac{N}{N+1} r_N(y)\right] P(x, dy; u). \end{aligned}$$

Thus the recurrence relationship

$$r_{N+1}(x) = \inf_u \int \left[\frac{1}{N+1} f(x, y, u) + \frac{N}{N+1} r_N(y)\right] P(x, dy; u) \tag{1.79}$$

is valid.

Assume that $r_N(x) \to s$ where s is a constant; moreover $[r_N(x) - s]N \to c(x)$, where $c(x)$ is a certain function. Then rewriting (1.79) in the form

$$(N + 1)s + c_{N+1}(x) = \inf_u \int [f(x, y, u) + Ns + c_N(y)]P(x, dy; u),$$

and approaching the limit we obtain

$$s + c(x) = \inf_u \int [f(x, y, u) + c(y)]P(x, dy; u). \tag{1.80}$$

Theorem 1.14. *Let the number s and a bounded function $c(x)$ satisfy* (1.80). *Then*

$$s \le \underline{\lim}_{N\to\infty} r_N(x).$$

If the Borel function $g_\varepsilon(x)$ satisfies the relation

$$s + \varepsilon + c(x) \geq \int [f(x, y, g_\varepsilon(x)) + c(y)]P(x, dy; g_\varepsilon(x)) \tag{1.81}$$

for some $\varepsilon \geq 0$, then the stationary Markov control v_ε defined by the functions $\{u_k = g_\varepsilon(x_k), k = 0, \ldots\}$ will be ε-optimal:

$$\overline{\lim_{N\to\infty}} \mathbf{E}_{v_\varepsilon}\left[\frac{1}{N}\sum_{k=0}^{N-1} f(\xi_k, \xi_{k+1}, \eta_k)/\xi_0 = x\right] < s + \varepsilon. \tag{1.82}$$

PROOF. Let v be an arbitrary admissible randomized control, $\mathfrak{F}_n$ be the σ-algebra generated by the random variables $\eta_0, \xi_1, \ldots, \eta_{n-1}, \xi_n$. Denote by $\varphi(du/\mathfrak{F}_n)$ the conditional distribution of η_n given $\mathfrak{F}_n$. Equality (1.80) implies that

$$\begin{aligned} c(\xi_n) + s &= \min_u \int [c(y) + f(\xi_n, y, u)]P(\xi_n, dy; u) \\ &\leq \int [c(u) + f(\xi_n, y, \eta_n)]P(\xi_n, dy; \eta_n), \end{aligned}$$

or denoting by $(\mathfrak{F}_n, \eta_n)$ the σ-algebra generated by $\mathfrak{F}_n$ and the random variable η_n, we have

$$c(\xi_n) + s \leq \mathbf{E}_v\{c(\xi_{n+1}) + f(\xi_n, \xi_{n+1}, \eta_n)/(\mathfrak{F}_n, \eta_n)\}.$$

Averaging over the conditional distribution $\varphi(d\eta_n/\mathfrak{F}_n)$ we obtain

$$c(\xi_n) + s \leq \mathbf{E}_v\{c(\xi_{n+1}) + f(\xi_n, \xi_{n+1}, \eta_n)/\mathfrak{F}_n\}.$$

Evaluating the conditional mathematical expectation given $\xi_0 = x$ for both sides of the last inequality and summing the relations obtained over n from 0 to N we have

$$\begin{aligned} c(x) + (N + 1)s &\leq \mathbf{E}_v\{c(\xi_{N+1})/\xi_0 = x\} \\ &\quad + \mathbf{E}_v\left\{\sum_{n=0}^{N} f(\xi_n, \xi_{n+1}, \eta_n)/\xi_0 = x\right\}. \end{aligned}$$

Whence (in view of the boundedness of $c(x)$)

$$s \leq r_N(x) + O\left(\frac{1}{N}\right).$$

Hence $s \leq \underline{\lim}_{N\to\infty} r_N(x)$.

Let relation (1.81) be fulfilled. Then

$$s + \varepsilon + c(\xi_k) \geq \mathbf{E}_{v_\varepsilon}[f(\xi_k, \xi_{k+1}, \eta_k)/\xi_k] + \mathbf{E}_{v_\varepsilon}(c(\xi_{k+1})/\xi_k).$$

Consequently,

$$s + \varepsilon + \mathbf{E}_{v_\varepsilon}[c(\xi_k)/\xi_0 = x] \geq \mathbf{E}_{v_\varepsilon}[c(\xi_{k+1}) + f(\xi_k, \xi_{k+1}, \eta_k)/\xi_0 = x].$$

Summing up over k from 0 to $N - 1$ we obtain

$$s + \varepsilon + \frac{c(x)}{N} \geq \mathbf{E}_{\nu_\varepsilon} \frac{1}{N} \sum_{k=0}^{N-1} f(\xi_k, \xi_{k+1}, \eta_k) + \frac{1}{N} \mathbf{E}_{\nu_\varepsilon}(c(\xi_N)/\xi_0).$$

The last inequality yields (1.82). □

Consider now the problem of the existence of functions $g_\varepsilon(x)$ which satisfy condition (1.81). The following lemma will be required for this purpose.

Lemma 1.11. *Let the function* $\Phi(x, u)$ *on* $X \times U$ *be a Baire function of the* 1st *class and let* $\inf_u \Phi(x, u) = 0$ *for all* x. *Then there exists for any* $\varepsilon > 0$ *a Borel function* $u = g(x)$ *such that* $\Phi(x, g(x)) < \varepsilon$.

PROOF. Let $\Phi_n(x, u)$ be a sequence of continuous functions such that $\Phi(x, u) = \lim_{n \to \infty} \Phi_n(x, u)$. Then $G_k = \bigcap_{n=k}^{\infty} \{(x, u) : \Phi_n(x, u) \leq \varepsilon/2\}$ is a closed set in $X \times U$ and

$$\left\{(x, u) : \Phi(x, u) < \frac{\varepsilon}{2}\right\} \subset \bigcup_k G_k \subset \{(x, u) : \Phi(x, u) < \varepsilon\}. \tag{1.83}$$

The compactness of U implies that the projection of G_k on X:

$$G_k' = \{x : (x, u) \in G_k\}$$

is a closed set (moreover in view of (1.83) $\bigcup_k G_k' = X$). The function $1 - \chi_{G_k}(x, u)$ on $G_k' \times U$ is lower semi-continuous and furthermore for $x \in G_k'$ $\inf_u[1 - \chi_{G_k}(x, u)] = 0$. Therefore in view of lemma 1.5 there exists a Borel function $g_k(x)$ defined on G_k' such that $1 - \chi_{G_k}(x, g_k(x)) = 0$. Setting $g(x) = g_1(x)$ for $x \in G_1'$, $g(x) = g_k(x)$ for $x \in G_k' \backslash G_{k-1}'$ we obtain the required function.

Corollary. *If the function* $c(x)$ *in* (1.80) *is a bounded Baire function of the first class then for any* $\varepsilon > 0$ *there exists a function* $g_\varepsilon(x)$ *satisfying* (1.81).

To prove this, observe that $\int c(y)P(x, dy; u)$ is also a Baire function of the first class; therefore we obtain the existence of function $g_\varepsilon(x)$ by applying lemma 1.11 to the function

$$\int [f(x, y; u) + c(y)]P(x, dy; u) - c(x) - s.$$

Thus if $c(x)$ is a bounded Baire function of the first class then the number s satisfying (1.80) will be the optimal average cost of controlling:

$$s = \lim_{N \to \infty} r_N(x).$$

In particular, it follows from here that the number s satisfying (1.80) for some bounded Baire function of the first class $c(x)$, is uniquely determined by relation (1.80).

We shall now investigate the problem of uniqueness of function $c(x)$ satisfying (1.80). Clearly in addition to $c(x)$ the function $c_1(x) = c(x) + c_0$, where c_0 is an arbitrary constant, also satisfies relation (1.80).

Lemma 1.12. *Denote by* $\mathbf{E}^g$ *the mathematical expectation corresponding to the control* $u_n = g(x_n)$. *Assume that:*

1. *for any Borel function* $u = g(x)$ *there exists a probability measure* π_g on X *and a sequence* $\rho_n \downarrow 0$ *which is independent of* g *and such that for any bounded measurable function* f *the inequality*
$$\left|\mathbf{E}^g\left(\frac{1}{n}\sum_{k=0}^{n-1} f(\xi_k)/\xi_0\right) - \int f(y)\pi_g(dy)\right| \le \rho_n \sup_x |f(x)| \tag{1.84}$$
is fulfilled;
2. *there exists a finite measure* π *on* X *such that* π *is absolutely continuous with respect to* π_g *and the function*
$$\lambda_g(x) = \frac{d\pi}{d\pi_g}(x)$$
is positive and satisfies the condition
$$\lim_{N\to\infty} \sup_g \pi(\{x : \lambda_g(x) > N\}) = 0. \tag{1.85}$$

Then if $c(x)$ *and* $c_1(x)$ *are two solutions for* (1.80), *it follows that* $c(x) - c_1(x)$ *is constant almost everywhere in measure* π.

Proof. Let control $u = g(x)$ be chosen in such a manner that
$$\int [f(x, y, g(x)) + c(y)]P(x, dy; g(x)) \le s + c(x) + \varepsilon. \tag{1.86}$$

Since also
$$\int [f(x, y, g(x)) + c_1(y)]P(x, dy; g(x)) \ge s + c_1(x), \tag{1.87}$$

putting $c(y) - c_1(y) = \Delta(y)$ we obtain
$$\Delta(x) + \varepsilon \ge \int \Delta(y)P(x, dy; g(x)).$$

If $\{\xi_n, n = 1, 2, \ldots\}$ is a Markov chain with the transition probability $P(\cdot, \cdot, g)$ then (1.84) implies that
$$\mathbf{E}(\Delta(\xi_n)/\xi_{n-1}) \le \Delta(\xi_{n-1}) + \varepsilon.$$

Therefore
$$\mathbf{E}(\Delta(\xi_n)/\xi_0 = x) \le \Delta(x) + n\varepsilon,$$
$$\mathbf{E}\left(\frac{1}{n}\sum_{k=0}^{n-1}\Delta(\xi_n)/\xi_0 = x\right) \le \Delta(x) + n\varepsilon.$$

Taking condition 1 and inequality (1.84) into account we obtain

$$\int \Delta(y)\lambda_g^{-1}(y)\pi(dy) \le \Delta(x) + n\varepsilon + \rho_n.$$

If $\Delta = \inf_x \Delta(x)$, then

$$\int \Delta(y)\lambda_g^{-1}(y)\pi(dy) \le \Delta + n\varepsilon + \rho_n.$$

Hence,

$$\int [\Delta(y) - \Delta]\frac{1}{\lambda_g(y)}\pi(dy) \le n\varepsilon + \rho_n.$$

Let $E_N^g = \{y : \lambda_g(y) \le N\}$. Then

$$\int [\Delta(y) - \Delta]\chi_{E_N^g}(y)\pi(dy) \le [n\varepsilon + \rho_n]N.$$

Taking (1.85) into account we may assert that for any sequence g_n and N_n

$$\chi_{E_{N_n}^{g_n}}(x) \to 1$$

in measure π. Choose $\varepsilon_n \downarrow 0$ so that $n\varepsilon_n \to 0$ and N_n so that $[n\varepsilon_n + \rho_n]N_n \to 0$. Then denoting by g_n a function for which (1.86) is fulfilled for $\varepsilon = \varepsilon_n$, we have

$$\int [\Delta(y) - \Delta]\chi_{E_{N_n}^{g_n}}(y)\pi(dy) \le [n\varepsilon_n + \rho_n]N_n.$$

Approaching the limit we obtain

$$\int [\Delta(y) - \Delta]\pi(dy) \le 0.$$

Hence $\Delta(y) = \Delta$ almost for all y in measure π. □

8 Optimal Stopping of Markov Chains

The problem of optimal stopping for a Markov chain is a particular case of the problem of optimal stopping for a random sequence studied in Section 5. However, a more specific class of sequences allows us to utilize more specific methods for solving this problem; these methods are different from those used in Section 5.

Let $\xi_0, \xi_1, \ldots$ be a Markov chain in a separable complete metric space X. Let $P_k(x, A) = P\{\xi_k \in A \mid \xi_{k-1} = x\}$ be its k-th step transition probability. We shall assume that the transition probability satisfies the following measurability condition: for any bounded measurable function $f(x)$ the function $\int f(y)P_k(x, dy)$ is measurable for all k.

Next, let a sequence of measurable functions $f(k, x)$ which defines the gain obtained after stopping at the k-th step when $\xi_k = x$ be given. The problem is to find the stopping time τ for which $\mathbf{E}f(\tau, \xi_\tau)$ is maximal. The optimal time τ is sought among stopping times relative to the σ-algebras $\tilde{\sigma}_n$ generated by sequence $\{\xi_0, \xi_1, \ldots\}$ (see Section 5). These stopping times are called *strategies*.

Denote

$$s(x) = \sup_\tau \mathbf{E}_x f(\tau, \xi_\tau),$$

where the supremum is taken over stopping times ($\mathbf{E}_x$ denotes the conditional mathematical expectation given $\xi_0 = x$); $s(x)$ which is the optimal gain is also called the *value*. The problem is to choose a stopping time τ for which the expression

$$s(x) = \mathbf{E}_x f(\tau, \xi_\tau)$$

is valid. If such a time does not exist, it is then desirable to be able to find ε-optimal strategies, i.e. Markov times τ_ε such that

$$\mathbf{E}_x f(\tau_\varepsilon, \xi_{\tau_\varepsilon}) > s(x) - \varepsilon.$$

A stopping time τ is called *truncated* if $\mathbf{P}\{\tau \le n\} = 1$ for some n. Observe that an ε-optimal strategy for a given x can be found among truncated stopping times provided $s(x)$ is finite. Indeed if τ_ε is such that

$$\mathbf{E}_x f(\tau_\varepsilon, \xi_{\tau_\varepsilon}) > s(x) - \frac{\varepsilon}{2},$$

and $\tau_\varepsilon^n = \tau_\varepsilon \wedge n$ then the relation

$$s(x) - \frac{\varepsilon}{2} < \mathbf{E}_x f(\tau_\varepsilon, \xi_{\tau\varepsilon}) = \lim_{n\to\infty} \mathbf{E}_x f(\tau_\varepsilon, \xi_{\tau_\varepsilon})\chi_{\{\tau_\varepsilon \le n\}}$$
$$\le \lim_{n\to\infty} \mathbf{E}_x f(\tau_\varepsilon^n, \xi_{\tau_\varepsilon^n})$$

implies that for n sufficiently large $\mathbf{E}_x f(\tau_\varepsilon^n, \xi_{\tau_\varepsilon^n}) > s(x) - \varepsilon$.

The problem at hand may be reduced to a general optimization problem for a controlled Markov chain. We shall assume that the Markov controlled object does not depend on the control and is determined by the transition probabilities $P_k(x, A)$. The phase space of controlling consists of two points: 0 and 1. The choice of the control $u = 0$ at the k-th step means that the stopping did not occur as yet at this step ($\tau > k$); if $\tau = k$ then $u = 1$. Hence the function which should be minimized (the control cost) is determined as follows:

$$F(\mathbf{x}, \mathbf{u}) = -f_k(x_k), \quad \text{if } u_0 = 0, \ldots, u_{k-1} = 0,\ u_k = 1,$$

or

$$F(\mathbf{x}, \mathbf{u}) = -\sum_{k=0}^{\infty} \left(\prod_0^{k-1} (1 - u_i) \right) u_k f(k, x_k). \tag{1.88}$$

This is a functional of the form (1.51) ($g_l(u) = 1 - u$, $f_k(x_k, u_k) = -uf(k, x)$). Since in this case U is finite one can utilize Remark 1.10. The functional $F_m(\mathbf{x}, \mathbf{u})$ in this case is of the form

$$F_m(\mathbf{x}, \mathbf{u}) = -\sum_{k=m}^{\infty}\left(\prod_{i=m}^{k-1}(1 - u_i)\right)u_k f(k, x_k).$$

The function $\rho_m(x)$ is of the form

$$\rho_m(x) = -\sup_{\tau} \mathbf{E}(F_m(\xi, \eta)/\xi_m = x),$$

and the relation connecting functions $\rho_m(x)$ and $\rho_{m+1}(x)$ (cf. (1.24)) is

$$\rho_m(x) = \inf_u \left[-uf(m, x) + (1 - u)\int \rho_{m+1}(y)P_{m+1}(x; dy)\right]. \tag{1.89}$$

Set $s_m(x) = -\rho_m(x)$. Then the sequence $s_m(x)$ satisfies the recurrent system of equalities

$$s_m(x) = \sup_u \left[uf(m, x) + (1 - u)\int s_{m+1}(y)P_{m+1}(x, dy)\right]$$
$$= \max\left[f(m, x), \int s_{m+1}(y)P_{m+1}(x, dy)\right].$$

Moreover $s_0(x) = s(x)$. If $\sup_x f(m, x) \to 0, f(m, x) \geq 0$, then $\sup_x s_m(x) \to 0$. Therefore the functional $F(\mathbf{x}, \mathbf{u})$ will be regular in this case (cf. Section 6). Hence the following theorem is valid.

Theorem 1.15. *Let $f(k, x) \to 0$ uniformly in x as $k \to \infty, f(k, x) \geq 0$ measurable in x, and $\xi_0, \ldots, \xi_n, \ldots$ be a Markov chain with the k-th step transition probability $P_k(x, A)$. Then there exists a stopping time $\bar{\tau}$ such that*

$$\mathbf{E}_x f(\bar{\tau}, \xi_{\bar{\tau}}) = \sup_{\tau} \mathbf{E}_x f(\tau, \xi_\tau)$$

(the supremum is taken over all the stopping times). Also there exists a unique sequence of measurable non-negative functions satisfying

$$s_m(x) = \max\left[f(m, x), \int s_{m+1}(y)P_{m+1}(x, dy)\right] \tag{1.90}$$

and the condition $\lim_{k\to\infty} s_k(x) = 0$. The stopping time $\bar{\tau}$ is defined as the first time at which $s_k(\xi_k) = f(k, \xi_k)$ and

$$\sup_{\tau} \mathbf{E}_x f(\tau, \xi_\tau) = s_0(x).$$

Only the last assertion of the theorem concerning the natüre of the optimal control requires a proof. Theorem 1.11 implies that the optimal control is determined by functions $u_k = \varphi_k(x_k)$ satisfying relation

$$\inf_u \left[-uf(k, x) + (1-u)\int \rho_{k+1}(y)P_{k+1}(x, dy)\right]$$

$$= -\varphi_k(x) + (1 - \varphi_k(x))\int \rho_{k+1}(y)P_{k+1}(x, dy).$$

Hence

$$\varphi_k(x) = \begin{cases} 1, & f(k, x) < \int s_{k+1}(y)P_{k+1}(x, dy); \\ 0, & f(k, x) \geq \int s_{k+1}(y)P_{k+1}(x, dy). \end{cases}$$

Let $\varphi_0(\xi_0) = 1, \ldots, \varphi_{k-1}(\xi_{k-1}) = 1$, $\varphi_k(\xi_k) = 0$. If τ is an optimal time then $\tau = k$, moreover $f(0, \xi_0) < s_0(\xi_0), \ldots, f(k-1, \xi_{k-1}) < s_{k-1}(\xi_{k-1})$, $f(k, \xi_k) = s_k(\xi_k)$. This shows that $\tau = \bar{\tau}$ with probability 1. □

Remark 1.13. Functions $s(x)$ under the conditions of theorem 1.15 are measurable.

Indeed,

$$s(x) = \mathbf{E}_x \sum_{k=0}^{\infty} \left(\prod_{i=0}^{k-1}(1 - \varphi_i(\xi_i))\right)\varphi_k(\xi_k)f(k, \xi_k)$$

$$= \sum_{k=0}^{\infty} \mathbf{E}_x\left(\prod_{i=0}^{k-1}(1 - \varphi_i(\xi_i))\right)\varphi_k(\xi_k)f(k, \xi_k)$$

in view of the uniform convergence on the right-hand-side. The measurability of functions φ_k and $f(k, \xi_k)$ implies the measurability of each one of the summands in the last sum. This yields the measurability of $s(x)$.

Below the following lemma will be used.

Lemma 1.13. *Let $f_n(k, x)$ be a sequence of non-negative measurable functions and let for all x and k*

$$f_n(k, x) \uparrow f(k, x)$$

$(n \to \infty)$. Set

$$s^{(n)}(x) = \sup_\tau \mathbf{E}_x f_n(\tau, \xi_\tau), \qquad s(x) = \mathbf{E}_x f(\tau, \xi_\tau).$$

Then

$$s^{(n)}(x) \uparrow s(x)$$

as $n \to \infty$.

PROOF. Relations $f_n(k, x) \le f_{n+1}(k, x) \le f(k, x)$ imply inequality

$$s^{(n)}(x) \le s^{(n+1)}(x) \le s(x).$$

Let for a given x and $c < s(x)$, τ be a stopping time such that $\mathbf{E}_x f(\tau, \xi_\tau) > c$. Then

$$\lim_{n\to\infty} s^{(n)}(x) \ge \lim_{n\to\infty} \mathbf{E}_x f_n(\tau, \xi_\tau) = \mathbf{E}_x f(\tau, \xi_\tau) > c. \qquad \square$$

Corollary 1. *For any non-negative function $f(k, x)$, measurable in x,*

$$s(x) = \sup_\tau \mathbf{E}_x f(\tau, \xi_\tau)$$

is measurable ($s(x)$ may admit the value $+\infty$).

Indeed, one can always find a sequence of functions $f_n(k, x)$ such that $f_n(k, x)$ are bounded and

$$\lim_{n\to\infty} f_n(k, x) = 0, \qquad f_n(k, x) \uparrow f(k, x).$$

Functions $s^{(n)}(x)$ are measurable in view of the Remark following theorem 1.15. Thus $s(x)$ is also measurable.

Corollary 2. *If $f(x)$ is a measurable non-negative bounded function and*

$$s(x) = \sup_\tau \mathbf{E}_x f(\xi_\tau),$$

then for any $\varepsilon > 0$ there exists an ε-optimal control.

To verify this we choose a sequence $q_n \uparrow 1$ and set

$$f_n(k, x) = q_n^k f(x).$$

If $s^{(n)}(x) = \sup_\tau \mathbf{E}_x f_n(\tau, \xi_\tau)$, then in view of theorem 1.15 there exists a stopping time τ_n such that

$$s^{(n)}(x) = \mathbf{E}_x f_n(\tau_n, \xi_{\tau_n}).$$

Let $\tilde{A}_n = \{x : s^{(n)}(x) > s(x) - \varepsilon\}$, $A_n = \tilde{A}_n \setminus \bigcup_1^{n-1} \tilde{A}_k$. According to lemma 1.10

$$\bigcup_{n=1}^{\infty} A_n = \bigcup_{n=1}^{\infty} \tilde{A}_n = X.$$

Since A_n is measurable, $\tau_\varepsilon = \sum_n \chi_{A_n}(\xi_0)\tau_n$ will clearly be a stopping time and

$$\begin{aligned} \mathbf{E}_x f(\xi_{\tau_\varepsilon}) &\ge \mathbf{E}_x \sum \chi_{A_n}(\xi_0) f_n(\tau_\varepsilon, \xi_{\tau_\varepsilon}) \\ &= \sum \chi_{A_n}(x) \mathbf{E}_x f_n(\tau_n, \xi_{\tau_n}) = \sum \chi_{A_n}(x) s^{(n)}(x) > s(x) - \varepsilon. \end{aligned}$$

Corollary 3. *Let $f(x)$ be measurable and non-negative and $0 \le g(x) < s(x)$, where $g(x)$ is a measurable finite function. Then a stopping time τ exists such that for all x*

$$\mathbf{E}_x f(\xi_\tau) > g(x) - \varepsilon.$$

Indeed, let f_n be bounded and $f_n \uparrow f$. Then s_n are bounded and $s_n \uparrow s$. In view of corollary 2 for each n there exists a time (stopping rule) τ_n such that $\mathbf{E}_x f(\xi_{\tau_n}) > s^{(n)}(x) - \varepsilon$. If

$$\tilde{A}_n = \{x : s^{(n)}(x) > g(x)\}, \qquad A_n = \tilde{A}_n \Big\backslash \bigcup_{k=1}^{n-1} \tilde{A}_k, \qquad \tau = \sum \chi_{A_n}(\xi_0)\tau_n,$$

then τ is the required stopping time.

Consider the case when the Markov chain $\xi_0, \xi_1, \ldots$ is homogeneous with one-step transition probability $P(x, A)$. Denote by Pf the operator

$$Pf(x) = \int f(y)P(x, dy).$$

We shall investigate the properties of the function $s(x)$.

I. $s(x)$ satisfies the inequality

$$s(x) \le \max[f(x), Ps(x)]. \tag{1.91}$$

Indeed for any stopping time τ the expression

$$\mathbf{E}_x \chi_{\{\tau > 0\}} f(\xi_\tau) = \mathbf{E}_x f(\xi_\tau) - \mathbf{P}_x\{\tau = 0\} f(x)$$

is valid. We define a stopping time τ_1 on the set $\{\tau > 0\}$ such that $\xi(\tau) = \theta\xi(\tau_1)$ where θ is the shift operator in the Markov chain defined by equation $\theta g(\xi_0, \ldots, \xi_n) = g(\xi_1, \ldots, \xi_{n+1})$ for any Borel function g and extended by continuity. (Cf. [20], Vol II, p. 87). Then

$$\begin{aligned} \mathbf{E}_x f(\xi_\tau) &= \mathbf{P}_x\{\tau = 0\} f(x) + \mathbf{E}_x \chi_{\{\tau > 0\}} \mathbf{E}_{\xi_1} f(\xi_{\tau_1}) \\ &\le \mathbf{P}_x\{\tau = 0\} f(x) + \mathbf{E}_x \chi_{\{\tau > 0\}} s(\xi_1) \\ &= \mathbf{P}_x\{\tau = 0\} f(x) + \mathbf{P}_x\{\tau > 0\} Ps(x). \end{aligned}$$

Hence

$$\mathbf{E}_x f(\xi_\tau) \le \max[f(x), Ps(x)].$$

Taking the supremum over τ we arrive at (1.91).

II. The function $s(x)$ is an excessive function for the chain $\xi_0, \xi_1, \ldots$. Recall that a function $g(x)$ is called *excessive* for a chain if it satisfies the inequality

$$Pg(x) \le g(x).$$

Let τ be an arbitrary stopping time. Set $\tau' = 1 + \theta\tau$, where $\theta\tau$ is defined in the same manner as τ, but for the shifted sequence $\xi_1, \xi_2, \ldots$. Then

$$s(x) \ge \mathbf{E} f(\xi_{\tau'}) = \mathbf{E}_x \mathbf{E}_x(f(\xi_{\tau'})/\xi_1) = P[\mathbf{E}_x f(\xi_\tau)].$$

Corollary 3 implies that for any $\varepsilon > 0$, $l > 0$ one can construct a stopping time τ such that the inequality

$$\mathbf{E}_x f(\xi_\tau) \geq \min[l, s(x)] - \varepsilon$$

is fulfilled. Consequently,

$$s(x) \geq Ps_l(x) - \varepsilon,$$

where $s_l = \min\{l, s\}$. Approaching the limit as $\varepsilon \downarrow 0$ and $l \uparrow \infty$ we arrive at $s(x) \geq Ps(x)$.

III. $s(x) \geq f(x)$. Indeed, $f(x) = \mathbf{E}_x f(\xi_\tau)$ if $\tau = 0$ with probability 1.

IV. It follows from properties I–III that $s(x)$ is an excessive function satisfying equation

$$s(x) = \max[f(x), Ps(x)]. \tag{1.91'}$$

V. Let $s(x)$ be finite. Then it is the smallest excessive function such that $s(x) \geq f(x)$. Indeed, let $s(x) \geq v(x) \geq f(x)$ where $v(x)$ is excessive. Then for any stopping time τ such that $\tau \leq N$ we have with probability 1

$$\mathbf{E}_x v(\xi_\tau) \geq \mathbf{E}_x f(\xi_\tau).$$

However it is easy to verify that $v_0(\xi_0), v(\xi_1), \ldots, v(\xi_N)$ form a supermartingale, therefore

$$\mathbf{E}_x v(\xi_\tau) \leq v(x).$$

Thus $\mathbf{E}_x f(\xi_\tau) \leq v(x)$. Taking the supremum over τ we obtain $s(x) \leq v(x)$, i.e. $s(x) = v(x)$. Our assertion is thus proved.

VI. Let $A_\infty = \{x : s(x) = +\infty\}$. Then for all $x \in X \backslash A_\infty$ $P(x, A_\infty) = 0$. Indeed, if $P(x, A_\infty) > 0$ then $\int s(y)P(x, dy) = +\infty$. However for $x \in X \backslash A_\infty$, $s(x) < \infty$

$$s(x) \geq \int s(y)P(x, dy).$$

Thus $X \backslash A_\infty$ is an invariant set for a given chain and the chain may be considered to be defined on this set. Therefore the assumption that $s(x)$ is finite is not restrictive.

We shall now investigate under the assumption of the finiteness of $s(x)$ the form of ε-optimal controls. Corollary 2 implies that these controls do exist. Let τ_ε^* be such a stopping time. It may be assumed to be truncated (by a quantity which depends on the initial state x). Then

$$0 \leq \mathbf{E}_x[s(\xi_{\tau_\varepsilon^*}) - f(\xi_{\tau_\varepsilon^*})] \leq s(x) - (s(x) - \varepsilon) < \varepsilon,$$

since $s(\xi_0), s(\xi_1), \ldots$ is a supermartingale and τ_ε^* is an ε-optimal control. The variable $s(\xi_{\tau_\varepsilon^*}) - f(\xi_{\tau_\varepsilon^*})$ is non-negative. Hence

$$\mathbf{P}_x\{s(\xi_{\tau_\varepsilon^*}) - f(\xi_{\tau_\varepsilon^*}) < \varepsilon_1\} > 1 - \frac{\varepsilon}{\varepsilon_1}.$$

Therefore

$$\mathbf{P}_x\{\inf(s(\xi_k) - f(\xi_k)) < \varepsilon_1\} = 1 - \frac{\varepsilon}{\varepsilon_1}.$$

Because $\varepsilon > 0$ is arbitrary we have for any ε_1

$$\mathbf{P}_x\left\{\inf_k (s(\xi_k) - f(\xi_k)) < \varepsilon_1\right\} = 1,$$

and hence for any ε the stopping time

$$\zeta_\varepsilon = \min[k : s(\xi_k) - f(\xi_k) < \varepsilon] \tag{1.92}$$

is finite with probability 1.

Let $C_\varepsilon = \{x : s(x) - f(x) \geq \varepsilon\}$. Then setting $\zeta_\varepsilon^n = n \wedge \zeta_\varepsilon$ we obtain

$$s(\xi_{\zeta_\varepsilon^n}) = \sum_{k=0}^{n} \prod_{i=0}^{k-1} \chi_{C_\varepsilon}(\xi_i)[s(\xi_k) - s(\xi_{k-1})] + s(\xi_0).$$

Observe that

$$\mathbf{E}_{\zeta_i}\chi_{C_\varepsilon}(\xi_i)[s(\xi_{i+1}) - s(\xi_i)] = \mathbf{E}_{\xi_i}\chi_{C_\varepsilon}(\xi_i)[Ps(\xi_i) - s(\xi_i)] = 0,$$

since $Ps(x) = s(x)$ for $f(x) < s(x)$ in view of (1.91). Therefore

$$\mathbf{E}_x s(\xi_{\zeta_\varepsilon^n}) = s(x).$$

It follows from relation

$$\begin{aligned}\varepsilon &> \mathbf{E}_x[s(\xi_{\zeta_\varepsilon}) - \mathbf{E}_x f(\xi_{\zeta_\varepsilon})] \\ &= \lim_{n\to\infty} \mathbf{E}_x s(\xi_{\zeta_\varepsilon^n}) - \mathbf{E}_x f(\xi_{\zeta_\varepsilon}) = s(x) - \mathbf{E}_x f(\xi_{\zeta_\varepsilon})\end{aligned}$$

that ζ_ε is an ε-optimal control. We have thus proved the following theorem.

Theorem 1.16. *If the function $s(x) = \sup_\tau \mathbf{E} f(\xi_\tau)$, where $\xi_0, \xi_1, \ldots$ is a homogeneous Markov chain and f is non-negative, measurable and everywhere finite, then $s(x)$ is the smallest excessive function satisfying the condition $s(x) \geq f(x)$ and equation* (1.91). *For any $\varepsilon > 0$ the Markov time ζ_ε defined by* (1.92) *is finite with probability* 1 *and is an ε-optimal control.*

When do we have an optimal strategy? Assume that $\bar{\tau}$ is an optimal strategy. Then setting $\bar{\tau}_n = \bar{\tau} \wedge n$ we have

$$0 \leq \mathbf{E}_x[s(\xi_{\bar{\tau}_n}) - f(\xi_{\bar{\tau}_n})] \leq s(x) - \mathbf{E}_x f(\xi_{\bar{\tau}_n}).$$

Approaching the limit as $n \to \infty$ we verify that

$$\mathbf{E}_x[s(\xi_{\bar{\tau}}) - f(\xi_{\bar{\tau}})] = 0,$$

and hence with probability 1

$$s(\xi_{\bar{\tau}}) - f(\xi_{\bar{\tau}}) = 0$$

and the time

$$\zeta_0 = \min[k : s(\xi_k) = f(\xi_k)] \tag{1.93}$$

is finite.

Analogously to the proof of theorem 1.16 we verify that if the time ζ_0 is finite with probability 1 then $\mathbf{E}_x f(\xi_{\zeta_0}) = s(x)$. Hence the following theorem is valid.

Theorem 1.17. *If f is a non-negative measurable finite function then the optimal control exists if and only if the time ζ_0 defined by* (1.93) *is finite with probability* 1. *This time is also an optimal control.*

Consider now the problem of how to define the function $s(x)$. We shall use the fact that $s(x)$ is the smallest excessive function such that $s(x) \geq f(x)$.

Define the operator $Qg(x)$ by equality

$$Qg(x) = \max[g(x); Pg(x)]. \tag{1.94}$$

If Q^n is the n-th power of the operator, then there exists

$$v(x) = \lim_{n \to \infty} Q^n f(x). \tag{1.95}$$

Clearly, $f(x) \leq v(x)$. We show that $v(x)$ is an excessive function. We have

$$v(x) \geq Q^n f(x) \geq Q[Q^{n-1} f(x)] \geq PQ^{n-1} f(x).$$

Approaching the limit as $n \to \infty$ we obtain

$$v(x) \geq Pv(x).$$

Let $f(x) \leq v_1(x) \leq v(x)$ and v_1 be excessive. Then

$$Qv_1(x) = \max[v_1(x), Pv_1(x)] = v_1(x),$$

since $Pv_1 \leq v_1$. Therefore $Q^n v_1 = v_1$ for all n. However,

$$Q^n f \leq Q^n v_1 = v_1.$$

Consequently $v \leq v_1$, $v = v_1$. Thus the function v defined by equality (1.95) where operator Q is given by (1.94) is the smallest excessive function such that $v(x) \geq f(x)$.

Theorem 1.18. *The value $s(x)$ is determined by equality*

$$s(x) = \lim_{n \to \infty} Q^n f(x).$$

PROOF. If $A_0 = \{x; v(x) < \infty\}$, then A_0 is an invariant set, and on this set in view of property V, $s(x) = v(x)$. Let $v(x) = +\infty$. Relations $\max[f, Ps] \geq \max[f, Pf] = Qf$ and $Qs = s$ yield that $Q^n f \leq s$. Hence $s \geq v$ and $s(x) = +\infty$ for $v(x) = +\infty$. □

Continuous-Time Control Processes 2

1 General Definitions

The definition of a controlled object and that of a control (or strategy) for the continuous time case can be directly carried over from the discrete-time case. The latter was given in Section 1 of Chapter 1.

Let $(X, \mathfrak{A})$ and $(U, \mathfrak{B})$ be two measurable spaces; the first space is the *phase space of the basic process* while the second is the *phase space of the controls*. We shall consider processes defined on the interval $[0, T]$.

Denote by $X^{[0, T]}$ the space of all functions defined on $[0, T]$ and taking on values in X; let $\mathfrak{A}^{[0, T]}$ be the minimal σ-algebra containing all the cylinders on $X^{[0, T]}$; $U^{[0, T]}$ and $\mathfrak{B}^{[0, T]}$ are defined analogously over the measurable space $(U, \mathfrak{B})$. Let $\mathfrak{A}_t$ be a σ-algebra in $X^{[0, T]}$ generated by the cylinders with bases over $[0, T]$; $\mathfrak{A}_{t-0} = \bigcup_{s<t} \mathfrak{A}_s$, $\mathfrak{B}_t$ and $\mathfrak{B}_{t-0}$ are defined analogously.

A *controlled object* $\mu(A \mid u(\cdot))$ is, by definition, a family of probability measures defined for $A \in \mathfrak{A}^{[0, T]}$ and $u(\cdot) \in U^{[0, T]}$ satisfying the following measurability condition: for all $t \in [0, T]$ and $A \in \mathfrak{U}$ $\mu(A \mid u(\cdot))$ is a $\mathfrak{B}_{t-0}$-measurable function of $u(\cdot)$.

A *control* $\nu(B \mid u(\cdot))$ is, by definition, a family of probability measures defined for $B \in \mathfrak{B}^{[0, T]}$, $x(\cdot) \in X^{[0, T]}$ satisfying the following measurability condition: for all $t \in [0, T]$ and $B \in \mathfrak{B}_t$ $\nu(B \mid x(\cdot))$ is a $\mathfrak{A}_t$-measurable function of $x(\cdot)$.

In the case of discrete time one can construct, using measures $\mu(\cdot/\cdot)$ and $\nu(\cdot/\cdot)$ a unique distribution on $X^N \times U^N$ which actually defines the distribution of the controlled sequence $\{(\xi_n, \eta_n), n = 0, 1, \ldots\}$. In the continuous case, the problem concerning the existence of a control process with a given controlled object and a given control and the uniqueness of such a process (more precisely the uniqueness of its distributions) is not quite so simple. Moreover, even if the compound process $(\xi(t), \eta(t))$ exists in $X \times U$ we cannot verify whether it is indeed a controlled process with a given controlled object $\mu(\cdot/\cdot)$ and control $v(\cdot/\cdot)$. Observe that the equality

$$\mathbf{P}\{\xi(\cdot) \in A/\eta(t), t \in [0, T]\} = \mu(A/\eta(\cdot))$$

is not valid. To see this it is sufficient to consider a discrete time controlled process.

Processes with a *non-randomized control* determined by a family of $\mathfrak{A}_t$-measurable functionals $\varphi_t(x(\cdot))$ defined for $t \in [0, T]$ and such that for all $A \in \mathfrak{B}_t$ the expression

$$v(A/x(\cdot)) = \chi_A(\varphi_t(x(\cdot)))$$

is valid, are an exception to the above. In this case the controlled process $(\xi(t), \eta(t))$ satisfies equality

$$\eta(t) = \varphi_t(\xi(\cdot)), \qquad 0 \leq t \leq T,$$

with probability 1 and hence, in this case a control can be defined (however, the form of the controlled object *cannot* be determined all the same). This is due to the following fact. In order to construct the basic process on the interval $[0, t]$ one needs to define the control on $[0, t)$ which is, in turn, defined if the basic process is defined on $[0, t)$. In the case of a discrete time one can sequentially define the process and its control. Here, however, attempting such a sequential definition of the basic process and the control we find ourselves in a vicious circle. One can present forcible arguments which explain that this vicious circle arises due to the essence of the problem since the knowledge of the control object and control are not sufficient for constructing the joint distribution of a pair of processes (the basic one and the controlled). Indeed, assume that a random function $\xi(\cdot, u(\cdot))$ with values on $X^{[0, T]}$ whose one dimensional marginals coincide with $\mu(\cdot \mid u(\cdot))$:

$$\mathbf{P}\{\xi(\cdot, u(\cdot)) \in A\} = \mu(A/u(\cdot)), \qquad A \in \mathfrak{A}^{[0, T]}, \tag{2.1}$$

is constructed on $U^{[0, T]}$. Such a function can be interpreted as the value of the basic processes when a non-random control $u(\cdot)$ is selected. It is natural to consider a solution of equation

$$\eta(\cdot) = \xi(\cdot, \varphi(\eta(\cdot))) \tag{2.2}$$

as a *value of the basic process* when the non-randomized control $u = \varphi(\xi)$ is applied. The nature of this equation depends on the choice of function $\xi(\cdot, u(\cdot))$ and its finite-dimensional marginal distributions are non-uniquely determined by relation (2.1). However, the existence and the uniqueness of the solution of equation (2.2) can not always be established.

In spite of the above, it would seem intuitively that knowing the behavior of the controlled object under non-random control and the nature of the dependence of the control on the basic process one could determine the nature of the behavior of the basic process under a random control as well. It is indeed the case if the control is somewhat delayed in relation to the process, i.e. if the control up to time t allows us to determine the process on an interval which is larger than $[0, t]$. A control $u(t)$ is called a *step control* if $u(t)$ is piecewise constant on $[0, T]$, i.e. if for some n and $0 = t_0 < t_1 < \cdots < t_n = T$, $u(t) = u(t_k) = u_k$, $t \in [t_k, t_{k+1})$.

A control $\nu(\cdot/\cdot)$ is called a *step control* if the measure $\nu(\cdot/x(\cdot))$ is concentrated for all $x(\cdot) \in X^{[0, T]}$ on step functions.

Denote by $t_1, t_2, \ldots$ the times at which the control changes its value. If $\nu(\cdot/x(\cdot))$ is defined, then the measure

$$\nu(\{u(\cdot): u(0) \in B_0\}/x(\cdot)) = \nu_0(B_0/x(0))$$

is also defined. Moreover, for any t

$$\nu(\{u(\cdot): u(0) \in B_0, u(s) = u(0), s \leq t\}/x(\cdot))$$

is a $\mathfrak{A}_t$-measurable function. Utilizing the absolute continuity of this measure (as a function of B_0) with respect to measure $\nu_0(B_0/x(0))$ we verify the existence of the function

$$\lambda_1(t/x(\cdot), u(0)),$$

such that

$$\nu(\{u(\cdot): u(0) \in B_0, u(s) = u(0), s \leq t\}/x(\cdot))$$

$$= \int_{B_0} \lambda_0(t/x(\cdot), u_0)\nu_0(du_0/x(0));$$

$\nu_0(B_0/x(0))$ determines the distribution of the control at the initial time and $\lambda_0(t/x(\cdot), u_0)$ determines the distribution of the first jump of the control, namely it gives us the conditional probability that $t_1 > t$.

Furthermore, define

$$\nu_1(B_1/x(\cdot), u_0, s_1)$$

to be the conditional probability that $u_1(s_1) \in B_1$ for $u(0) = u_0$, $t_1 = s_1$. Analogously we define, for all k the functions,

$$\lambda_k(t/x(\cdot), u_0, \ldots, u_{k-1}, s_1, \ldots, s_{k-1}) \quad (s_1 < \cdots < s_{k-1} < t),$$

which give us the conditional distributions of t_k under the condition that $t_1 = s_1, \ldots, t_{k-1} = s_{k-1}$,

$$u(0) = u_0, \ldots, u(t_{k-1}) = u_{k-1},$$

and

$$\nu_k(B_k/x(\cdot), u_0, \ldots, u_{k-1}, s_1, \ldots, s_k) \quad (s_1 < \cdots < s_k)$$

which gives the conditional distributions of $u(t_k)$ under the condition that $t_1 = s_1, \ldots, t_k = s_k$, $u(0) = u_0, \ldots, u(t_{k-1}) = u_{k-1}$. The functions λ_k and ν_k are measurable in all their arguments; moreover $\lambda_k(t/\cdot)$ is measurable in $x(\cdot)$

with respect to $\mathfrak{A}_t$ while $\nu_k(\cdot/\cdot)$ is measurable with respect to $\mathfrak{A}_{s_k}$. We now show how one can, using functions λ_k and ν_k, determine the distribution of the controlled process $(\xi(t), \eta(t))$ for a given controlled object.

We also introduce conditional measures $\mu_t(A/x(\cdot), u(\cdot))$ defined on the σ-algebra $\mathfrak{A}_T^t$ generated by cylinders with bases over $[t, T]$, measurable in $x(\cdot)$ and $u(\cdot)$ with respect to $\mathfrak{A}_t \times \mathfrak{B}^{[0, T]}$ and such that for any $A' \in \mathfrak{A}_t$

$$\int_{A'} \mu_t(A/x(\cdot), u(\cdot))\mu(dx/u(\cdot)) = \mu(A \cap A').$$

In other words, $\mu_t(A/x(\cdot), u(\cdot))$ defines the conditional distribution of processes with the corresponding measure $\mu(\cdot/u(\cdot))$ on $[t, T]$ provided its value on the interval $[0, t]$ is known.

We construct now a sequence of processes (more precisely their distributions)

$$(\xi^n(t), \eta^n(t))$$

as follows. Let $\eta^0(0)$ possess for a given $\xi(0)$ the conditional distribution which coincides with $\nu_0(B_0/\xi_0)$ and let for all $t \in [0, T]$ $\eta^0(t) = \eta^0(0)$. Define $\xi^0(t)$ in such a manner that $\xi^0(0) = \xi(0)$ and for all $A \in \mathfrak{U}^{[0, T]}$

$$\mathbf{P}\{\xi^0(\cdot) \in A/\eta^0(\cdot)\} = \mu(A/\eta^0(0)).$$

Now define variables t_1 and $\eta^1(t_1)$ in such a manner that

$$\mathbf{P}\{t_1 > t/\xi^0(\cdot), \eta^0(\cdot)\} = \lambda_1(t/\xi^0(\cdot), \eta^0(\cdot)),$$

$$\mathbf{P}\{\eta^1(t_1) \in B_1/\xi^0(\cdot), \eta^0(\cdot)\} = \nu_1(B_1/\xi^0(\cdot), \eta^0(\cdot), t_1).$$

Set

$$\eta^1(t) = \begin{cases} \eta^0(t) & \text{for } t < t_1; \\ \eta^1(t_1) & \text{for } t \geq t_1. \end{cases}$$

Next we construct a process $\xi^1(t)$ satisfying the conditions $\xi^1(t) = \xi^0(t)$ for $t < t_1$ and if $A \in \mathfrak{U}_T^t$

$$\mathbf{P}\{\xi^1(\cdot) \in A/\xi^1(s), s < t, \eta^1(\cdot)\} = \mu_t(A/\xi^1(\cdot), \eta^1(\cdot))$$

for $t_1 \leq t$.
Continuing this construction one can define a sequence of processes $(\xi^k(t), \eta^k(t))$ such that $\eta^k(t)$ possesses exactly k jumps on $[0, T]$, $\eta^k(t) = \eta^{k-1}(t)$ for all k and $t < t_{k-1}$, where $0 < t_1 < \cdots < t_k$ are the jump points of $\eta^k(t)$ and $\xi^k(t) = \xi^{k-1}(t)$ for $t < t_k$. After $\xi^k(t)$ and $\eta^k(t)$ are constructed we first determine the time t_{k+1} and a value of $\eta^{k+1}(t_{k+1})$ such that

$$\mathbf{P}\{t_{k+1} > t/\xi^k(\cdot), \eta^k(\cdot)\} = \lambda_{k+1}(t/\xi^k(\cdot), \eta^k(0), \ldots, \eta^k(t_k), t_1, \ldots, t_k),$$

$$\mathbf{P}\{\eta^{k+1}(t_{k+1}) \in B_{k+1}/\xi^k(\cdot), \eta^k(\cdot)\}$$
$$= \nu_{k+1}(B_{k+1}/\xi^k(\cdot), \eta^k(0), \ldots, \eta^k(t_k), t_1, \ldots, t_k).$$

Next we set $\eta^{k+1}(t) = \eta^k(t)$ for $t < t_{k+1}$ and $\eta^{k+1}(t) = \eta^{k+1}(t_{k+1})$ for $t \geq t_{k+1}$. After the process $\eta^{k+1}(t)$ is constructed we define the process $\xi^{k+1}(t)$ by setting it equal to $\xi^k(t)$ on the interval $[0, t_{k+1}]$ and continue it on the interval $[t_{k+1}, T]$ in such a manner that for all $A \in \mathfrak{A}_T^t$ and $t_{k+1} \le t$

$$\mathbf{P}\{\xi^1(\cdot) \in A/\xi^1(s), s < t, \eta^{k+1}(\cdot)\} = \mu_t(A/\xi^{k+1}(\cdot), \eta^{k+1}(\cdot)).$$

For $t_{k+1} \ge T$ the process $(\xi^{k+1}(t), \eta^{k+1}(t))$ is the required process $(\xi(t), \eta(t))$.

One can find sufficiently simple restrictions on a control such that the procedure described above will terminate with probability 1, i.e. a k can be found such that $t_k = T$. One such restriction is as follows:

Let a function $\alpha(\delta)$, $\alpha(\delta) \downarrow 0$ as $\delta \downarrow 0$, exist such that for all k, $x(\cdot)$, $u_0, \ldots, u_{k-1}, s_1, \ldots, s_{k-1} < s < s + \delta < T$ the inequality

$$\lambda_k(s/x(\cdot), u_0, \ldots, u_{k-1}, s_1, \ldots, s_{k-1}) - \lambda_k(s + \delta/x(\cdot), u_0, \ldots, u_{k-1}, s_1, \ldots, s_{k-1}) \le \alpha(\delta)$$

is satisfied. Then for any controlled object $\mu(\cdot/\cdot)$

$$\mathbf{P}\left\{\eta(\cdot) \in \bigcup_n D_n\right\} = 1,$$

where D_n is the set of step functions in $U^{[0, T]}$ possessing n discontinuities (this is equivalent to the existence (with probability 1) of a k such that $t_k = T$). Indeed define $\eta(t)$ on the interval $(\max_k t_k, T)$ (provided $\max_k t_k < T$) to be equal to some $\bar{u} \in U$. Then for $0 < s_2 - s_1 < \delta$, $s_2 < T$

$$\mathbf{P}\{\eta(s_2) \ne \eta(s_1)/\eta(s), s \le s_1\} \le \alpha(\delta).$$

We introduce on U a metric $r(\cdot/\cdot)$ such that

$$r(u_1, u_2) = 1 \quad \text{for } u_1 \ne u_2.$$

Then for the process $\eta(t)$ for $\varepsilon > 0$ the relation

$$\mathbf{P}\{r(\eta(s + h), \eta(s)) > \varepsilon/\eta(t), t \le s\} \le \alpha(\delta)$$

is valid. Moreover the process $\eta(t)$ is separable by construction. Hence in view of theorem 2 in Section 4 of Chapter 3 of [20] the process $\eta(t)$ possesses with probability 1 no discontinuities of the second kind.

The procedure for constructing a controlled process using a controlled object and a step control presented above can be succinctly described in the following manner.

Given the value of $\xi(0)$ we determine the control at the initial time $\eta(0)$. Since this is a step control $\eta(t) = \eta(0)$ on some interval $[0, t_1]$. The time t_1 and the value $\eta(t_1)$ depend on the course of the process on the interval $[0, t_1]$. If one defines $\eta(t)$ for $t > t_1$ in an arbitrary manner, then the process $\xi(t)$ is not changed on the interval $[0, t_1]$. The only way to define $\eta(t)$ so that $\eta(t) = \eta(0)$ on $[0, t_1]$, (since t_1 is unknown) is to set $\hat{\eta}(t) = \eta(0)$ for all $t \in [0, T]$. Using this control one can construct the process $\hat{\xi}(t)$ and determine time t_1; moreover, for $t \in [0, t_1]$ and $\hat{\xi}(t) = \xi(t)$, $t \le t_1$, $\hat{\eta}(t) = \eta(t)$, $t \le t_1$, where $(\xi(t), \eta(t))$ is the required controlled process. Knowing $\xi(t)$ on

$[0, t]$ one can determine $\eta(t_1)$. Thereupon the preceding construction is simply applied to the interval $[t_1, T]$.

Utilizing the fact that a constant control uniquely determines the joint distributions of $(\xi(t), \eta(t))$ one can verify that the controlled object and the step control uniquely determine the joint distributions of the controlled process.

When solving certain problems for controlled processes with a step control it is useful to apply the following representation of these processes as discrete-time controlled processes. Let us view $(X^{[0,T]}, \mathfrak{A}^{[0,T]})$ and $(\tilde{U}^{[0,T]}, \mathfrak{B}^{[0,T]})$ where $\tilde{U}^{[0,T]}$ is a subset of step functions in $U^{[0,T]}$ as new phase spaces for the basic process and control respectively. Define the control object by means of a sequence of measures

$$P_k(A/x^0(\cdot), \ldots, x^{k-1}(\cdot); u^0(\cdot), \ldots, u^{k-1}(\cdot)), \; A \in \mathfrak{A}^{[0,T]},$$

defined by equality

$$\begin{aligned} &P_k(A/x^0(\cdot), \ldots, x^{k-1}(\cdot); u^0(\cdot), \ldots, u^{k-1}(\cdot)) \\ &\quad = \chi_{A_1}(x^{k-1}(\cdot))\mu_{t_k}(A^2/x^{k-1}(\cdot); u^{k-1}(\cdot)), \end{aligned}$$

with $A = A^1 \cap A^2$, where $A^1 \in \mathfrak{U}_{t_{k-1}}$, $A^2 \in \mathfrak{U}_T^{t_{k-1}}$ and $u^{k-1}(\cdot)$ is a step function possessing $(k-1)$ jumps at the points $t_1 < \cdots < t_{k-1}$. The control is given by the sequence of measures

$$q_k(B/x^0(\cdot), \ldots, x^{k-1}(\cdot); u^0(\cdot), \ldots, u^{k-1}(\cdot)),$$

defined as follows: if $u^{k-1}(\cdot)$ is a step function and $\bar{t}$ is its maximal discontinuity point then the measure q_k is concentrated on the step functions $u(\cdot)$ satisfying the equality $u(t) = u^{k-1}(t)$ for $t \le \bar{t}$ and possessing at most one jump on $(\bar{t}, T)$; moreover,

$$\begin{aligned} &q_k(\{u(\cdot)\colon u(t) = u^{k-1}(t), t \le T\}/x^0(\cdot), \ldots, x^k(\cdot); u^0(\cdot), \ldots, u^{k-1}(\cdot)) \\ &\quad = \lambda_{n+1}(T/x^k(\cdot), u^{k-1}(t_0), \ldots, u^{k-1}(t_n); t_1, \ldots, t_n). \qquad (2.3) \end{aligned}$$

Here $n \ge k - 1$, $0 = t_0$, $t_1 < t_2 < \cdots < t_n < T$ are all points of discontinuity of $u^{k-1}(\cdot)$; for $s > \bar{t}$, $B \in \mathfrak{B}$,

$$\begin{aligned} &q_k(\{u(\cdot)\colon u(t) = u^{k-1}(t), t < s, u(T) \in B\}/x^0(\cdot), \ldots, x^k(\cdot), \\ &u^0(\cdot), \ldots, u^{k-1}(\cdot)) = -\int_s^T v_{n+1}(B/x^k(\cdot), u^{k-1}(t_0), \ldots, u^{k-1}(t_n); \qquad (2.4) \\ &\quad t_1, \ldots, t_n, t')\, d\lambda(t'/x^k(\cdot), u^{k-1}(t_0), \ldots, u^{k-1}(t_n); t_1, \ldots, t_n). \end{aligned}$$

For $k = 0$ the formulas (2.3) and (2.4) have no meaning.

Let the measure

$$q_0(B/x^0(\cdot))$$

be concentrated on the functions $u(\cdot)$ constant on $[0, T]$; and let for $B \in \mathfrak{B}$ the equality

$$q_0(\{u(\cdot): u(t) = u(0), t \in [0, T], u(0) \in B\}/x^0(\cdot)) = \nu(\{u(\cdot): u(0) \in B\}/x(0))$$

be valid. It follows from the properties of a control that the sequence of controls is such that $u^k(\cdot)$ possesses at most k discontinuities and up to the last discontinuity coincides with $u^{k-1}(\cdot)$; if for some k, $u^k(\cdot)$ possesses less than k discontinuities, then for all $l > k$, $u^k(\cdot) = u^l(\cdot)$. The random sequence obtained under this choice of the controlled object and control coincides with the above distributed processes $\xi^k(t)$ and $\eta^k(t)$. In the same manner one may construct a controlled random process with a step controlled object, i.e. with a controlled object $\mu(\cdot/\cdot)$ such that $\mu(\tilde{X}^{[0,T]}, u(\cdot)) = 1$, where $\tilde{X}^{[0,T]}$ is the set of all step functions on $X^{[0,T]}$. In this case we set $\xi^{(0)}(t) = \xi(0)$, $t \in [0, T]$ and construct the control $\eta^{(0)}(t)$ by means of $\xi^{(0)}(t)$. Knowing the control $\eta^{(0)}(t)$ we then construct the process $\tilde{\xi}^{(1)}(t)$ by means of the control $\eta^{(0)}$. If t_1 is the time of the first jump of the process $\tilde{\xi}^{(1)}(t)$ we set $\xi^{(1)}(t) = \xi(0)$ for $t < t_1$, $\xi^{(1)}(t) = \tilde{\xi}^{(1)}(t + 0)$ for $t \geq t_1$ and so on.

2 Representation of the Controlled Objects and Construction of Controlled Processes

In this section we shall discuss in more detail the possibility (as indicated in Section 1) of constructing a controlled process using a random function $\xi(\cdot, u(\cdot))$ which determines the controlled object defined on a probability space.

Now let $\{\Omega, \mathfrak{S}, \mathbf{P}\}$ be a probability space. We say that a family of random processes $\xi(t, \omega; u(\cdot))$, $t \in [0, T]$, $\omega \in \Omega$, $u(\cdot) \in U^{[0,T]}$ is a *representation of a controlled object* $\mu(\cdot/\cdot)$ if the following conditions are satisfied:

1. for all $A \in \mathfrak{A}^{[0,T]}$ $\mathbf{P}\{\xi(\cdot, \omega; u(\cdot)) \in A\} = \mu(A/u(\cdot))$;
2. for $u_1(t) = u_2(t)$ with $t \leq t_1$, $\xi(t, \omega, u_1(\cdot)) = \xi(t, \omega, u_2(\cdot))$ with $t \leq t_1$;
3. $\xi(\cdot, \omega; u(\cdot))$ is a measurable random function defined on $U^{[0,T]}$ with values in $X^{[0,T]}$, i.e. for any $t \in [0, T]$, $A \in \mathfrak{U}_t$, $\{(\omega; u(\cdot)) : \xi(\cdot, \omega; u(\cdot)) \in A\} \in \mathfrak{S} \times \mathfrak{B}_t$.

The first condition is evidently necessary in order that the process $\xi(\cdot, \omega; u(\cdot))$ will possess for a fixed $u(\cdot)$ the same distributions as the controlled process with the controlled object $\mu(\cdot/\cdot)$ and fixed control $u(\cdot)$. The second condition is the compatability conditions for the control and the basic process: to define the basic process on $[0, t]$ one must define a control on $[0, t]$. The third condition is necessary in order that it will be permissible to replace $u(\cdot)$ in $\xi(\cdot, \omega; u(\cdot))$ by the random process $\eta(\cdot)$.

Indeed the following assertion is valid.

Lemma 2.1. *If $f(u, \omega)$ is a measurable mapping of $(U \times \Omega, \mathfrak{B} \times \mathfrak{S})$ into $(X, \mathfrak{A})$ and $\varphi(\omega)$ is a measurable mapping of $(\Omega, \mathfrak{S})$ into $(U, \mathfrak{B})$, then $f(\varphi(\omega), \omega)$ is a measurable mapping of $(\Omega, \mathfrak{S})$ into $(X, \mathfrak{A})$.*

PROOF. Let $A \in \mathfrak{A}$. It is necessary to prove that

$$\{\omega : f(\varphi(\omega), \omega) \in A\} \in \mathfrak{S}.$$

Since $\{(u, \omega) : f(u, \omega) \in A\} \in \mathfrak{B} \times \mathfrak{S}$, it is sufficient to show that for any set $\Gamma \in \mathfrak{B} \times \mathfrak{S}$ $\{\omega: (\varphi(\omega), \omega) \in \Gamma\} \in \mathfrak{S}$. Evidently the set of such Γ forms a σ-algebra. This σ-algebra contains all the rectangles $B \times \Delta$, where $B \in \mathfrak{B}$ and $\Delta \in \mathfrak{S}$ since

$$\{\omega : (\varphi(\omega), \omega) \in B \times \Delta\} = \{\omega : \varphi(\omega) \in B\} \cap \Delta.$$

Hence this algebra coincides with $\mathfrak{B} \times \mathfrak{S}$. □

Denote by $\mathfrak{F}_t$ the current of σ-algebras in $\mathfrak{S}$ generated by the variables $\xi(s, \omega, u(\cdot))$, $s \le t$, for all possible $u(\cdot)$. Any process $\eta(t)$ with values in U for all t, measurable with respect to $\mathfrak{F}_{t-0} = \bigcup_{s<t} \mathfrak{F}_s$ will be referred to as a *generalized control*. It follows from lemma 2.1 that for any generalized control $\eta(t)$, $(\xi(t, \omega), \eta(\cdot))$ is a random process on the probability space $\{\Omega, \mathfrak{S}, \mathbf{P}\}$. We shall call it the *controlled process under the control* $\eta(t)$. If in place of condition 3) the following condition 3′) is stipulated: for any $t \in [0, T]$

$$A \in \mathfrak{A}_t \{(\omega, u(\cdot)) : \xi(\cdot, \omega, u(\cdot)) \in A\} \in \mathfrak{F}_t \times \mathfrak{B}_t,$$

then the process $\xi(t, \omega, \eta(t))$ becomes $\mathfrak{F}_t$-measurable.

We now examine representations of some controlled objects. First we assume that the controlled object is stepwise, i.e. the probability measure $\mu(\cdot/u(\cdot))$ is such that for all $u(\cdot) \in U^{[0, T]}$, $\mu(\tilde{X}^{[0, T]}/u(\cdot)) = 1$. A step controlled object can be conveniently defined by the following collection of conditional distribution functions.

$$P(dx_0),\ \lambda_1(ds/x_0, u(\cdot)),\ P_1(dx_1/x_0, \tau_1, u(\cdot)), \ldots,$$

$$\lambda_k(ds/x_0, \ldots, x_{k-1}, \tau_1, \ldots, \tau_{k-1}; u(\cdot)),$$

$$P_k(dx_k/x_0, \ldots, x_{k-1}, \tau_1, \ldots, \tau_k; u(\cdot)), \ldots.$$

Here $P(dx_0)$ is the distribution of $x(0)$ which does not depend on $u(\cdot)$; $\lambda_1(ds/\cdots)$ is the conditional distribution of time of the first jump of the process; $P_1(dx_1/x_0, \tau_1)$ is the conditional distribution of the state of the process after the first jump for the given time of the jump and the initial state and so on. These conditional distributions may be chosen so that they satisfy the conditions:

a. *the measures $\lambda_k(ds/x_0, \ldots, x_{k-1}, \tau_1, \ldots, \tau_{k-1}; u(\cdot))$, $P_k(dx_k/x_0, \ldots, x_{k-1}, \tau_1, \ldots, \tau_k; u(\cdot))$ be measurable with respect to $\mathfrak{A}^k \times \mathfrak{A}_{[0,T]}^{k-1} \times \mathfrak{B}^{[0, T]}$ and $\mathfrak{A}^{k-1} \times \mathfrak{A}_{[0, T]}^k \times \mathfrak{B}^{[0, T]}$ respectively (here $\mathfrak{A}_{[0, T]}$ is the σ-algebra of Borel sets on $[0, T]$);*

b. *for* any $\Gamma \in \mathfrak{A}_{[0,T]}$, $\Gamma \subset [t, t+h]$, $\lambda_k(\Gamma/x_0, \ldots, x_{k-1}, \tau_1, \ldots, \tau_{k-1}; u(\cdot))$ *for* $\tau_{k-1} < t$ *depends only on the values of* $u(s)$ *on* $[0, t+h)$, and $P_k(dx_k/x_0, \ldots, x_{k-1}, \tau_1, \ldots, \tau_k, u(\cdot))$ *depends only on* $u(s)$ *on the interval* $[0, \tau_k)$.

The following auxiliary assertion will be required for the construction of a representation of controlled objects.

Lemma 2.2. *Let X be a complete separable metric space; $\mathfrak{A}$ be the σ-algebra of its Borel sets; U be a topological space with σ-algebra $\mathfrak{B}$ containing all the open sets; $\{\mu_u(\cdot), u \in U\}$ be a family of measures on $\mathfrak{A}$ such that $\mu_u(A)$ is $\mathfrak{B}$-measurable for all $A \in \mathfrak{A}$. Then there exists a function $f(\alpha, u)$ from $[0, 1] \times U$ into X measurable with respect to $\mathfrak{A}_{[0,1]} \times \mathfrak{B}$ and satisfying the conditions*:

1. *if* $\mu_{u_1} = \mu_{u_2}$, *then* $f(\alpha, u_1) = f(\alpha, u_2)$;
2. *if m is the Lebesgue measure on* $[0, 1]$ *then for all* $A \subset \mathfrak{A}$ *and* $u \in U$

$$\mu_u(A) = m(\{\alpha : f(\alpha, u) \in A\}), \tag{2.5}$$

PROOF. Under the conditions of the lemma one can construct a system of sets $A_{k_1, \ldots, k_r}$ defined for $r = 1, 2, \ldots, k_i = 1, 2, \ldots$ and satisfying the conditions:

1. $A_{k_1, \ldots, k_{r-1}, i} \cap A_{k_1, \ldots, k_{r-1}, j} = \varnothing$ for $i \neq j$;
2. the diameter of $A_{k_1, \ldots, k_r}$ is at most 2^{-r};
3. $\bigcup_{k_r} A_{k_1, \ldots, k_{r-1}, k_r} = A_{k_1, \ldots, k_{r-1}}$, $\bigcup_{k_1} A_{k_1} = X$.

Select in each one of the sets $A_{k_1, \ldots, k_r}$ a point $x_{k_1, \ldots, k_r}$. Let $\Delta^{(u)}_{k_1, \ldots, k_r}$ be a system of intervals on $[0, 1]$ open on the right defined by the conditions:

A. $\Delta^{(u)}_{k_1, \ldots, k_{r-1}, i} \cap \Delta^{(u)}_{k_1, \ldots, k_{r-1}, j} = \varnothing$ for $i < j$ and $\Delta^{(u)}_{k_1, \ldots, k_{r-1}, i}$ is situated to the left of $\Delta^{(u)}_{k_1, \ldots, k_r}$;
B. the length of $\Delta^{(u)}_{k_1, \ldots, k_r}$ equals $\mu_u(A_{k_1, \ldots, k_r})$;
C. $\bigcup_{k_r} \Delta^{(u)}_{k_1, \ldots, k_{r-1}} = \Delta^{(u)}_{k_1, \ldots, k_{r-1}}$ $\bigcup_k \Delta^{(u)}_k = [0, 1)$.

Set $f_n(\alpha, u) = x_{k_1, \ldots, k_n}$ for $\alpha \in \Delta^{(u)}_{k_1, \ldots, k_n}$. Then

$$r(f_{n+1}(\alpha, u), \quad f_n(\alpha, u)) \leq 2^{-n},$$

since $f_{n+1}(\alpha, u)$ and $f_n(\alpha, u)$ belong to the same set $A_{k_1, \ldots, k_n}$. Hence $f_n(\alpha, u)$ converges uniformly to a function $f(\alpha, u)$ with values in X. Let $g(x)$ be a function belonging to $C(X)$. Then

$$\begin{aligned}\int_0^1 g(f(\alpha, u))m(d\alpha) &= \lim_{n\to\infty} \sum g(x_{k_1, \ldots, k_n})m(\Delta^{(u)}_{k_1, \ldots, k_n}) \\ &= \lim_{n\to\infty} \sum g(x_{k_1, \ldots, k_n})\mu_u(A_{k_1, \ldots, k_n}) = \lim_{n\to\infty} \int g_n(x)\mu_u(dx) \\ &= \int g(x)\mu_u(dx),\end{aligned}$$

where $g_n(x) = g(x_{k_1, \ldots, k_n})$ for $x \in A_{k_1, \ldots, k_n}$, $g_n(x) \to g(x)$ in view of the continuity of g and condition 1)); moreover $|g_n(x)|$ are jointly bounded by the bound $\sup_x |g(x)|$. Hence the equality (2.5) is valid. If, for a pair u_1 and u_2 $\mu_{u_1} = \mu_{u_2}$, then $\Delta^{(u_1)}_{k_1, \ldots, k_n} = \Delta^{(u_2)}_{k_1, \ldots, k_n}$, for any n and $k_1, k_2, \ldots, k_n$. Therefore $f_n(\alpha, u_1) = f_n(\alpha, u_2)$ and $f(\alpha, u_1) = f(\alpha, u_2)$. The measurability of $f(\alpha, u)$ with respect to $\mathfrak{U}_{[0, 1]} \times \mathfrak{B}$ follows from the measurability of $f_n(\alpha, u)$; the latter property is a corollary from equality

$$\{(\alpha, u) : f_n(\alpha, u) = x_{k_1, k_2, \ldots, k_n}\}$$
$$= \left\{(\alpha, u) : \sum_{i=1}^{k_1 - 1} \mu_u(A_i) + \sum_{i=1}^{k_2 - 1} \mu_u(A_{k_1, i}) + \cdots \right.$$
$$\left. + \sum_{i=1}^{k_n} \mu_u(A_{k_1, \ldots, k_n, i}) \le \alpha < \mu_u(A_{k_1, \ldots, k_n})\right\}$$

and measurability of functions $\mu_u(A)$ for all $A \in \mathfrak{B}$. □

Remark. Assume that all measures μ_u are absolutely continuous with respect to a measure μ. Then the function $f(\alpha, u)$ can be chosen in such a manner that for any point u_0 such that for all $g \in C_x$ the limit

$$\lim_{u \to u_0} \int g(x) \mu_u(dx) = \int g(x) \mu_{u_0}(dx)$$

(i.e. u_0 is a point of weak continuity for μ_u), $f(\alpha, u)$ is continuous in u for almost all α. Indeed, in this case one can choose $A_{k_1, \ldots, k_n}$ in such a manner that for all u, $\mu_u(A_{k_1, \ldots, k_n}) = 0$ (A' is the boundary of A). Then $m(\Delta_{k_1, \ldots, k_n})$ is continuous in u at point u_0.

Therefore for all points α located in the interior of an interval $\Delta^{(u_0)}_{k_1, \ldots, k_n}$ the equality $\lim_{u \to u_0} f(\alpha, u) = f(\alpha, u_0)$ is satisfied. Hence if α does not coincide with either one of the end-points of the intervals $\Delta^{(u_0)}_{k_1, \ldots, k_n}$, we have

$$\overline{\lim_{u \to u_0}} \; r(f(\alpha, u), f(\alpha, u_0)) \le \overline{\lim_{u \to u_0}} \; [r(f_n(\alpha, u), f(\alpha, u))$$
$$+ r(f_n(\alpha, u), f(\alpha, u_0))] \le 2^{-n+2}.$$

Thus for all α, except possibly for a countable number of them, $\lim_{u \to u_0} r(f(\alpha, u), f(\alpha, u_0)) = 0$.

Lemma 2.3. *Let the family of distributions $\mu(\cdot/v, u(\cdot))$ defined on the σ-algebra $\mathfrak{A}_{[0, T]}$, where $v \in (V, \mathfrak{C})$ is a measurable set and $u(\cdot) \in U^{[0, T]}$, satisfy the following conditions:*

1. *$\mu(\cdot/v, u(\cdot))$ is measurable with respect to $\mathfrak{C} \times \mathfrak{B}^{[0, T]}$;*
2. *if $\Gamma \in \mathfrak{A}_{[0, T]}$ and $u_1(s) = u_2(s)$ for $s \le t$, then*

$$\mu(\Gamma/v, u_1(\cdot)) = \mu(\Gamma/v, u_2(\cdot)).$$

Then there exists a function $\delta(\alpha, v, u(\cdot))$ defined on $[0, T] \times V \times U^{[0, T]}$ and measurable with respect to $\mathfrak{A}_{[0, T]} \times \mathfrak{C} \times \mathfrak{B}^{[0, T]}$ possessing the following properties:

A. $m(\{\alpha : \delta(\alpha, v, u(\cdot)) \in \Gamma\}) = \mu(\Gamma/v, u(\cdot))$ *for all Borel sets* $\Gamma \in [0, T]$;
B. *if* $u_1(s) = u_2(s)$, $s < t$, *then* $\delta(\alpha, v, u_1(\cdot)) = \delta(\alpha, v, u_2(\cdot))$ *for all* α *such that* $\delta(\alpha, v, u_1(\cdot)) \le t$.

Proof. For a given distribution function $F(\cdot)$ define $F^{-1}(\cdot)$ to be the function which is equal to t on the interval $[F(t-0), F(t)]$. Denote

$$F(t, v, u(\cdot)) = \mu([0, t)/v, u(\cdot)).$$

In view of assumption 2 $F(s, v, u_1(\cdot)) = F(s, v, u_2(\cdot))$, $s \le t$, provided $u_1(s) = u_2(s)$, $s < t$. Setting

$$\delta(\alpha, v, u(\cdot)) = F^{-1}(\alpha, v, u_1(\cdot)),$$

we obtain a function which is measurable jointly in the variables. Clearly assertion A is valid for this function. Let $u_1(s) = u_2(s)$ for $s < t$. Then $F(s, v, u_1(\cdot)) = F(s, v, u_2(\cdot))$ for $s \le t$ and hence $F^{-1}(\alpha, v, u_1(\cdot)) = F^{-1}(\alpha, v, u_2(\cdot))$ if $\alpha \in [0, F(t, v, u_1(\cdot))]$ i.e. $\delta(\alpha, v, u_2(\cdot)) < t$. □

We now proceed to construct a step controlled object defined by the conditional distributions P_k and λ_k. Utilizing lemmas 2.2 and 2.3 one can construct functions $\delta_k(\alpha, x_0, \ldots, x_{k-1}, \tau_1, \ldots, \tau_{k-1}, u(\cdot))$, $f_k(\alpha, x_0, x_1, \ldots, x_{k-1}, \tau_1, \ldots, \tau_{k-1}, u(\cdot))$ defined on $[0, 1] \times X^k \times [0, T]^{k-1} \times U^{[0, T]}$ and $[0, 1] \times X^k \times [0, 1]^k \times U^{[0, T]}$ taking on values in $[0, T]$ and X respectively, measurable jointly in the variables, and such that the following conditions are satisfied:

1. if Γ is a Borel set in $[0, T]$, then

$$m(\{\alpha : \delta_k(\alpha, x_0, \ldots, x_{k-1}, \tau_1, \ldots, \tau_{k-1}, u(\cdot)) \in \Gamma\}) = \lambda_k(\Gamma/x_0, \ldots, x_{k-1}, \tau_1, \ldots, \tau_{k-1}, u(\cdot));$$

   if $A \in \mathfrak{A}$, then

$$m(\{\alpha : f_k(\alpha, x_0, \ldots, x_{k-1}, \tau_1, \ldots, \tau_k, u(\cdot)) \in A\}) = P_k(A/x_0, \ldots, x_{k-1}, \tau_1, \ldots, \tau_k, u(\cdot));$$

2. if $\tau_1 < \cdots < \tau_{k-1} < t$, $u_1(s) = u_2(s)$ for $s < t$, then

$$\delta_k(\alpha, x_0, \ldots, x_{k-1}, \tau_1, \ldots, \tau_{k-1}, u_1(\cdot)) = \delta_k(\alpha, x_0, \ldots, x_{k-1}, \tau_1, \ldots, \tau_{k-1}, u_2(\cdot))$$

   for all α such that $\delta_k(\alpha, x_0, \ldots, x_{k-1}, \tau_1, \ldots, \tau_{k-1}, u_1(\cdot)) \le t$.

Now let $\beta_0, \alpha_1, \beta_1, \alpha_2, \beta_2, \ldots$—be a sequence of independent uniformly distributed variables on a probability space. Set

$$\left.\begin{aligned}
&\hat{x}_0(\omega) = f_0(\beta_0), \ \hat{\tau}_1(\omega, u(\cdot)) = \delta_1(\alpha_1, f_0(\beta_0), u(\cdot)),\\
&\hat{x}_1(\omega, u(\cdot)) = f_1(\beta_1, \hat{x}_0(\omega), \hat{\tau}_1(\omega, u(\cdot)), u(\cdot)), \ldots,\\
&\hat{\tau}_k(\omega, u(\cdot)) = \delta_k(\alpha_k, \hat{x}_0(\omega), \ldots, \hat{x}_k(\omega, u(\cdot)),\\
&\hat{\tau}_1(\omega, u(\cdot)), \ldots, \hat{\tau}_{k-1}(\omega, u(\cdot)), u(\cdot)),\\
&\hat{x}_k(\omega, u(\cdot)) = f_k(\beta_k, \hat{x}_0(\omega), \ldots, \hat{x}_{k-1}(\omega)u(\cdot)),\\
&\hat{\tau}_1(\omega, u(\cdot)), \ldots, \hat{\tau}_k(\omega, u(\cdot)), u(\cdot)), \ldots
\end{aligned}\right\} \tag{2.6}$$

Conditions 1 and 2 imply the following properties of the random variables constructed above: if $u_1(s) = u_2(s)$ for $s < t$, then for all ω such that $\hat{\tau}_k(\omega, u_1(\cdot)) \leq t$ the equalities

$$\hat{\tau}_k(\omega, u_1(\cdot)) = \hat{\tau}_k(\omega, u_2(\cdot)),$$
$$\hat{x}_k(\omega, u_1(\cdot)) = \hat{x}_k(\omega, u_2(\cdot))$$

are satisfied (these equalities are verified using the induction argument and taking into account that $\hat{\tau}_k$ increases with k).

The joint distribution of the variables $\hat{x}_0(\omega), \hat{\tau}_1(\omega, u(\cdot)), \ldots, \hat{\tau}_k(\omega, u(\cdot)), \hat{x}_k(\omega, u(\cdot))$ coincides with the joint distribution of the variables $x(0), \tau_1, \ldots, \tau_k, x(\tau_k)$ with respect to the probability measure $\mu(\cdot/u)$ since the conditional distributions of each one of the variables in the chain given the previous one are the same. Therefore the function

$$\xi(t, \omega, u(\cdot)) = \hat{x}_k(\omega, u(\cdot))$$
$$\hat{\tau}_{k-1}(\omega, u(\cdot)) \leq t < \hat{\tau}_k(\omega, u(\cdot))$$

for $(\tau_0 = 0)$ is the required representation of the controlled object. Conditions 1 and 2 are satisfied by construction. Condition 3 is a corollary of lemma 2.1.

Two representations of the controlled object $\xi(t, \omega, u(\cdot))$ and $\xi'(t, \omega, u(\cdot))$ (possibly even on different probability spaces) are called *stochastically equivalent* if the joint distributions

$$\xi(t_1, \omega, u_1(\cdot)), \ldots, \xi(t_n, \omega, u_n(\cdot)) \quad \text{and}$$
$$\xi'(t_1, \omega, u_1(\cdot)), = \ldots, \xi'(t_n, \omega, u_n(\cdot))$$

coincide for any $t_1, \ldots, t_n \in [0, T]$, $u_1(\cdot), \ldots, u_n(\cdot) \in U^{[0, T]}$. Clearly the procedure presented above uniquely determines (up to stochastic equivalence) the representation of a step controlled object provided only the subdivision $\{A_{i_1, \ldots, i_n}\}$ of the space X which was utilized in lemma 2.2 is fixed.

Consider now a controlled object $\mu(\cdot/\cdot)$ for which $\mu(D_{[0, T]}(X)/u(\cdot)) = 1$

for all $u(\cdot) \in U^{[0,T]}$, where $D_{[0,T]}(X)$ is the space of functions $x(t)$ without discontinuities of the second kind, continuous from the right and continuous at $t = T$. Denote by ρ_D a metric on $D_{[0,T]}(X)$. Introduce in $D_{[0,T]}(X)$ a family of operators S_ε:

$$[S_\varepsilon x](t) = x(\tau_k) \quad \text{for} \quad \tau_k \leq t < \tau_{k+1},$$

where $0 = \tau_0 < \tau_1 < \cdots$ are the points which are recursively defined by the relation $\tau_{i+1} = \inf[t > \tau_i, \ r(x(t), \ x(\tau_i)) > \varepsilon]$; if however, $\sup_{t > \tau_i} r(x(t), x(\tau_i)) \leq \varepsilon$ we then set $\tau_{i+1} = T$, $x(T) = x(\tau_{i+1})$. The inequality

$$\sup_t r([S_\varepsilon x](t), x(t)) \leq \varepsilon$$

is valid by the very definition of $[S_\varepsilon x](t)$.

Denote by $\mathfrak{A}^{[0,T]}(D)$ the σ-algebra of Borel sets in $D_{[0,T]}$. It is easy to see that the mapping $[S_\varepsilon x]$ measurably maps $D_{[0,T]}(X)$ into $D_{[0,T]}(X)$. Next denote by $\mu_\varepsilon(A/u(\cdot))$ the controlled object defined by the equality

$$\mu_\varepsilon(A/u(\cdot)) = \mu_\varepsilon(\{x(\cdot) : [S_\varepsilon x](\cdot) \in A\}/u(\cdot)).$$

It is easy to verify that this is indeed a controlled object since for any cylinder A with basis on $[0, t]$ $\{x(\cdot) : [S_\varepsilon x](\cdot) \in A\} \in \mathfrak{A}_t(D)$, where $\mathfrak{A}_t(D)$ is the σ-algebra in $\mathfrak{A}^{[0,T]}(D)$, generated by cylinders with their basis on $[0, t]$.

Denote by $\xi_\varepsilon(t, \omega, u(\cdot))$ a representation of the probability object $\mu_\varepsilon(\cdot/\cdot)$. It is natural to expect that if the limit of $\xi_\varepsilon(t, \omega, u(\cdot))$ exists (in a certain sense) as $\varepsilon \to 0$, this limit is then a representation of the probability object $\mu(\cdot/\cdot)$. We present some necessary conditions for the existence of such a limit. The representation constructed above may turn out to be stochastically continuous in $u(\cdot)$. For this it is sufficient that the following conditions be satisfied: measures $\mu(dx/\cdot)$ be absolutely continuous with respect to a measure $\pi(dx)$ on $\mathfrak{A}$ and $A_{k_1, \ldots, k_n}$ in lemma 2.2 be chosen in such a manner that $\pi(A'_{k_1, \ldots, k_n}) = 0$ where A' is the boundary of A.

Theorem 2.1. *Let X be a locally compact space and let the controlled object $\mu(\cdot/u(\cdot)$ satisfy the conditions:*

1. $\mu(D_{[0,T]}(X)/u(\cdot)) = 1$ *for all* $u(\cdot) \in D_{[0,T]}(U)$;
2. $\lim_{R \to \infty} \sup_{u(\cdot) \in D_{[0,T](u)}} \mu(D_{[0,T]}(X) \cap \{x(\cdot);$

$$\rho_D(x(\cdot), x_0(\cdot)) \geq R/u(\cdot)) = 0;$$

3. *there exist representations of controlled objects* $\mu_\varepsilon(\cdot/u(\cdot))$, $\xi_\varepsilon(t, \omega, u(\cdot))$, *such that* $\xi_\varepsilon(t, \omega, u(\cdot))$ *is stochastically continuous jointly in the variables* $t \in [0, T]$ *and* $u(\cdot) \in D_{[0,T]}(U)$ *uniformly in* ε.

Then there exists a representation of the controlled object $\mu(\cdot/u(\cdot))$ *for* $u(\cdot) \in D_{[0,T]}(U)$, $\xi(t, \omega, u(\cdot))$, *which is stochastically continuous jointly in the variables.*

PROOF. Utilizing the separability of $D_{[0,T]}(U)$ one can choose a countable set $\Lambda \subset D_{[0,T]}(U)$ which is dense in $D_{[0,T]}(U)$. Condition 2 of the theorem implies that for any $t_1, \ldots, t_n \in [0, T]$ and $u_1(\cdot), \ldots, u_n(\cdot) \in \Lambda$ the joint distribution of the variables

$$\xi_\varepsilon(t_1, \omega, u_1(\cdot)), \ldots, \xi_\varepsilon(t_n, \omega, u_n(\cdot))$$

for different $\varepsilon > 0$ form a compact set since

$$\begin{aligned}\mathbf{P}\{r(\xi_\varepsilon(t_n, \omega, u_n(\cdot)), x_0(\cdot)) \le R\} \\ \le \mu(\{x(\cdot): \rho_D(x(\cdot), x_0(\cdot)) \le R + \varepsilon\}/u(\cdot)).\end{aligned}$$

Therefore if $\{t_k\}$ is a sequence dense in $[0, T]$ and $\Lambda = \{u_1(\cdot), u_2(\cdot), \ldots\}$, one can choose a subsequence $\varepsilon_k \downarrow 0$ such that for all n and $i_1, \ldots, i_n$ the joint distribution of the variables

$$\xi_{\varepsilon_k}(t_{i_1}, \omega, u_1(\cdot)), \ldots, \xi_{\varepsilon_k}(t_{i_n}, \omega, u_n(\cdot))$$

converge to a limit distribution

$$\begin{aligned}P(t_{i_1}, \ldots, t_{i_n}, u_1(\cdot), \ldots, u_n(\cdot); A_1, \ldots, A_n) \\ = \lim_{n\to\infty} \mathbf{P}\{\xi_{\varepsilon_k}(t_{i_j}, \omega, u_j(\cdot)) \in A_j, \quad j = 1, \ldots, n\} \qquad (2.7) \\ (A_i \in \mathfrak{A}).\end{aligned}$$

Utilizing the uniform stochastic continuity of random functions $\xi_\varepsilon(t, \omega, u(\cdot))$ we verify that the limits

$$\lim_{\substack{t_{i_1}\to t^{(1)}, \ldots, t_{i_n}\to t^{(n)} \\ \rho_D(u_{j_1}, u^{(1)})\to 0 \\ \cdots\cdots\cdots\cdots \\ \rho_D(u_{j_n}, u^{(n)})\to 0}} \mathbf{E} f_1(\xi_{\varepsilon_k}(t_{i_1}, \omega, u_{j_1}(\cdot))) \cdots f_n(\xi_{\varepsilon_k}(t_{i_n}, \omega, u_{j_n}(\cdot)))$$

exist for $f_i \in C_X$ and that these limits can be represented in the form

$$\int \cdots \int f_1(x_1) \cdots f_n(x_n) P(t^{(1)}, \ldots, t^{(n)}, u^1(\cdot), \ldots, u^n(\cdot), dx_1, \ldots, dx_n),$$

where $P(t^{(1)}, \ldots, t^{(n)}, u^{(1)}, \ldots, u^{(n)}, A_1, \ldots, A_n)$ is the continuation in weak continuity of the distributions (2.7). Therefore these are limit distributions for finite-dimensional marginal distributions of the random functions $\xi_{\varepsilon_k}(t, \omega, u(\cdot))$. A random function $\xi(t, \omega, u(\cdot))$ such that

$$\begin{aligned}\mathbf{P}\{\xi(t^{(1)}, \omega, u^{(1)}(\cdot)) \in A_1, \ldots, \xi(t^{(n)}, \omega, u^{(n)}(\cdot)) \in A_n\} \\ = P(t^{(1)}, \ldots, t^{(n)}, u^{(1)}(\cdot), \ldots, u^{(n)}(\cdot), A_1, \ldots, A_n)\end{aligned}$$

is the required representation. The existence of such a function follows from Kolmogorov's theorem (see, e.g. [20], Vol I, p. 46 theorem 2). Indeed, for $f_k \in C_X$ and $t_k \in [0, T]$ we have

$$\mathbf{E} \prod_{j=1}^{n} f_j(\xi(t_i, \omega, u(\cdot))) = \lim_{k\to\infty} \mathbf{E} \prod_{j=1}^{n} f_j(\xi_{\varepsilon_k}(t_j, \omega, u(\cdot)))$$

$$= \lim_{\varepsilon\to 0} \int f_1(x_1) \cdots f_n(x_n)\mu_\varepsilon(\{x(\cdot): x(t_1) \in dx_1, \ldots, x(t_n) \in dx_n\}/u(\cdot))$$

$$= \int f_1(x_1) \cdots f_n(x_n)\mu(\{x(\cdot): x(t_1) \in dx_1, \ldots, x(t_n) \in dx_n\}/u(\cdot)).$$

Therefore condition 1 is satisfied. Let $u_1(s) = u_2(s)$, $s \le t$. Then for $t_1 < \cdots < t_n \le t$

$$\mathbf{E} \exp\left\{- \sum_{j=1}^{n} r(\xi(t_j, \omega, u_1(\cdot)), \xi(t_j, \omega, u_2(\cdot)))\right\}$$

$$= \lim_{k\to\infty} \mathbf{E} \exp\left\{- \sum_{j=1}^{n} r(\xi_{\varepsilon_k}(t_j, \omega, u_1(\cdot)), \xi(t_j, \omega, u_2(\cdot)))\right\} = 1.$$

Hence,

$$\mathbf{P}\{\xi(t_j, \omega, u_1(\cdot)) = \xi(t_j, \omega, u_2(\cdot)), \quad j = 1, \ldots, n\} = 1.$$

The last relation and the expressions $\mathbf{P}\{\xi(\cdot, \omega, u(\cdot)) \in D_{[0, T]}(X)\} = \mu(D_{[0, T]}(X)/u(\cdot)) = 1$ imply that

$$\mathbf{P}\{\xi(s, \omega, u_1(\cdot)) = \xi(s, \omega, u_2(\cdot)), s \le t\} = 1.$$

Finally, the measurability of $\xi(s, \omega, u_1(\cdot))$ jointly in the variables may be obtained by passing to a stochastically equivalent process (cf [20] Vol I, p. 171, theorem 1). □

If the measures $\mu(\cdot/u(\cdot))$ for all $u(\cdot)$ correspond to a class of stochastic processes which is effectively described by means of certain characteristics, then the construction of the controlled object reduces to the construction of a stochastic process with the given characteristics. Consider, for example, the cases of controlled objects with independent increments and a Gaussian controlled object.

Let $\mu(\cdot/u(\cdot))$ for any $u(\cdot)$ correspond to a stochastically continuous stochastic process with independent increments in R^m. Denote its characteristic function:

$$\varphi(t, z; u(\cdot)) = \exp\left\{i(z, \gamma(t, u(\cdot))) - \frac{1}{2}(B(t, u(\cdot))z, z) + \int \left(e^{i(z, x)} - 1 - \frac{i(z, x)}{1 + |x|^2}\right) M_t(dx; u(\cdot))\right\}.$$

where $\gamma(t, u(\cdot))$ is a function continuous in t with values in R^m; $B(t, u(\cdot))$ is a continuous function in t whose values are non-negative symmetric operators in R^m and $M_t(A; u(\cdot))$ is a measure on R^m, continuously dependent on t, non-decreasing in t, such that $M_t(\{0\}, u(\cdot)) = 0$ and

$$\int \frac{(x, x)}{1 + (x, x)} M_t(dx, u(\cdot)) < \infty.$$

Assume that $B(t, u(\cdot))$ and $M_t(A, u(\cdot))$ are absolutely continuous in t:

$$B(t; u(\cdot)) = \int_0^t V(s, u(\cdot))ds, \qquad M_t(A, u(\cdot)) = \int_0^t \Pi_s(A, u(\cdot))\, ds.$$

Let $V^{1/2}$ be the positive square root of the positive operator V and the function $f(s, y, u(\cdot))$ with values in R^m be such that

$$\Pi_s(A; u(\cdot)) = \int_{\{f(s,\, y,\, u(\cdot)) \in A\}} \frac{dy}{|y|^{m+1}} \quad \text{and} \quad \int_{|y| \leq 1} |f|^2 \frac{dy}{|y|^{m+1}} < \infty$$

(the existence of such a function follows from Lemma 2.2).

Denote, furthermore,

$$\begin{aligned}\xi(t, \omega, u(\cdot)) &= \gamma_1(t_1, u(\cdot)) + \int_0^t V^{1/2}(s, u(\cdot))\, dw(s)\\ &\quad + \int_0^t \int f(s, y, u(\cdot)) \left[\nu(ds \times dy) - \chi_{\{|y| \leq 1\}} \frac{ds\, dy}{|y|^{m+1}} \right],\end{aligned} \tag{2.8}$$

where

$$\begin{aligned}\gamma_1(t, u(\cdot)) &= \gamma(t, u(\cdot)) + \int_0^t \int_{|y| \leq 1} \frac{|f|^2}{1 + |f|^2} f(s, y, u(\cdot)) \frac{ds\, dy}{|y|^{m+1}}\\ &\quad - \int_0^t \int_{|y| > 1} \frac{1}{1 + |f|^2} f(s, y, u(\cdot)) \frac{ds\, dy}{|y|^{m+1}},\end{aligned}$$

$w(t)$ is a Wiener process in R^m, $\nu(A)$ is a Poisson measure with independent values in $[0, T] \times R^m$ such that $\mathbf{E}\nu(ds\, dy) = ds\, dy/|y|^{m+1}$.

If the characteristics of the process γ, B and M depend measurably on $u(\cdot)$, then (2.8) is a representation of a controlled object.

A controlled object $\mu(\cdot/u(\cdot))$ is called *Gaussian* if for all $u(\cdot)$ the measure $\mu(\cdot/u(\cdot))$ corresponds to a Gaussian process in R^m. Such a process is characterized by its mean value $a(t, u(\cdot))$—which is a function with values in R^m—and the correlation function $B(t, s, u(\cdot))$ defined on $[0, T] \times [0, T]$ whose values are linear operators in R^m; moreover, $B(t, s, u(\cdot)) = B^*(s, t, u(\cdot))$ (B^* is the conjugate of B). Clearly, $a(t, u(\cdot))$ depends on values of $u(s)$ for $s < t$, and $B(t, s, u(\cdot))$ on values of $u(\tau)$ for $\tau < s \vee t$. In the case when a Gaussian process is stochastically continuous, the functions $a(t, \cdot)$ and $B(t, s, \cdot)$ are continuous jointly in variables t and s; however, if these functions define a Gaussian controlled object they are measurable with respect to $u(\cdot)$. Assume that the operator function $B(t, s, u(\cdot))$ admits factorization:

$$B(t, s, u(\cdot)) = \int_0^s V(t, \tau)V^*(s, \tau)\, d\tau, \qquad (s < t),$$

where $V(t, \tau)$ is an operator function which depends on the value of $u(\cdot)$ on $[0, t]$. Then

$$\xi(t, u(\cdot)) = a(t, u(\cdot)) + \int_0^t V(t, \tau, u(\cdot))\, dw(\tau), \tag{2.9}$$

where $w(\tau)$ is a Wiener process in R^m, will be a representation of a Gaussian controlled object.

Consider the possibility of "attaching" a given non-randomized control $\{u(t) = \varphi_t(x(\cdot))\}$, where $\varphi_t(x(\cdot))$ is a $\mathfrak{A}_t$-measurable function for all t, to a controlled object defined by its representation $\xi(t, \omega, u(\cdot))$. It is natural to assume that $(\xi(t), \eta(t))$ is a controlled process with a given representation of the controlled object $\xi(t, \omega, u(\cdot))$ and control $\varphi_t(x(\cdot))$ provided the equalities

$$\begin{cases} \xi(t) = \xi(t, \omega, \eta(\cdot)), \\ \eta(t) = \varphi_t(\xi(\cdot)) \end{cases}$$

are satisfied.

Thus the basic process $\xi(t)$ must satisfy the relation

$$\xi(t) = \xi(t, \omega, \varphi(\xi(\cdot))), \tag{2.10}$$

which should be viewed as an equation in $\xi(\cdot)$.

To begin with, we shall assume that $\varphi_t(\xi(\cdot))$ is a step function with jumps at the points $0 < t_1 < t_2 \cdots < t_r < T$ (t_i are non-random). Then

$$\begin{aligned} \varphi_s(\xi(\cdot)) &= \varphi_0(\xi(0)) \quad \text{for } s < t_1; \\ \varphi_s(\xi(\cdot)) &= \varphi_{t_k}(\xi(\cdot)) \quad \text{for } t_k \le s < t_{k+1},\ k < r; \\ \varphi_s(\xi(\cdot)) &= \varphi_{t_r}(\xi(\cdot)) \quad \text{for } t_r \le s \le T, \end{aligned}$$

and the solution of (2.10) exists and is unique:

$$\begin{aligned} \xi(t) &= \xi(t, \omega, \varphi_0(\xi(0))) \quad \text{for } s < t_1; \\ \xi(t) &= \xi(t, \omega, \varphi_{t_1}(\xi(\cdot))) \quad \text{for } t_1 \le s < t_2 \end{aligned}$$

(the variable $\varphi_{t_1}(\xi(\cdot))$ is determined by the values of $\xi(\cdot)$ for $s \le t_1$) and so on. If no restrictions are imposed on the control it is unlikely that the existence of equation (2.10) could be established. Below we present a general theorem concerning the existence of a solution for (2.10) under specified restrictions on the representation of the probabilistic object and the control.

We shall assume that the probability space $(\Omega, \mathfrak{S}, \mathbf{P})$ is of the form $\{C_{[0,T]}(Z), \mathfrak{C}, \mathbf{P}\}$ where $C_{[0,T]}(Z)$ is the space of continuous functions on

$[0, T]$ taking on values in a complete metric space Z; $\mathfrak{C}$ is the σ-algebra generated by the cylinders in $C_{[0,T]}(Z)$ and $\mathbf{P}$ is a probability measure on $C_{[0,T]}(Z)$. We shall denote by $\mathfrak{C}_t$ the σ-algebra generated by cylinders with bases on $[0, t]$. Concerning the representation of the probabilistic object $\xi(t, z(\cdot), u(\cdot))$ we shall assume in addition to the measurability jointly in the variables also the measurability for each t with respect to the σ-algebra $\mathfrak{C}_t \times \mathfrak{B}_t$.

We shall assume that X and U are complete metric separable spaces. $D_{[0,T]}(X)$ and $D_{[0,T]}(U)$ are spaces of functions defined on $[0, T]$ without discontinuities of the second kind and with values in X and U respectively.

Theorem 2.2. *Let the following conditions be satisfied:*

a. $\xi(t, z(\cdot), u(\cdot))$ *maps* $C_{[0,T]}(Z) \times D_{[0,T]}(U)$ *continuously into* $D_{[0,T]}(X)$;
b. *for any* $\varepsilon > 0$ *one can find a compact set* $K_\varepsilon \subset D_{[0,T]}(X)$ *and a set* $C \subset C_{[0,T]}(Z)$ *such that* $\mathbf{P}(C) > 1 - \varepsilon$ *and*

$$\xi(\cdot, z(\cdot), u(\cdot)) \in K_\varepsilon,$$

provided $z(\cdot) \in C$ *and* $u(\cdot) \in D_{[0,T]}(U)$;
c. $\varphi_t(x(\cdot))$ *maps* $D_{[0,T]}(X)$ *continuously into* $D_{[0,T]}(U)$.

Then there exists a solution of equation (2.10).

Proof. Let $\varphi_t^{(n)}(x(\cdot))$ be a sequence of continuous functions such that the following conditions are satisfied:

1. $\varphi_t^{(n)}(x(\cdot))$ are constant for $t \in (kT/n, (k+1)T/n)$, $k = 1, \ldots, n-1$;
2. On any compact set $K \subset D_{[0,T]}(X)$,

$$\lim_{n\to\infty} \sup_{x(\cdot)\in K} \rho_D(\varphi^{(n)}(x(\cdot)), \quad \varphi(x(\cdot))) = 0.$$

Denote by $\xi^{(n)}(t)$ the solution of the equation

$$\xi^{(n)}(t) = \xi(t, z(\cdot), \varphi^{(n)}(\xi^{(n)}(\cdot)))$$

(it was shown above that this solution exists and is unique). Let

$$\eta^{(n)}(t) = \varphi_t^{(n)}(\xi^{(n)}(\cdot)).$$

Furthermore, denote by π_n a measure in the space $C_{[0,T]}(Z) \times D_{[0,T]}(X) \times D_{[0,T]}(U)$ which corresponds to the compound process $(z(t), \xi^{(n)}(t), \eta^{(n)}(t))$ (the measure π_n is defined on the product of σ-algebras of Borel sets of the corresponding metric spaces).

Let C be a compact set in $C_{[0,T]}(Z)$ such that $\mathbf{P}(C) > 1 - \varepsilon$ and let for $z(\cdot) \in C$

$$\xi(\cdot, z(\cdot), u(\cdot)) \in K_\varepsilon$$

for all $u(\cdot)$ where K_ε is a compact in $D_{[0,T]}(X)$ whose existence is guaranteed by the condition "b" of the theorem. It follows from assumptions 1 and 2 that

$$K^U = \varphi(K_\varepsilon) \cup \left(\bigcup_n \varphi^{(n)}(K_\varepsilon)\right)$$

is also a compact in $D_{[0,T]}(U)$ ($\varphi(K_\varepsilon)$ and $\varphi^{(n)}(K_\varepsilon)$ are the images of the set K_ε under the mappings φ and $\varphi^{(n)}$). Since for $z(\cdot) \in C$ $\xi^{(n)} \in K_\varepsilon$ and $\eta^{(n)} \in K^U$, it follows that

$$\pi_n(C \times K_\varepsilon \times K^U) \geq 1 - \varepsilon.$$

Consequently the sequence of measures π_n is weakly compact. Without loss of generality one can assume that π_n is weakly convergent to a measure π.

Let

$$\psi(z(\cdot), x(\cdot), u(\cdot)) = \exp\{-\rho_D(x(\cdot), \xi(\cdot, z(\cdot), u(\cdot))) - \rho_D(u(\cdot), \varphi(x(\cdot)))\}.$$

This is a continuous bounded functional on $C_{[0,T]}(Z) \times D_{[0,T]}(X) \times D_{[0,T]}(U)$. Thefore

$$\lim_{n\to\infty} \psi(z(\cdot), x(\cdot), u(\cdot))\pi_n(dz \times dx \times du) = \int \psi(z(\cdot), x(\cdot), u(\cdot))\pi(dz \times dx \times du).$$

However,

$$\begin{aligned}\int \psi(z(\ ), x(\ ), u(\ ))\pi_n(dz \times dx \times du) &= \mathbf{E} \exp\{-\rho_D(\xi^{(n)}(\cdot), \xi(\cdot, z(\cdot), \eta^{(n)}(\cdot)) - \rho_D(\eta^{(n)}(\cdot), \varphi(\xi^{(n)}(\cdot)))\} \\ &= \mathbf{E} \exp\{-\rho_D(\eta^{(n)}(\cdot), \varphi(\xi^{(n)}(\cdot)))\} \\ &\geq \mathbf{E} \exp\{-\rho_D(\varphi_0^{(n)}(\xi^{(n)}(\cdot)), \varphi_0(\xi^{(n)}(\cdot))\}\end{aligned}$$

(since $\rho_D(\xi^{(n)}(\cdot), \xi(\cdot, z(\cdot), \eta^{(n)}(\cdot)) = 0$, $\rho_D(\eta^{(n)}(\cdot), \varphi_0^{(n)}(\xi^{(n)}(\cdot))) = 0$). Since for $\xi^{(n)}(\cdot) \in K_\varepsilon$ $\rho_D(\varphi(\xi^{(n)}(\cdot)), \varphi^{(n)}(\xi^{(n)}(\cdot))) \to 0$ in view of condition 2, we have

$$\lim_{n\to\infty} \mathbf{E} \exp\{-\rho_D(\varphi_0^{(n)}(\xi^{(n)}(\cdot)), \varphi_0(\xi^{(n)}(\cdot)))\} \geq 1 - \varepsilon$$

for any $\varepsilon > 0$. Hence

$$\int \psi(z(\cdot), x(\cdot), u(\cdot))\pi(dz \times dx \times du) \geq 1.$$

It follows from the form of the functional ψ that

$$x(t) = \xi(t, z(\cdot), u(\cdot)), \qquad u(t) = \varphi_t(x(\cdot))$$

almost everywhere in measure π. □

3 Optimization Problem; Approximation Theorem

Assume that a controlled object is defined by means of its representation $\xi(t, \omega, u(\cdot))$ on a given probability space $(\Omega, \mathfrak{S}, \mathbf{P})$. We shall consider only those controls $u(\cdot)$ which belong to $D_{[0,T]}(U)$. We shall assume that under this condition $\xi(\cdot, \omega, u(\cdot)) \in D_{[0,T]}(X)$. Furthermore, let a function $F(x(\cdot), u(\cdot))$ which represents the cost of control be defined on $D_{[0,T]}(X) \times D_{[0,T]}(U)$. The problem consists in determining a control for which the quantity $\mathbf{E}_v F(\xi(\cdot), \eta(\cdot))$ is minimized. Here v is a control v which can be "attached" to the controlled object, $\mathbf{E}_v$ is the mathematical expectation with respect to the measure which corresponds to the pair $(\xi(t), \eta(t))$ where $\xi(t)$ is the basic process and $\eta(t)$ is the control.

Consider a somewhat modified problem. Denote by $\mathfrak{F}_t$ the current of σ-algebras generated by the variables $\xi(s, \omega, u(\cdot))$ for $s \le t$ and $u(\cdot) \in D_{[0,T]}(U)$. In Section 2, generalized controls were considered—i.e. processes $\eta(t)$ on the probability space $\{\Omega, \mathfrak{S}, P\}$ with values in U and measurable with respect to $\mathfrak{F}_t$. We shall consider such controls without discontinuities of the second kind. One can pose the problem of determining a generalized control $\eta(t)$ at which the infimum

$$\inf \mathbf{E}F(\xi(\cdot, \omega, \eta(\cdot)), \eta(\cdot))$$

is attained. It turns out that under very general assumptions equality

$$\inf_v \mathbf{E}_v F(\xi(\cdot), \eta(\cdot)) = \inf_\eta \mathbf{E}(\xi(\cdot, \omega, \eta(\cdot)), \eta(\cdot))$$

is valid. Denote by $\mathcal{N}$ the set of all controls, by $\mathcal{N}_0$ the set of step controls, by $\mathcal{H}$ the set of all generalized controls and by $\mathcal{H}_0$ the set of all generalized step controls.

Theorem 2.3. *Let the function $F(x(\cdot), u(\cdot))$ be bounded and continuous jointly in the variables $x(\cdot)$ and $u(\cdot)$ in metric ρ_D and the representation of the controlled object $\xi(t, \omega, u(\cdot))$ satisfy the condition:*

(A) *for any $\varepsilon > 0$ and a sequence of random processes $\eta_n(t) \in \mathcal{H}$ for which $\rho_D(\eta_n(\cdot), \eta_0(\cdot)) \to 0$ in probability,*

$$\rho_D(S_\varepsilon[\xi(\cdot, \omega, \eta_n(\cdot))], \quad S_\varepsilon[\xi(\cdot, \omega, \eta_0(\cdot))]) \to 0$$

in probability (operators S_ε were introduced in Section 2). Then

$$\begin{aligned}\inf_{v \in \mathcal{N}} \mathbf{E}_v F(\xi(\cdot), \eta(\cdot)) &= \inf_{v \in \mathcal{N}_0} \mathbf{E}_v F(\xi(\cdot), \eta(\cdot)) \\ &= \inf_{\eta \in \mathcal{H}} \mathbf{E}F(\xi(\cdot, \omega, \eta(\cdot)), \eta(\cdot)) = \inf_{\eta \in \mathcal{H}_0} \mathbf{E}F(\xi(\cdot, \omega, \eta(\cdot)), \eta(\cdot)).\end{aligned} \tag{2.11}$$

PROOF. The following inequalities are self-evident:

$$\inf_{\eta \in \mathscr{H}} \mathbf{E}F(\xi(\cdot, \omega, \eta(\cdot)), \eta(\cdot)) \leq \inf_{\eta \in \mathscr{H}_0} \mathbf{E}F(\xi(\cdot, \omega, \eta(\cdot)), \eta(\cdot)),$$

$$\inf_{\eta \in \mathscr{H}} \mathbf{E}F(\xi(\cdot, \omega, \eta(\cdot)), \eta(\cdot)) \leq \inf_{\nu \in \mathscr{N}} \mathbf{E}_\nu F(\xi(\cdot), \eta(\cdot))$$
$$\leq \inf_{\nu \in \mathscr{N}_0} \mathbf{E}F(\xi(\cdot), \eta(\cdot)).$$

Therefore to prove the theorem it is sufficient to verify two equalities:

$$\inf_{\eta \in \mathscr{H}} \mathbf{E}F(\xi(\cdot, \omega, \eta(\cdot)), \eta(\cdot)) = \inf_{\eta \in \mathscr{H}_0} \mathbf{E}F(\xi(\cdot, \omega, \eta(\cdot)), \eta(\cdot))$$
$$= \inf_{\nu \in \mathscr{N}_0} \mathbf{E}F(\xi(\cdot), \eta(\cdot)).$$

To begin with we shall prove the first one. Let $\eta(\cdot) \in \mathscr{H}$. Construct a sequence $\eta_n(\cdot) \in \mathscr{H}_0$ such that $\rho_D(\eta(\cdot), \eta_n(\cdot)) \to 0$ (setting for example $\eta_n(t) = S_{\varepsilon_n}[\eta_n](t)$, where $\varepsilon_n \downarrow 0$). Then also

$$\rho_D(S_\varepsilon[\xi(\cdot, \omega, \eta_n(\cdot))], S_\varepsilon[\xi(\cdot, \omega, \eta(\cdot))]) \to 0 \quad \text{in probability.}$$

Furthermore,

$$\rho_D(\xi(\cdot, \omega, \eta_n(\cdot)), S_\varepsilon[\xi(\cdot, \omega, \eta_n(\cdot))]) \leq \varepsilon,$$
$$\rho_D(\xi(\cdot, \omega, \eta(\cdot)), S_\varepsilon[\xi(\cdot, \omega, \eta(\cdot))]) \leq \varepsilon.$$

Since

$$\rho_D(\xi(\cdot, \omega, \eta_n(\cdot)), \xi(\cdot, \omega, \eta(\cdot))) \leq \rho_D(S_\varepsilon[\xi(\cdot, \omega, \eta_n(\cdot))],$$
$$S_\varepsilon[\xi(\cdot, \omega, \eta(\cdot))]) + \rho_D(S_\varepsilon[\xi(\cdot, \omega, \eta(\cdot))], \xi(\cdot, \omega, \eta(\cdot)))$$
$$+ \rho_D(S_\varepsilon[\xi(\cdot, \omega, \eta_n(\cdot))], \xi(\cdot, \omega, \eta_n(\cdot)))$$
$$\leq 2\varepsilon + \rho_D(S_\varepsilon[\xi(\cdot, \omega, \eta_n(\cdot))], S_\varepsilon[\xi(\cdot, \omega, \eta'\cdot))]),$$

it follows that $\rho_D(\xi(\cdot, \omega, \eta_n(\cdot)), \xi(\cdot, \omega, \eta(\cdot))) \to 0$ in probability. Hence

$$F(\xi(\cdot, \omega, \eta_n(\cdot)), \eta(\cdot)) \to F(\xi(\cdot, \omega, \eta(\cdot)), \eta(\cdot))$$

in probability. Therefore,

$$\mathbf{E}F(\xi(\cdot, \omega, \eta(\cdot)), \eta(\cdot)) = \lim_{n\to\infty} \mathbf{E}F(\xi(\cdot, \omega, \eta_n(\cdot)), \eta_n(\cdot))$$
$$\geq \inf_{\eta_0 \in \mathscr{H}_0} \mathbf{E}F(\xi(\cdot, \omega, \eta_0(\cdot)), \eta_0(\cdot)).$$

Since the last inequality is valid for all $\eta(\cdot)$,

$$\inf_{\eta \in \mathscr{H}} \mathbf{E}F(\xi(\cdot, \omega, \eta(\cdot)), \eta(\cdot)) \geq \inf_{\eta_0 \in \mathscr{H}_0} \mathbf{E}F(\xi(\cdot, \omega, \eta_0(\cdot)), \eta_0(\cdot)).$$

Thus the first of the equalities (2.11) is verified. We now prove the second one.

First we observe that

$$\inf_{\eta \in \mathscr{H}_0} \mathbf{E}F(\xi(\cdot, \omega, \eta(\cdot)), \eta(\cdot)) \leq \inf_{\nu \in \mathscr{N}_0} \mathbf{E}_\nu F(\xi(\cdot), \eta(\cdot)). \tag{2.12}$$

Hence it is sufficient to prove the opposite inequality. For this purpose we show that for any $\varepsilon > 0$ and $\eta \in \mathscr{H}_0$ a $\nu \in \mathscr{N}_0$ can be found such that

$$\mathbf{E}F(S_\varepsilon[\xi(\cdot, \omega, \eta(\cdot))], \eta(\cdot)) \geq \mathbf{E}_\nu F(S_\varepsilon[\xi(\cdot, \omega, \eta(\cdot))], \eta(\cdot)). \tag{2.13}$$

The process $S_\varepsilon[\xi](t)$—with the chosen control ν—can be obtained if the control ν and the controlled object $\mu_\varepsilon(A/\cdot) = \mu(\{x(\cdot) : S_\varepsilon[x] \in A\}/\cdot)$ is utilized. Consider the left-hand-side of (2.13). Let $\tau_k(\omega, u(\cdot))$ be the instants of the jumps of the process $S[\xi(\cdot, \omega, u(\cdot)]$, $x_k(\omega, u(\cdot))$ be the state of the process at time $\tau_k(\omega, u(\cdot))$ and $x_0(\omega)$ be the initial state. Let the step generalized control $\eta(t)$ possess jumps at points σ_j, $u_0 = \eta(0)$, $u_j = \eta(\sigma_j)$. Consider successively the conditional distributions of the variables u_j and σ_j:

$$q_0(A/x_0) = P\{u_0 \in A/x_0\},$$

$$\theta_1(ds/u_0, \xi(\cdot)) = P\{\sigma_1 \in ds/u_0; \xi(t), t \leq s\},$$

$$q_j(A/u_0, \ldots, u_{j-1}, \sigma_1, \ldots, \sigma_{j-1}, \sigma_j; \xi(t), t \leq \sigma_j),$$

$$\theta_j(ds/u_0, \ldots, u_{j-1}, \sigma_1, \ldots, \sigma_{j-1}; \xi(t), t \leq s) \qquad (s > \sigma_{j-1}).$$

Here $S_\varepsilon[\xi(\cdot, \omega, \eta(\cdot))]$ is denoted for brevity by $\xi(s)$. Taking the form of $\xi(\cdot)$ into account one can write these conditional probabilities as:

$$\theta_1(ds/u_0, \xi(t), t \leq s) = \hat{\theta}_1(ds/u_0, \tau_1(\omega, u_0), \ldots, \tau_k(\omega, u_0),$$
$$x_0(\omega, u_0), \ldots, x_k(\omega, u_0)),$$

provided $\tau_k(\omega, u_0) < s$, $\tau_{k+1}(\omega, u_0) \geq s$;

$$\theta_j(ds/u_0, \ldots, u_{j-1}, \sigma_1, \ldots, \sigma_{j-1}; \xi(t), t \leq s)$$
$$= \hat{\theta}_j(ds/u_0, \ldots, u_{j-1}, \sigma_1, \ldots, \sigma_{j-1}, \tau_1(\omega, u_0), \ldots, \tau_{k_1}(\omega, u_0),$$
$$x_1(\omega, u_0), \ldots,$$
$$x_{k_1}(\omega, u_0), \ldots, \tau_{k_{j-1}+1}(\omega, u_{j-1}(\cdot)), \ldots, \tau_{k_j}(\omega, u_{j-1}(\cdot)),$$
$$\ldots, x_{k_{j-1}+1}(\omega, u_{j-1}(\cdot)), \ldots, x_k(\omega, u_{j-1}(\cdot))),$$

provided $\tau_1 < \cdots < \tau_{k_1} < \sigma_1 \leq \tau_{k_1+1} < \cdots < \sigma_{j-1} \leq \tau_{k_{j-1}+1} < \cdots < \tau_{k_j} < s \leq \tau_{k_j+1}$, where $u_0(t) = u_0$, $u_{j-1}(t)$ is a step function with jumps at points $\sigma_1, \ldots, \sigma_{j-1}$ and with values $u_0, u_1, \ldots, u_{j-1}$;

$$q_j(A/u_0, \ldots, u_{j-1}, \sigma_1, \ldots, \sigma_j; \xi(t), t \leq \sigma_j)$$
$$= \hat{q}_j(A/u_0, \ldots, u_{j-1}, \sigma_1, \ldots, \sigma_j, \tau_1(\omega, u_0), \ldots, \tau_{k_1}(\omega, u_0),$$
$$x_1(\omega, u_0), \ldots, x_{k_1}(\omega, u_0), \ldots, \tau_{k_{j-1}+1}(\omega, u_{j-1}(\cdot)), \ldots,$$
$$\tau_{k_j}(\omega, u_{j-1}(\cdot)), x_{k_{j-1}+1}(\omega, u_{j-1}(\cdot)), \ldots, x_{k_j}(\omega, u_{j-1}(\cdot)),$$

provided $\tau_1 < \cdots < \tau_{k_1} < \sigma_1 \le \tau_{k_1+1} < \cdots < \sigma_{j-1} \le \tau_{k_{j-1}+1} < \cdots < \tau_{k_j} < \sigma_j \le \tau_{k_j+1}$.

Let $F(x(\cdot), u(\cdot))$ be a functional. Then

$$F(S_\varepsilon[\xi(\cdot, \omega, \eta(\cdot))], \eta(\cdot)) = \Phi(x_0, \tau_1, x_1, \ldots; u_0, \sigma_1, u_1, u_2, \sigma_2, \ldots),$$

where Φ is a measurable function in its variables. Assume that Φ is nonvanishing only under the condition that $\tau_1 < \cdots < \tau_{k_1} < \sigma_1 \le \tau_{k_2} < \cdots < \sigma_n \le \tau_{k_n+1} < \cdots < \tau_{k_{n+1}} \le T$. Then

$$\begin{aligned}\mathbf{E}F(S_\varepsilon[\xi(\cdot, \omega, \eta(\cdot))], \eta(\cdot)) &= \mathbf{E}\Phi(x_0(\omega, u_0), \tau_1(\omega, u_0), \\ &\ldots, \tau_{k_1}(\omega, u_0), x_{k_1}(\omega, u_0), \ldots, \tau_{k_n+1}(\omega, u_n(\cdot)), x_{k_n+1}(\omega, u_n(\cdot)), \\ &\ldots, \tau_{k_{n+1}}(\omega, u_n(\cdot)), x_{k_{n+1}}(\omega, u_n(\cdot)), \sigma_1, \ldots, \sigma_n, u_n) \\ &= \mathbf{E} \int \Phi(x_0(\omega), \ldots, \tau_{k_n}(\omega, u_n(\cdot)), x_{k_n}(\omega, u_n(\cdot)), s_1, \ldots, \\ &s_{k_{n+1}-k_n}, y_1, \ldots, y_{k_{n+1}-k_n}, \sigma_1, \ldots, \sigma_n; u_k)p_{k_n+1}(dy_1/x_0, \ldots, \\ &x_{k_n}, \tau_0, \ldots, s_1; u_n(\cdot)) \cdots p_{k_{n+1}}(dy_{k_{n+1}-k_n}/x_0, \ldots, x_{k_n}, \\ &y_1, \ldots, y_{k_{n+1}-k_n-1}, \tau_1, \ldots, \tau_{k_n}, \\ &s_1, \ldots, s_{k_{n+1}-k_n}; u_n(\cdot))\lambda_{k_n+1}(ds_1/x_0, \ldots, x_{k_n}, \tau_0, \ldots, \tau_{k_n}; u_n(\cdot)) \\ &\cdots \lambda_{k_{n+1}}(ds_{k_{n+1}-k_n}/x_0, \ldots, x_{k_n}, y_1, \ldots, y_{k_{n+1}-k_n-1}, \\ &\tau_1, \ldots, \tau_{k_n}, s_1, \ldots, s_{k_{n+1}-k_n-1}; u_n(\cdot)).\end{aligned}$$

Here p_j, λ_j are the conditional distributions which define the probabilistic object $\mu_\varepsilon(\cdot/\cdot)$.

Denote the integral appearing under the sign of the mathematical expectation by $\bar{\Phi}_{n-1}(x_0(\omega), \ldots, \tau_{k_n}(\omega, u_{n-1}), x_{k_n}(\omega, u_{n-1}), \sigma_1, \ldots, \sigma_n, u_n)$.

Set

$$\begin{aligned}&\Phi_{n-1}(x_0(\omega), \ldots, \tau_{k_n}(\omega, u_n), x_{k_n}(\omega, u_{n-1}), \sigma_1, \ldots, \sigma_n, u_0, \ldots, u_n) \\ &\quad = \mathbf{E}[\bar{\Phi}_{n-1}(x_0(\omega), \ldots, \tau_{k_n}(\omega, u_{n-1}(\cdot)), x_{k_n}(\omega, u_{n-1}(\cdot)), \\ &\quad \sigma_1, \ldots, \sigma_n, u_0, \ldots, u_n/x_0(\omega), \ldots, \tau_{k_n}(\omega, u_n), x_{k_n}(\omega, u_n), \\ &\quad \sigma_1, \ldots, \sigma_{n-1}, u_0, \ldots, u_n] = \int \bar{\Phi}_n(x_0(\omega), \ldots, \tau_{k_n}(\omega, u_{n-1}(\cdot)), \\ &\quad x_{k_n}(\omega, u_{n-1}(\cdot)), \sigma_1, \ldots, \sigma_{n-1}, s, u_0, \ldots, u_{n-1}, u)\hat{\theta}_n(ds/u_0, \\ &\quad \ldots, u_{n-1}, \sigma_1, \ldots, \sigma_{n-1}, x_0(\omega), \tau_1(\omega, u_0(\cdot)), \ldots, x_{k_n}(\omega, u_{n-1}(\cdot)) \\ &\quad \times \hat{q}_n(du/u_0, \ldots, u_{n-1}, \sigma_1, \ldots, \sigma_{n-1}, x_0(\omega), \tau_1(\omega, u_0(\cdot)), \\ &\quad \ldots, x_{k_n}(\omega, u_{n-1}(\cdot))).\end{aligned}$$

Defining $\bar{\Phi}_{n-2}$ and Φ_{n-2} and so on analogously, we verify that $\mathbf{E}F(S_\varepsilon[\xi(\cdot, \omega, \eta(\cdot))], \eta(\cdot))$ coincides with the expression for $\mathbf{E}_\nu(S_\varepsilon[\xi(\cdot)], \eta(\cdot))$, provided ν is a randomized step control defined by the conditional probabilities q_j and θ_j.

Thus the inequality (2.13) is verified. Utilizing the fact that $\rho_D(S_\varepsilon[\xi(\cdot, \omega, \eta(\cdot)], \xi(\cdot)) \leq \varepsilon$ and hence the distribution of $S_\varepsilon[\xi(\cdot, \omega, \eta(\cdot))]$ converges weakly in $D_{[0,T]}(X)$ to the distribution of $\xi(\cdot, \omega, \eta(\cdot))$ one can find a compact set $K_X \subset D_{[0,T]}(X)$, such that for $\varepsilon > 0$ sufficiently small

$$\mathbf{P}\{S_\varepsilon[\xi(\cdot, \omega, \eta(\cdot)] \in K_X\} \geq 1 - \delta.$$

Let $K_U \subset D_{[0,T]}(X)$ be a compact set such that

$$\mathbf{P}\{\eta(\cdot) \in K_U\} \geq 1 - \delta.$$

The functional $F(x(\cdot), u(\cdot))$ is uniformly continuous on $K_X \times K_U$. Let F^δ be its uniformly continuous extension onto $D_{[0,T]}(X) \times D_{[0,T]}(U)$ such that $\sup|F| = \sup|F^\delta|$ and $F^\delta > F - \delta$. Then

$$\begin{aligned}
&\mathbf{E}|F(S_\varepsilon[\xi(\cdot, \omega, \eta(\cdot))], \eta(\cdot)) - F^\delta(S_\varepsilon[\xi(\cdot, \omega, \eta(\cdot))], \eta(\cdot))| \\
&\quad \leq (\sup|F| + \sup|F^\delta|)[\mathbf{P}\{S_\varepsilon[\xi(\cdot, \omega, \eta(\cdot))] \notin K_X\} \\
&\quad\quad + \mathbf{P}\{\eta(\cdot) \notin K_U\} = O(\delta).
\end{aligned}$$

It follows from the uniform continuity of F^δ that

$$\begin{aligned}
&\lim_{\varepsilon \to 0} \sup_{v \in \mathcal{N}_0} \mathbf{E}|F^\delta(S_\varepsilon[\xi(\cdot)], \eta(\cdot)) - F^\delta(\xi(\cdot), \eta(\cdot))| \\
&\quad \leq \lim_{\varepsilon \to 0} \sup_{\substack{u(\cdot) \\ x(\cdot)}} |F^\delta(S_\varepsilon[x(\cdot)], u(\cdot)) - F^\delta(x(\cdot), u(\cdot))| = 0.
\end{aligned}$$

Hence

$$\begin{aligned}
\inf_{v \in \mathcal{N}_0} \mathbf{E}_v F(\xi(\cdot), \eta(\cdot)) &\leq \inf_{v \in \mathcal{N}_0} \mathbf{E}_v F^\delta(\xi(\cdot), \eta(\cdot)) + \delta \\
&\leq \inf_{v \in \mathcal{N}_0} \mathbf{E}_v F^\delta(S_\varepsilon[\xi(\cdot)], \eta(\cdot)) + \delta + \sup_{v \in \mathcal{N}_0} \mathbf{E}_v |F^\delta(S_\varepsilon[\xi(\cdot)], \eta(\cdot)) \\
&\quad - F^\delta(\xi(\cdot), \eta(\cdot))| \\
&\leq \mathbf{E}_v F(S_\varepsilon[\xi(\cdot, \omega, \eta(\cdot))], \eta(\cdot)) + O(\delta) + \sup_{v \in \mathcal{N}_0} \mathbf{E}_v |F^\delta(S_\varepsilon[\xi(\cdot)], \eta(\cdot)) \\
&\quad - F^\delta(\xi(\cdot), \eta(\cdot))|. \text{ (We have used inequality (2.13) for } F^\delta.)
\end{aligned}$$

Approaching the limit as $\varepsilon \to 0$ and then as $\delta \to 0$ we obtain

$$\inf_{v \in \mathcal{N}_0} F(\xi(\cdot), \eta(\cdot)) \leq \mathbf{E}F(\xi(\cdot, \omega, \eta(\cdot)), \eta(\cdot))$$

for all $\eta \in \mathcal{H}_0$. Together with (2.12) it gives us the second of equations (2.11). □

Remark 2.1. Effective conditions for the validity of condition A) of the theorem can be obtained in terms of conditional probabilities:

$$p_k(\cdot/x_0, \ldots, x_{k-1}, \tau_1, \ldots, \tau_{k-1}, \tau_k; u(\cdot))$$

and

$$\lambda_k(\cdot/x_0, \ldots, x_{k-1}, \tau_0, \ldots, \tau_{k-1}; u(\cdot)),$$

which define the controlled object $\mu_\varepsilon(\cdot/\cdot)$; in particular, these conditions are satisfied if the conditional distributions are continuous jointly in the variables and measures $p_k(dx/\ldots)$ are absolutely continuous with respect to the same measure $\pi(dx)$ for all values of the arguments appearing to the right of the dash in the expression for $p_k(dx/\ldots)$. This follows from Lemma 2.2

Remark 2.2. Let $\mathscr{H}'_0$ denote the set of step generalized controls which are constant for some n on the interval $[Tk/n, T(k+1)/n)$. Since for any control $\eta(t) \in \mathscr{H}$ the control $\eta_n(t) = \eta(Tk/n)$ for $t \in [Tk/n, T(k+1)/n)$ belongs to $\mathscr{H}'_0$, one can verify analogously to the proof of theorem 2.3 that

$$\inf_{\eta \in \mathscr{H}'_0} \mathbf{E}F(\xi(\cdot, \omega, \eta(\cdot)), \eta(\cdot)) = \inf_{\eta \in \mathscr{H}} \mathbf{E}F(\xi(\cdot, \omega, \eta(\cdot)), \eta(\cdot)).$$

Let $\mathscr{N}'_0$ be the set of non-randomized step controls constant on the intervals $[Tk/n, T(k+1)/n)$ for some n. Once again repeating almost verbatim the argument of theorem 2.3 we verify that

$$\inf_{\eta \in \mathscr{H}'_0} \mathbf{E}F(\xi(\cdot, \omega, \eta(\cdot)), \eta(\cdot)) = \inf_{\nu \in \mathscr{N}'_0} \mathbf{E}_\nu F(\xi(\cdot, \omega, \eta(\cdot)), \eta(\cdot)).$$

Denote by $\mathscr{N}_0^{(n)}$ the set of step non-randomized controls which are constant on the interval $[kT2^{-n}, (k+1)T2^{-n})$.

Theorem 2.4. *If F and the controlled object satisfy the conditions of theorem 2.3, and moreover, if*

$$\lim_{n\to\infty} \sup_{\eta} \mathbf{E}|F(\xi(\cdot, \omega, \eta(\cdot)), \eta(\cdot)) - F^{(n)}(\xi(\cdot, \omega, \eta(\cdot)), \eta(\cdot))| = 0,$$

then

$$\inf_{\eta \in \mathscr{H}} \mathbf{E}F(\xi(\cdot, \omega, \eta(\cdot)), \eta(\cdot)) = \lim_{n\to\infty} \inf_{\nu \in \mathscr{N}_0^{(n)}} \mathbf{E}_\nu F^{(n)}(\xi(\cdot), \eta(\cdot)).$$

Proof. Let n and ν_n be such, that

$$\mathbf{E}_{\nu_n} F(\xi(\cdot), \eta(\cdot)) \le \inf_{\eta \in H} \mathbf{E}F(\xi(\cdot), \eta(\cdot)) + \varepsilon.$$

Denote

$$F^{(n)}(x(\cdot), u(\cdot)) = F(\Gamma_n x(\cdot), u(\cdot)),$$

where

$$\Gamma_n x(t) = x\left(\frac{kT}{2^n}\right) \quad \text{for} \quad \frac{kT}{2^n} \le t < \frac{k+1}{2^n} T.$$

It follows from the continuity of F that

$$\lim_{m\to\infty} \mathbf{E}_{\nu_n} F(\Gamma_m \xi(\cdot), \eta(\cdot)) = \mathbf{E}_{\nu_n} F(\xi(\cdot), \eta(\cdot)).$$

For $m > n$, $\nu_n \in \mathcal{N}_0^{(m)}$. Therefore

$$\begin{aligned}\lim_{m\to\infty} \inf_{\nu \in \mathcal{N}_0^{(m)}} \mathbf{E}_\nu F^{(m)}(\xi(\cdot), \eta(\cdot)) &\le \lim_{m\to\infty} \mathbf{E}_{\nu_n} F^{(m)}(\xi(\cdot), \eta(\cdot)) = \mathbf{E}_{\nu_n} F(\xi(\cdot), \eta(\cdot)) \\ &\le \inf_{\eta\in\mathscr{H}} \mathbf{E}F(\xi(\cdot), \eta(\cdot)) + \varepsilon.\end{aligned}$$

Hence

$$\lim_{m\to\infty} \inf_{\nu \in \mathcal{N}_0^{(m)}} \mathbf{E}_\nu F^{(m)}(\xi(\cdot), \eta(\cdot)) \le \inf_{\eta\in\mathscr{H}} \mathbf{E}F(\xi(\cdot), \eta(\cdot)), \tag{2.15}$$

which implies the assertion of the theorem. □

Remark 2.3. If ν_n is a sequence of controls in $\mathcal{N}_0^{(n)}$ such that

$$\mathbf{E}_{\nu_n} F^{(n)}(\xi(\cdot), \eta(\cdot)) \le \inf_{\nu \in \mathcal{N}_0^{(n)}} F(\xi(\cdot), \eta(\cdot)) + \varepsilon_n,$$

where $\varepsilon_n \to 0$, then under the conditions of the theorem

$$\inf_{\eta\in\mathscr{H}} \mathbf{E}F(\xi(\cdot, \omega, \eta(\cdot)), \eta(\cdot)) = \lim_{n\to\infty} \mathbf{E}_{\nu_n} F(\xi(\cdot), \eta(\cdot)).$$

Thus for any $\varepsilon > 0$ a sequence of controls ν_n contains an ε-optimal control.

Remark 2.4. For any n we find a discrete-parameter (time) controlled object defined by the conditional probabilities

$$\begin{aligned} p_k^{(n)}(A/x_0, \ldots, x_{k-1}, u_0, \ldots, u_{k-1}) &= \mathbf{P}\left\{\xi\left(\frac{k}{2^n} T, \omega, u(\cdot)\right) \in A/\xi(0, \omega, u(\cdot))\right. \\ &\qquad \left. = x_0, \ldots, \xi\left(\frac{k-1}{2^{n-1}} T, \omega, u(\cdot)\right) = x_{k-1}\right\},\end{aligned}$$

where $u(\cdot) \in \mathscr{H}_0^{(n)}$ and $u(jT/2^n) = u_j$.

Furthermore, let $\bar{\nu}$ be a non-randomized control defined by the functions $u_k = \varphi_k(x_0, \ldots, x_k)$. Then denoting by (ξ_k, η_k) the controlled sequence with the indicated controlled object and control we obtain

$$\mathbf{E}_{\bar{\nu}} \Phi_n(\xi_0, \ldots, \xi_{2^n}, \eta_0, \ldots, \eta_{2^n}) = \mathbf{E}_{\nu_n} F^{(n)}(\xi(\cdot), \eta(\cdot)),$$

provided

$$F^{(n)}(x(\cdot), u(\cdot)) = \Phi_n\left(x(0), \ldots, x\left(\frac{2^n - 1}{2^n} T\right), u(0), \ldots, u\left(\frac{2^n - 1}{2^n} T\right)\right);$$

v_n is defined by relation

$$u(t) = \varphi_k\left(x(0), \ldots, x\left(\frac{k}{2^n}T\right)\right) \quad \text{for} \quad \frac{k}{2^n}T \le t < \frac{k+1}{2^n}T.$$

Therefore the determination of optimal and ε-optimal controls in the class $\mathcal{N}_0^{(n)}$ for functional $F^{(n)}$ can be carried out using the methods developed in Section 3 of Chapter 1.

4 Controlled Markov Processes

A *controlled Markov object* with phase space of the basic process $(X, \mathfrak{A})$ and phase space of controls $(U, \mathfrak{B})$ is defined by the set of transition probabilities $P(t, x, s, A; u(\cdot))$ defined for $0 \le t < s \le T$, $x \in X$, $A \in \mathfrak{A}$ and $u(\cdot) \in U^{[0, T]}$. This transition probability is assumed to be measurable for a fixed $t < s$ and A is measurable with respect to $\mathfrak{A} \times \mathfrak{B}_{[t, s)}$ where $\mathfrak{B}_{[t, s)}$ is a σ-algebra generated in $\mathfrak{B}^{[0, T]}$ by cylinders with bases in $[t, s)$. Transition probabilities $P(t, x, s, A; u(\cdot))$ allow us to construct a family of probabilistic objects $\mu_x(\cdot/\cdot)$ which depend on the initial value of the basic process x as on a parameter. For each $u(\cdot)$ $\mu_x(\cdot/u(\cdot))$ is a family of measures corresponding to a Markov process with transition probability $P(\cdot)$. We shall consider only the case when X is a complete separable metric space, U is a compact space and $\mathfrak{A}$ and $\mathfrak{B}$ are σ-algebras of Borel sets on these spaces. Observe that to define transition probabilities $P(t, x, s, A; u(\cdot))$ for step functions $u(\cdot)$ it is sufficient to have transition probability $P(t, x, s, A; u) = P(t, x, s, A; u(\cdot))$ where $u(\tau) = u$ for all τ. Indeed, if $u(t) = u_k$ for $t_k \le t < t_{k+1}$ where $0 = t_0 < t_1 < \cdots < t_n = T$, then

$$P(t, x, s, A; u(\cdot)) = \int \cdots \int P(t, x, t_j, dx_j; u_{j-1}) \cdots P(t_k, x_k, s, A; u_k)$$

for $t_{j-1} \le t < t_j < \cdots < t_k < s \le t_{k+1}$. Therefore if $P(t, x, s, A; u(\cdot))$ depends on $u(\cdot)$ in such a manner that it can be well defined and if it is known for step functions $u(\cdot)$ then to define a probabilistic object it is sufficient to prescribe values of $P(t, x, s, A; u)$. This is valid only when controls belonging to $D_{[0, T]}(X)$ are considered provided for all $f \in C_X$

$$\int P(t, x, s, dy; u(\cdot))f(y)$$

is continuous in $u(\cdot)$ with respect to metric ρ_D. In the general case the knowledge of function $P(t, x, s, A; u)$ is sufficient to define distributions of a probabilistic object under step controls.

It was established in the previous Section that under quite general assumptions $\inf \mathbf{E}_v F(\xi(\cdot), \eta(\cdot))$ remains unchanged if we confine ourselves to step controls only. Therefore when dealing below with an optimization problem we shall take infimum over step controls only. Evidently it is hardly possible to obtain an optimal control among the step controls, however ε-optimal control are to be found among them.

The Markovian nature of the controlled object does not simplify the investigation of the cost functions of a general type. However, as was the case in the discrete case, for functionals of the evolutional type the results obtained in Section 6 of Chapter 1 allow us to determine ε-optimal controls. In an analogy with the discrete case, a functional of the form

$$F(x(\cdot), u(\cdot)) = \int_0^T \exp\left\{\int_0^t g(s, x(s), u(s))\, ds\right\} f(t, x(t), u(t))\, dt \tag{2.16}$$

is called a *functional of the evolutional type*. Here the functions $g(s, x, u)$ and $f(s, x, u)$ are defined and measurable on $[0, T] \times X \times U$. The following approximation theorem will be required in the course of investigation of the problem of optimization of functional (2.16).

Theorem 2.5. *Let a controlled Markovian object defined by the collection of conditional probabilities $P(t, x, s, A; u)$ satisfy the condition:*

The function $\int \varphi(t, y, s, u)P(t, x, s, dy; u)$ is continuous jointly in the variables for $0 \le t < s \le T$, $x \in X$ and $u \in U$ provided only the function φ is continuous jointly in the variables.

Then if the functional $F(x(\cdot), u(\cdot))$ is of the form (2.16) *and the functions $g(s, x, u)$ and $f(s, x, u)$ are continuous, we have*

$$\inf_{v \in \mathcal{N}_0} \mathbf{E}_v F(\xi(\cdot), \eta(\cdot)) = \lim_{n \to \infty} \mathbf{E}_{\bar{v}_n} F_n(\xi(\cdot), \eta(\cdot)), \tag{2.17}$$

where $0 = t_{n0} < t_{n1} < \cdots < t_{nn} = T$, $\max \Delta t_{nk} \to 0$,

$$F_n(x(\cdot), u(\cdot)) = \sum_{k=0}^{n-1} \exp\left\{\sum_{i=0}^{k-1} (g(t_{ni}), x(t_{ni}), u(t_{ni}))\, \Delta t_{ni}\right\} f(t_{nk}, x(t_{nk}), u(t_{nk}))\, \Delta t_{nk}, \tag{2.18}$$

and $\bar{v}_n$ is a non-randomized Markovian control which is optimal for a Markovian controlled object with one-step transition probability $P(t_{nk}, x, t_{nk+1}, A; u)$ and functional (2.18).

Proof. Let v_ε be a step control such that

$$\mathbf{E}_{v_\varepsilon} F(\xi(\cdot), \eta(\cdot)) < \inf_v \mathbf{E}_v F(\xi(\cdot), \eta(\cdot)) + \varepsilon.$$

For all $x(\cdot) \in D_{[0, T]}(X)$, $u(\cdot) \in D_{[0, T]}(U)$ the equality

$$\lim_{n \to \infty} F_n(x(\cdot), u(\cdot)) = F(x(\cdot), u(\cdot))$$

is valid. Therefore

$$\lim_{n \to \infty} F_n(\xi(\cdot), \eta(\cdot)) = F(\xi(\cdot), \eta(\cdot)),$$

with probability 1. Hence we have for n sufficiently large

$$\mathbf{E}_{\nu_\varepsilon} F_n(\xi(\cdot), \eta(\cdot)) < \inf_{\nu} \mathbf{E}_\nu F(\xi(\cdot), \eta(\cdot)) + \varepsilon,$$

and more so

$$\mathbf{E}_{\bar{\nu}_n} F(\xi(\cdot), \eta(\cdot)) \le \mathbf{E}_{\nu_\varepsilon} F_n(\xi(\cdot), \eta(\cdot)) \le \inf_{\nu} \mathbf{E}_\nu F(\xi(\cdot), \eta(\cdot)) + \varepsilon.$$

Hence

$$\overline{\lim_{n\to\infty}} \mathbf{E}_{\bar{\nu}_n} F(\xi(\cdot), \eta(\cdot)) \le \inf_{\nu} \mathbf{E}_\nu F(\xi(\cdot), \eta(\cdot)). \qquad \square$$

Theorem 2.5 allows us to seek ε-optimal controls with the aid of discrete approximations of controlled objects.

5 Jump Markovian Controlled Processes

Consider a special class of controlled Markovian objects which by analogy with ordinary Markov processes is natural to call a class of jump controlled objects. Such controlled objects belong to the class of step controlled objects, however, what distinguishes them is the fact that the instants of jumps τ_k of a process with measure $\mu(\cdot/u(\cdot))$ possess bounded densities.

We give a general definition of a *jump Markovian object*. Assume that $P(t, x, s, A, u(\cdot))$ satisfies the following condition: for all t, x, s, A and $u(\cdot) \in D_{[0,T]}(U)$ there exists the limit

$$\lim_{s\downarrow t} \frac{1}{s-t}[P(t, x, s, A, u(\cdot)) - \chi_A(x)] = \Pi(t, x, u(t), A),$$

where the function $\Pi(t, x, u, A)$ is continuous in t, measurable jointly in t, x and u and is countably-additive on A while the quantity

$$\Pi(t, x, u, \{x\}) = -\Pi(t, x, u, X - \{x\})$$

is bounded. Let $\nu(\cdot/x(\cdot))$ be a given control. If $x(0) = x_0$ is the initial value of the process, we define the process $\eta_0(t)$ with distribution $\nu(\cdot/x_0(\cdot))$ where $x_0(t) = x_0$, $0 \le t \le T$. Furthermore, it will be convenient to assume that $T = +\infty$. Let $\xi_1(t)$ be a jump process such that $\xi_1(0) = x_0$ and the instant of the first jump τ_1 possesses the conditional distribution

$$\mathbf{P}\{\tau_1 > t/\eta_0(\cdot)\} = \exp\left\{\int_0^t \Pi(s, x_0, \eta_0, (s), \{x_0\})\, ds\right\}$$

and

$$\mathbf{P}\{\xi_1(\tau_1 + 0) \in A/\tau_1 = t, \eta(\cdot)\} = \frac{\Pi(t, x_0, \eta_0(t), A)}{\Pi(t, x_0, \eta_0(t), X - \{x_0\})},$$

with $\xi_1(t) = \xi_1(\tau_1 + 0)$ for $t > \tau_1$. Denote by $\eta_1(t)$ the process which coincides with $\eta_0(t)$ up to time τ_1 and the conditional distribution of $\eta_1(t)$ for given x_0, τ_1, x_1 and $\eta(\cdot)$ coincides with $\nu(\cdot/\xi_1(\cdot))$. We next define the time τ_2 whose conditional distribution for given x_0, τ_1, x_1 and $\eta_1(\cdot)$ is given by the formula

$$\mathbf{P}\{\tau_2 > t/\eta_1(\cdot), x_0, x_1 \tau_1 = s\} = \begin{cases} 1, & t \le s; \\ \exp\left\{\int_s^t \Pi(r, x_1, \eta_1(r), \{x_1\})\, dr\right\}, & s > t, \end{cases}$$

and let $\xi_2(\tau_2 + 0)$ possess the conditional distribution

$$\mathbf{P}\{\xi_2(\tau_2 + 0) \in A/\eta_1(\cdot), x_0, x_1, \tau_1 = s, \tau_2 = t\} = \frac{\Pi(t, x_1, \eta_1(t), A - \{x\})}{\Pi(t, x_1, \eta_1(t), X - \{x\})}.$$

Continuing a similar construction for $k = 2, 3, \ldots$ we verify that there exist sequences of processes $\xi_k(t), \eta_k(t)$ and instants of time τ_k and that the following conditions are fulfilled: (1) $\xi_k(t)$ is a step process continuous from the right which has exactly k jumps and $\tau_1, \ldots, \tau_k$ are the instants of these jumps. (2) $\xi_{k-1}(t) = \xi_k(t)$ for $t < \tau_k$; (3) $\eta_{k-1}(t) = \eta_k(t)$ for $t < \tau_k$ (4) let $\mathfrak{F}_t^k$ be a current of σ-algebras generated by the process $\eta_k(t)$ and $\mathfrak{M}_k$ be a σ-algebra generated by the process $\xi_k(t)$, then

$$\mathbf{P}\{\tau_{k+1} > t/\mathfrak{F}_\infty^k, \mathfrak{M}_k\} = \exp\left\{\int_{\tau_k}^{t \vee \tau_k} \Pi(s, \xi_k(\tau_k), \eta_k(s), \{\xi_k(\tau_k)\})\, ds\right\}, \tag{2.19}$$

$$\mathbf{P}\{\xi(\tau_{k+1}) \in A/\mathfrak{F}_\infty^k, \mathfrak{M}_k, \tau_{k+1}\} = \frac{\Pi(\tau_{k+1}, \xi_k(\tau_k), \eta_k(\tau_{k+1}), A - \{\xi_k(\tau_k)\})}{\Pi(\tau_{k+1}, \xi_k(\tau_k), \eta_k(\tau_{k+1}), X - \{\xi_k(\tau_k)\})} \tag{2.20}$$

(here $\mathbf{P}\{\cdot/\mathfrak{F}_\infty^k, \mathfrak{M}_k\}$ is the conditional probability with respect to the σ-algebra generated by $\mathfrak{F}_\infty^k$ and $\mathfrak{M}_k$; in the second case the σ-algebra is generated by $\mathfrak{F}_\infty^k$, $\mathfrak{M}_k$ and the quantity τ_{k+1}); 5) let $\mathfrak{N}_k$ be a σ-algebra generated by the events of the form

$$\{\tau_k > t\} \cap C_t \cap D,$$

where $C_t \in \mathfrak{F}_t^k$, $D \in \mathfrak{M}_k$ and $t > 0$; then for any set B belonging to $\mathfrak{B}^{[t, \infty]}$

$$\mathbf{P}\{\eta_{k+1}(\cdot) \in B/\mathfrak{N}_k\} = \nu(B/\xi_k(\cdot)) \tag{2.21}$$

for $\tau_k \le t$. It is easy to verify that formulas (2.19)–(2.21) and conditions 1, 2, and 3 uniquely determine the joint distributions of processes $\eta_k(t)$ and $\xi_k(t)$ provided only $\xi_0(0) = \xi_k(0)$ is given. Set $\xi(t) = \xi_k(t)$, $\eta(t) = \eta_k(t)$, $t \in [\tau_k, \tau_{k+1})$, $(\tau_0 = 0)$. Observe that $\mathbf{P}(\tau_k \to \infty) = 1$. Indeed

$$\mathbf{P}\{\tau_k \le T/\tau_1, \ldots, \tau_{k-1}\} \le 1 - e^{-cT},$$

where c is such that $|\Pi(t, x, u, \{x\})| \le c$. Therefore

$$\begin{aligned} \mathbf{P}\{\tau_k \le T\} &= \mathbf{E}\chi_{\{\tau_k \le T\}} \\ &= \mathbf{E}\chi_{\{\tau_1 \le T\}} \mathbf{E}(\chi_{\{\tau_2 \le T\}}/\tau_1) \cdots \mathbf{E}(\chi_{\{\tau_k \le T\}}/\tau_1, \ldots, \tau_{k-1}) \\ &\le (1 - e^{-cT})^k \to 0 \end{aligned} \tag{2.22}$$

as $k \to \infty$. Hence the processes $\xi(t)$ and $\eta(t)$ are defined on $[0, \infty)$. A pair of processes $\xi(t)$, $\eta(t)$ constructed in such a manner is a controlled process with a given controlled object and control. In the case when $\nu(\cdot/\cdot)$ is a non-randomized Markovian control defined by the function $u = \varphi(t, x)$ the construction is substantially simplified. As formulas (2.19) and (2.20) show $\xi(t)$ will then also be a jump Markovian process with the transition probability $\mathbf{P}^{\varphi}(t, x, s, A)$ satisfying the relation

$$\lim_{s \downarrow t} \frac{1}{s - t} [\mathbf{P}^{\varphi}(t, x, s, A) - \chi_A(x)] = \Pi(t, x, \varphi(t, x), A). \tag{2.23}$$

Consider now the problem of minimization of functionals of the evolutional type for jump controlled Markov processes. To simplify the notation we shall discuss in detail the additive functional

$$F_f(\xi(\cdot), \eta(\cdot)) = \int_0^T f(s, \xi(s), \eta(s))\, ds. \tag{2.24}$$

Let $f(s, x, u)$ be a non-negative function measurable jointly in the variables. Denote

$$Z(f) = \inf_{\nu} \mathbf{E}_\nu F_f(\xi(\cdot), \eta(\cdot)),$$

where the infimum is taken over all admissible controls. We note several self-evident properties of $Z(f)$.

I. If $f_n \uparrow f$, then $Z(f) = \lim_{n \to \infty} Z(f_n)$.

II. $Z(f) = \inf_{\nu \in \mathcal{N}_0} \mathbf{E}_\nu F_f(\xi(\cdot), \eta(\cdot))$, where $\mathcal{N}_0$ is a class of step controls (i.e. such controls for which the measure $\nu(\cdot/x(\cdot))$ is concentrated at step functions).

III. If $g \ge f$ then for all ν

$$\mathbf{E}_\nu F_g(\xi(\cdot), \eta(\cdot)) \ge \mathbf{E}_\nu F_f(\xi(\cdot), \eta(\cdot)).$$

Let $f(t, x, u)$ be a function continuous jointly in the variables. Let X and U be compact sets. For any $\varepsilon > 0$ there exists a subdivision of the interval $[0, T]: 0 = t_0 < t_1 < \cdots < t_n = T$ and control $\bar{\nu}_\varepsilon$ concentrated at the step functions with jumps only at points $t_1, \ldots, t_{n-1}$ such that

$$\mathbf{E}_{\bar{\nu}_\varepsilon} F_f(\xi(\cdot), \eta(\cdot)) \le Z(f) + \varepsilon.$$

Let $\bar{f}(t, x, u) = f(t_k, x, u)$ for $t \in [t_k, t_{k+1})$. If $\max_k(t_{k+1} - t_k)$ is sufficiently small we have

$$|\bar{f}(t, x, u) - f(t, x, u)| \leq \frac{\varepsilon}{T}.$$

Hence,

$$\mathbf{E}_{\bar{v}_\varepsilon} \int_0^T \bar{f}(s, \xi(s), \eta(s))\, ds \leq Z(f) + 2\varepsilon.$$

Furthermore,

$$\left| \int_0^T \bar{f}(s, \xi(s), \eta(s))\, ds - \sum_{k=0}^{n-1} f(t_k, \xi(t_k), \eta(t_k))\, \Delta t_k \right| \leq 2\zeta \|f\| \max_k \Delta t_k,$$

where ζ is the number of jumps of the process $\xi(s)$ on the interval $[0, T]$. The bound (2.22) implies that $\mathbf{E}_v \zeta$ is uniformly bounded over all controls. Consequently, for $\max_k \Delta t_k$ sufficiently small

$$\mathbf{E}_{\bar{v}_\varepsilon} \sum_{k=0}^{n-1} f(t_k, \xi(t_k), \eta(t_k))\, \Delta t_k \leq Z(f) + 3\varepsilon.$$

We now utilize theorem 1.11. Let $\bar{f}_\varepsilon(t, x)$ be defined by the recurrent system of equalities:

$$\bar{f}_\varepsilon(t_n, x) = 0,$$

$$\bar{f}_\varepsilon(t_k, x) = \inf_u \left[f(t_k, x, u)\, \Delta t_k + \int P(t_k, x, t_{k+1}, dy; u) \bar{f}_\varepsilon(t_{k+1}, y) \right]. \tag{2.25}$$

Then if $\varphi(t_k, x)$ is a measurable function such that

$$\bar{f}_\varepsilon(t_k, x) = f(t_k, x, \varphi(t_k, x))\, \Delta t_k + \int P(t_k, x, t_{k+1}, dy; \varphi(t_k, x)) \bar{f}_\varepsilon(t_{k+1}, y),$$

setting $\eta(t_k) = \varphi(t_k, \xi(t_k))$ one can find an optimal non-randomized control v_ε such that

$$\mathbf{E}_{v_\varepsilon} \sum_{k=0}^{n-1} f(t_k, \xi(t_k), \eta(t_k))\, \Delta t_k \leq Z(f) + 3\varepsilon.$$

The same arguments as above show that in this case

$$\mathbf{E}_{v_\varepsilon} F_f(\xi(\cdot), \eta(\cdot)) \leq Z(f) + 5\varepsilon.$$

This allows us to seek ε-optimal controls.

To determine an optimal control we shall formally pass to the limit in (2.25) as $\Delta t_k \to 0$. We write (2.25) in the form

$$\frac{1}{\Delta t_k}(\bar{f}_\varepsilon(t_k, x) - \bar{f}_\varepsilon(t_{k+1}, x))$$

$$= \inf_u \left\{ f(t_k, x, u) + \frac{1}{\Delta t_k} \int P(t_k, x, t_{k+1}, dy; u) \right\} [\bar{f}_\varepsilon(t_{k+1}, y) - \bar{f}_\varepsilon(t_{k+1}, x)].$$

After approaching the limit and taking the properties of $P(\cdot, \cdot, \cdot, \cdot; \cdot)$ into account we obtain

$$-\frac{\partial \bar{f}}{\partial t}(t, x) = \inf_u \left[f(t, x, u) + \int \bar{f}(t, y)\Pi(t, x, u, dy) \right]. \tag{2.26}$$

Equation (2.26) is called *Bellman's equation*. It is solved on the interval $(0, T)$ with the "initial" condition $\bar{f}(T, x) = 0$. Integrating (2.26) on the interval $[t, T]$ we obtain

$$\bar{f}(t, x) = \int_t^T \inf_u \left[f(s, x, u) + \int \bar{f}(s, y)\Pi(s, x, u, dy) \right] ds. \tag{2.27}$$

Denote by $Qg(t, x)$ the operator defined by equality

$$Qg(t, x) = \int_t^T \inf_u \left[f(s, x, u) + \int g(s, y)\Pi(s, x, u, dy) \right] ds.$$

Assume that the function $\Pi(s, x, u, dy)$ satisfies the following condition: for any continuous function $g(t, x)$ the integral

$$\int \Pi(s, x, u, dy) g(s, y)$$

is continuous jointly in the variables. Then Q maps $C_{[0,T] \times X}$ into itself and

$$|Qg_1(t, x) - Qg_2(t, x)| \le \int_t^T \left| \inf_u \left[f(s, x, u) + \int g_1(s, y)\Pi(s, x, u, dy) \right] \right.$$

$$\left. - \inf_u \left[f(s, x, u) + \int g_2(s, y)\Pi(s, x, u, dy) \right] \right| ds$$

$$\le \int_t^T \sup_u \left| \int g_1(s, y)\Pi(s, x, u, dy) - \int g_2(s, y)\Pi(s, x, u, dy) \right| ds$$

$$\le C \int_t^T \sup_y |g_1(s, y) - g_2(s, y)| \, ds,$$

where C is a constant dependent on Π. Hence

$$\begin{aligned}\int_0^T t^n \sup_x |Qg_1(t,x) - Qg_2(t,x)|\,dt &\le C\int_0^T t^n \int_t^T \sup_y |g_1(s,y) - g_2(s,y)|\,ds \\ &= C\int_0^T \frac{s^{n+1}}{n+1} \sup_y |g_1(s,y) - g_2(s,y)|\,ds \\ &\le \frac{CT}{n+1}\int_0^T s^n \sup_y |g_1(s,y) - g_2(s,y)|\,ds.\end{aligned}$$

If we introduce on $C_{[0,T]\times X}$ the norm

$$\|g(\cdot,\cdot)\| = \int_0^T t^n \sup_x |g(t,x)|\,dt,$$

then the operator Q becomes a contractible operator for $n \ge CT$. This implies that (2.27) possesses a unique solution (the continuity of the solution in t follows from the continuity of the r.h.s. of (2.27) in t). Furthermore, let $u = \varphi(t, x)$ be a Borel function satisfying

$$\begin{aligned}&\inf_u\left[f(t,x,u) + \int \bar{f}(t,y)\Pi(t,x,u,dy)\right] \\ &\quad = f(t,x,\varphi(t,x)) + \int \bar{f}(t,y)\Pi(t,x,\varphi(t,x),dy).\end{aligned} \tag{2.28}$$

We shall show that $u = \varphi(t, x)$ defines an optimal control.

Moreover under this control $\xi(t)$ will be a Markov process with transition probability P^φ satisfying relation (2.23). If

$$\begin{aligned}f^*(t,x) &= \mathbf{E}^\varphi\left(\int_t^T f(s,\xi(s),\eta(s))\,ds/\xi(t) = x\right) \\ &= \mathbf{E}^\varphi_{t,x}\int_t^T f(s,\xi(s),\varphi(s,\xi(s)))\,ds,\end{aligned}$$

where $\mathbf{E}^\varphi_{t,x}$ is the mathematical expectation which corresponds to the process with transition probability P^φ and the initial condition $\xi(t) = x$, then

$$\begin{aligned}&\lim_{h\downarrow 0}\frac{1}{h}(f^*(t+h,x) - f^*(t,x)) \\ &\quad = -\lim_{h\downarrow 0}\frac{1}{h}\mathbf{E}^\varphi_{t,x}\int_t^{t+h} f(s,\xi(s),\varphi(s,\xi(s)))\,ds \\ &\qquad + \lim_{h\downarrow 0}\frac{1}{h}\int P(t,x,t+h,dy;\varphi(t,x))[f^*(t+h,x) - f^*(t+h,y)] \\ &\quad = -f(t,x,\varphi(t,x)) - \int f^*(t,y)\Pi(t,x,\varphi(t,x),dy).\end{aligned}$$

Since the r.h.s. of this relation is continuous in t, we have

$$-\frac{\partial f^*(t, x)}{\partial t} = f(t, x, \varphi(t, x)) + \int f^*(t, y)\Pi(t, x, \varphi(t, x), dy).$$

Equations (2.26) and (2.28) imply that $\bar{f}(t, x)$ also satisfies equation

$$-\frac{\partial \bar{f}(t, x)}{\partial t} = f(t, x, \varphi(t, x)) + \int \bar{f}(t, y)\Pi(t, x, \varphi(t, x), dy).$$

Since $f^*(T, x) = \bar{f}(T, x) = 0$ the uniqueness of the last equation (which is a consequence of the uniqueness of the solution of Bellman's equation) yields $f^*(t, x) = \bar{f}(t, x)$. Thus

$$\mathbf{E}^{\varphi}_{0, x} \int_0^T f(s, \xi(s), \eta(s))\, ds = \bar{f}(0, x).$$

We now show that for any other control v

$$\mathbf{E}_v \left[\int_0^T f(s, \xi(s), \eta(s))\, ds / \xi(0) = x \right] \geq \bar{f}(0, x).$$

The r.h.s. of (2.28) is continuous, therefore for max Δt_k sufficiently small we have

$$\begin{aligned}
&\bar{f}(t_k, x) - \bar{f}(t_{k+1}, x) \\
&= \int_{t_k}^{t_{k+1}} \left[f(t, x, \varphi(t, x)) + \int \bar{f}(t, y)\Pi(t, x, \varphi(t, x), dy) \right] dt \\
&= \left[f(t_{k+1}, x, \varphi(t_{k+1}, x)) + \int \bar{f}(t_{k+1}, y)\Pi(t_{k+1}, x, \varphi(t_{k+1}, x), dy) + \alpha_k \right] \Delta t_k \\
&= \inf_u \left[f(t_{k+1}, x, u) + \int \bar{f}(t_{k+1}, y)\Pi(t_{k+1}, x, u, dy) \right] \Delta t_k + \alpha_k \Delta t_k,
\end{aligned}$$

where $|\alpha_k| < \varepsilon$. Thus,

$$\begin{aligned}
\bar{f}(t_k, x) = \inf_u \Big[& f(t_{k+1}, x, u)\, \Delta t_k + \bar{f}(t_{k+1}, x) \\
& + \Delta t_k \int \bar{f}(t_{k+1}, y)\Pi(t_{k+1}, x, u, dy) \Big] + \alpha_k \Delta t_k.
\end{aligned}$$

Condition (2.23) implies that

$$\begin{aligned}
\Big| & \bar{f}(t_{k+1}, x) + \Delta t_k \int \bar{f}(t_{k+1}, y)\Pi(t_{k+1}, x, u, dy) \\
& - \int \bar{f}(t_{k+1}, y)P(t_k, x, t_{k+1}, dy; u) \Big| \leq \varepsilon \Delta t_k
\end{aligned}$$

for Δt_k sufficiently small. Therefore

$$\bar{f}(t_k, x) = \inf_u \left[f(t_{k+1}, x, u)\, \Delta t_k + \int \bar{f}(t_{k+1}, y) P(t_k, x, t_{k+1}, dy; u) \right] + \beta_k\, \Delta t_k,$$

where $|\beta_k| \le 2\varepsilon$. Finally, utilizing the continuity of $f(t, x, u)$ we verify that for $\max_k \Delta t_k$ sufficiently small the relation

$$\bar{f}(t_k, x) = \inf_u \left[f(t_k, x, u)\, \Delta t_k + \int \bar{f}(t_{k+1}, y) P(t_k, x, t_{k+1}, dy; u) \right] + \gamma_k\, \Delta t_k \tag{2.29}$$

is valid with $|\gamma_k| \le 3\varepsilon$. Let $f_\varepsilon(t_k, x)$ be defined by the system of equalities (2.25). Then

$$\begin{aligned} |\bar{f}_\varepsilon(t_k, x) - \bar{f}(t_k, x)| &\le \sup_u \int |\bar{f}_\varepsilon(t_{k+1}, y) - \bar{f}(t_{k+1}, y)| P(t_k, x, t_{k+1}, dy; u) \\ &\quad + |\gamma_k|\, \Delta t_k \le \sup_y |\bar{f}_\varepsilon(t_{k+1}, y) - \bar{f}(t_{k+1}, y)| + 3\varepsilon\, \Delta t_k. \end{aligned}$$

Since $\bar{f}_\varepsilon(t_n, x) = \bar{f}(t_n, x) = 0$, we have

$$\sup_x |\bar{f}_\varepsilon(0, x) - \bar{f}(0, x)| \le 3\varepsilon \Sigma\, \Delta t_k = 3T\varepsilon.$$

As it was mentioned above one can choose control v_ε concentrated only on step functions with jumps at the points $t_1, \ldots, t_{n-1}$ such that

$$\bar{f}_\varepsilon(0, x) = \mathbf{E}_{v_\varepsilon} \sum_{k=0}^{n-1} f(t_k, \xi(t_k), \eta(t_k))\, \Delta t_k \le Z(f) + 3\varepsilon.$$

Hence

$$\bar{f}(0, x) \le Z(f) + 3\varepsilon + 3T\varepsilon.$$

Finally noting that $\varepsilon > 0$ is arbitrary we obtain the following.

Theorem 2.6. *If X and U are compact spaces and a controlled Markovian object satisfies condition (2.23) where the function $\Pi(t, x, u, dy)$ is such that*

1. *$\Pi(t, x, u, \{x\})$ is bounded*
2. *$\int f(t, y)\Pi(t, x, u, dy)$ is continuous for any continuous function f and the function $f(t, x, u)$ is continuous*

then an optimal control for functional (2.24) *is a non-randomized Markov control defined by the function* $u = \varphi(t, x)$ *where* $\varphi(t, x)$ *is a measurable function satisfying* (2.28); $f(t, x)$ *is a solution of equation* (2.27). *Moreover,*

$$Z(f) = \inf_{v} \mathbf{E}_v F_j(\xi(\cdot), \eta(\cdot)) = \bar{f}(0, x),$$

where x *is the initial value of the process.*

Analogously one can obtain the following more general theorem.

Theorem 2.7. *Let the conditions of theorem* 2.7 *be satisfied and* $g(t, x, u)$ *be a continuous function. Set*

$$F(x(\cdot), u(\cdot)) = \int_0^T \exp\left\{\int_0^t g(s, x(s), u(s))\, ds\right\} f(t, x(t), u(t))\, dt. \quad (2.30)$$

Define $\bar{f}(t, x)$ *to be the solution of equation*

$$\bar{f}(t, x) = \int_t^T \inf_u \left[f(s, x, u) + g(s, x, u)\bar{f}(s, x) + \int \bar{f}(s, y)\Pi(s, x, u, dy)\right] ds. \quad (2.31)$$

then, if $u = \varphi(t, x)$ *is a measurable function satisfying*

$$\frac{\partial \bar{f}}{\partial t}(t, x) = f(t, x, \varphi(t, x)) + g(t, x, \varphi(t, x))\bar{f}(t, x) + \int \bar{f}(t, y)\Pi(t, x, \varphi(t, x), dy), \quad (2.32)$$

the non-randomized Markov control $\bar{v}$ *defined by the function* $u = \varphi(t, x)$ *is optimal and moreover*

$$\inf_v \mathbf{E}_v F(\xi(\cdot), \eta(\cdot)) = \mathbf{E}_{\bar{v}} F(\xi(\cdot), \eta(\cdot)) = \bar{f}(0, x),$$

where x *is the initial value of the process.*

Remark 2.5. The existence and uniqueness of the solution of equation (2.31) is established in exactly the same manner as that of equation (2.27). This solution may be derived using the method of successive approximations.

3 Controlled Stochastic Differential Equations

1 Some Preliminaries

In this Section we present definitions and results related to the theory of stochastic integration which will be used frequently in what follows. Proofs can be found in the books listed in the Bibliography.

Let $\{\Omega, \mathfrak{S}, \mathbf{P}\}$ be a probability space. A family of σ-algebras $\{\mathfrak{F}_t, t \geq 0\}$ (or $t \in [0, T]$) satisfying the conditions $\mathfrak{F}_t \subset \mathfrak{S}$, $\mathfrak{F}_{t_1} \subset \mathfrak{F}_{t_2}$ for $t_1 < t_2$ is called a *current* of σ-algebras. A stochastic process $\{\xi(t), t \geq 0\}$ $(t \in [0, T])$ is called *adapted to the current* $\{\mathfrak{F}_t\}$ if $\xi(t)$ is $\mathfrak{F}_t$-measurable for all $t \geq 0$ $(t \in [0, T])$. Sometimes an object consisting of a current of σ-algebras $\mathfrak{F}_t$, $t \in [0, T]$ $(t \in [0, \infty))$ and a function $\xi(t) = \xi(t, \omega)$ adapted to $\{\mathfrak{F}_t\}$ is called a *stochastic process* and is denoted by $\{\xi(t), \mathfrak{F}_t, t \in [0, T]\}$.

A process $\{\mu(t), \mathfrak{F}_t, t \geq 0\}$ with values in $\mathscr{R}^d$ is called a *process with independent increments* (or $\mu(t)$ is said to be a process with independent increments with respect to $\{\mathfrak{F}_t, t \geq 0\}$) if

a. the process $\{\mu(t), t \geq 0\}$ is adapted to the current of σ-algebras $\{\mathfrak{F}_t, t \geq 0\}$;
b. the variables $\mu(s) - \mu(t)$ do not depend on σ-algebra $\mathfrak{F}_t$ for all $s \geq t$ and $t \geq 0$ $(\mu(0) = 0)$.

Evidently such a process $\mu(t)$ will also be a process with independent increments with respect to any current $\{\hat{\mathfrak{F}}_t, t \geq 0\}$ provided $\mu(t)$ is adapted to this current and $\hat{\mathfrak{F}}_t \subset \mathfrak{F}_t$. Denote by $\mathfrak{F}_t^\mu$ the smallest σ-algebra with respect to which all the random variables $\mu(s)$, $s \in [0, t]$ are measurable. Then $\{\mu(t), \mathfrak{F}_t^\mu, t \geq 0\}$ is a process with independent increments. If $\{\xi_\alpha(t), t \geq 0\}$, $\alpha \in A$, is a collection of arbitrary stochastic processes which do not depend on $\mu(t)$, $t \geq 0$ and $\hat{\mathfrak{F}}_t$ is the smallest σ-algebra with respect to which all the variables $\mu(s)$ and $\zeta_\alpha(s)$, $\alpha \in A$, $s \in [0, t]$ are measurable, then $\{\mu(t), \tilde{\mathfrak{F}}_t, t \geq 0\}$ is also a process with independent increments. The process is called *homogeneous* if the distribution of the vector $\mu(t+h) - \mu(t)$ does not depend on t.

Special cases of homogeneous processes with independent increments are the Wiener (sometimes called Brownian motion) and Poisson processes.

A process $\{w(t), \mathfrak{F}_t, t \geq 0\}$ with values in $\mathscr{R}^d$, where $w(t) = (w^{(1)}(t), \ldots, w^{(d)}(t))$ is called a *Wiener process* if

a. for a fixed B $v(t, B)$ is a homogeneous Poisson process, $\mathbf{E}v(t, B) = tq(B)$ σ-algebras $\{\mathfrak{F}_t, t \geq 0\}$;
b. its components $w^{(k)}(t)$, $k = 1, \ldots, d$, are mutually independent processes with continuous sample functions, $w^{(k)}(0) = 0$, and the variables $\Delta w^{(k)}(t) = w^{(k)}(t + \Delta t) - w^{(k)}(t)$ possess the Gaussian distribution for all $t \geq 0$ with $\mathbf{E}\Delta w^{(k)}(t) = 0$ and $\mathbf{V}[\Delta w^{(k)}(t)] = \Delta t$ $(\Delta t > 0)$.

We now present a definition of a Poisson measure. Let $\mathfrak{B}^d$ be a σ-algebra of Borel sets in $\mathscr{R}^d$, $\mathfrak{B}_0^d$ be a subalgebra of $\mathfrak{B}^d$ consisting of Borel sets such that their closure does not contain point 0.

A *Poisson measure* $v(t, B)$, $t \geq 0$, $B \in \mathfrak{B}^d$ is a family of random variables possessing the following properties:

a. for a fixed B $v(t, B)$ is a homogeneous Poisson process, $\mathbf{E}v(t, B) = tq(B)$ where $q(B)$ is a measure on $\mathfrak{B}^d$ and $q(B) < \infty$ for $B \in \mathfrak{B}_0$;
b. if $B_1 \cap B_2 = \varnothing$, then

$$v(t, B_1 \cup B_2) = v(t, B_1) + v(t, B_2);$$

c. the family of random variables $\{v(s, A), s \in [0, t], A \in \mathfrak{B}_0^d\}$ and $\{v(s', B) - v(t, B); s' > t, B \in \mathfrak{B}_0^d\}$ are independent for all $t > 0$.

We say that a Poisson measure $v(t, A)$ is a *Poisson measure with respect to the current of σ-algebras* $\{\mathfrak{F}_t, t \geq 0\}$ or that $\{v(t, A), \mathfrak{F}_t, t \geq 0\}$ is a Poisson measure if the process $v(t, A)$, $t > 0$ is adapted to the current of σ-algebras $\{\mathfrak{F}_t\} \forall A \in \mathfrak{B}_0^d$ and the family of random variables $\{v(s', A) - v(t, A), s' > t, A \in \mathfrak{B}_0^d\}$ does not depend on the σ-algebra $\mathfrak{F}_t$, $\forall t > 0$.

The process $\tilde{v}(t, A) = v(t, A) - tq(A)$ is called a *centered Poisson measure*. For this measure

$$\mathbf{E}\tilde{v}(t, A) = 0,$$

$$\mathbf{V}\tilde{v}(t, A) = tq(A).$$

A process $\xi(t)$, $t \geq 0$, is called *stochastically continuous* if for all $t \geq 0$ and any $\varepsilon > 0$

$$\mathbf{P}\{|\zeta(t+h) - \zeta(t)| > \varepsilon\} \to 0 \quad \text{as } h \to 0.$$

If a process with independent increments is stochastically continuous one can assume that its sample functions possess for each t a left-hand limit and are continuous from the right with probability 1 (cf. [*20*, Vol II, p. 41]).

Denote by D (D^d or $D^d[0, T]$) the space of functions $f(t)$, $t \in [0, T]$ with values in $\mathscr{R}^d$ possessing for each $t \in (0, T]$ a left-hand limit and continuous from the right for all $t \in [0, T)$.

In what follows when discussing a process with independent increments it will be assumed that the process is stochastically continuous and that its sample functions belong to D^d with probability 1.

An arbitrary homogeneous stochastically continuous process with independent increments is representable as the sum

$$\mu(t) = at + Cw(t) + \mu_1(t),$$

where a is a constant (non-random) vector, C is a matrix with constant (non-random) entries, $w(t)$ is a Wiener process and $\mu_1(t)$ is a purely discontinuous homogeneous process with independent increments which does not depend on the process $w(t)$. The process $\mu_1(t)$ is representable in the form

$$\mu_1(t) = a_1 t + \int_{|u| \leq 1} u(v(t, du) - tq(du)) + \int_{|u| > 1} uv(t, du),$$

where $v(t, A)$ is a Poisson measure on $\mathfrak{B}_0^d$, a_1 is a constant and $tq(A) = \mathbf{E}v(t, A)$. The integrating measure $v(t, du) - tq(du)$ appears in the first of the integrals on the r.h.s. above because the integral $\int_{R^d} uv(t, du)$ is in general infinite. However if the process $\mu_1(t)$ possesses finite moments of the second order it can be represented in the form

$$\mu_1(t) = tc + \int_{R^d} u\tilde{v}(t, du),$$

where $\tilde{v}(t, A) = v(t, A) - tq(A)$.

In what follows we shall consider *homogeneous processes with independent increments* and finite moments of the second order such that $\mathbf{E}\mu(t) = 0$. Such a process admits representation

$$\mu(t) = Cw(t) + \mu_1(t),$$

$$\mu(t) = \int_{R^d} u\tilde{v}(t, du),$$

$$\tilde{v}(t, A) = v(t, A) - tq(A),$$

where $tq(A) = \mathbf{E}v(t, A)$. The meaning of the measure $v(t, A)$ appearing in this representation is as follows: $v(t, A)$ is equal to the number of jumps of the process $\mu(t)$ with the values in the set A (i.e. the number of instances of time s, $s \leq t$, such that $\mu(s) - \mu(s-) \in A$). If the closure of the set A does not contain point 0, then $v(t, A)$ takes on a finite value with probability 1.

Sample functions of the process $w(t)$ possess with probability 1 an unbounded variation on any interval $[0, T]$; sample functions of the process

$\mu(t)$ in general possess the same property. Nevertheless it is possible to develop a theory of integrals of the form

$$\int_0^t \varphi(s)\mu(ds),$$

which is in several aspects analogous to the theory of the Stieltjes integral. These integrals are called *stochastic*.

We present the definition and point out several properties of stochastic integrals.

Let $\{\mu(t), \mathfrak{F}_t, t \geq 0\}$ be a homogeneous process with independent increments, $\mathbf{E}\mu(t) = 0$, $\mathbf{E}[\mu(t+h) - \mu(t)]^2 = h$. A function $\varphi(t)$, $t \in [0, T]$ is called *simple* if there exist $t_1, t_2, \ldots, t_n$, $0 = t_0 < t_1 < t_2 \cdots < t_n \leq T = t_{n+1}$, such that $\varphi(t) = \varphi_k$ for $t \in (t_k, t_{k+1}]$, $k = 0, \ldots, n$, where φ_k are bounded with probability 1 and $\mathfrak{F}_{t_k}$-measurable random variables. For simple functions $\varphi(\cdot)$ a *stochastic integral* over the process $\mu(t)$ is defined by the formula

$$\int_0^T \varphi(s)\mu(ds) = \sum_{k=0}^{n} \varphi_k[\mu(t_{k+1}) - \mu(t_k)].$$

Denote by $H_2 = H_2(\mathfrak{F}_t)$ the class of all the stochastic processes adapted to the current $\{\mathfrak{F}_t, t \in [0, T]\}$ such that there exists a sequence of simple functions $\varphi_n(t)$ satisfying

$$\int_0^T [\varphi(t) - \varphi_n(t)]^2 \, dt \to 0$$

with probability 1. By definition, set

$$\int_0^T \varphi(s)\mu(ds) = \mathbf{P}\text{-lim} \int_0^T \varphi_n(s)\mu(ds).$$

This limit does exist. Define for $t \in [0, T]$ a stochastic integral with a variable limit of integration:

$$\mathscr{T}(t) = \mathscr{T}(\varphi, t) = \int_0^t \varphi(s)\mu(ds) = \int_0^T \chi_t(s)\varphi(s)\mu(ds),$$

where $\chi_t(s) = 1$ for $s \leq t$ and $\chi_t(s) = 0$ for $s > t$. For each t the integral $\mathscr{T}(t)$ is defined only with probability 1. Utilizing this fact a process $\mathscr{T}(t)$ can be defined in such a manner that its sample functions with probability 1 possesses for each t a left-hand limit and are continuous from the right and in the case when $\mu(t) = w(t)$ the process $\mathscr{T}(t)$ can be so defined that its sample functions will be continuous for all values of t. In what follows we shall always assume that the stochastic integral $\mathscr{T}(\varphi, t)$ is a process whose sample functions possess the property stated above. We now list a number of properties of a stochastic integral.

1. The process $\mathcal{T}(\varphi, t)$ is adapted to the current $\{\mathfrak{F}_t, t \geq 0\}$ and its sample functions belong to D with probability 1.
2. For any two constants c_1 and c_2

$$\mathcal{T}(c_1\varphi_1 + c_2\varphi_2, t) = c_1\mathcal{T}(\varphi_1, t) + c_2\mathcal{T}(\varphi_2, t).$$

3. If

$$\mathbf{E}\int_0^T \varphi^2(t)\,dt < \infty,$$

then $\mathcal{T}(\varphi, t)$ possess finite moments of the second order and

$$\mathbf{E}\{\mathcal{T}(\varphi, t+h) - \mathcal{T}(\varphi, t) \mid \mathfrak{F}_t\} = 0,$$

$$\mathbf{E}\{[\mathcal{T}(\varphi, t+h) - \mathcal{T}(\varphi, t)]^2 \mid \mathfrak{F}_t\} = \mathbf{E}\left\{\int_t^{t+h} \varphi^2(s)\,ds \mid \mathfrak{F}_t\right\}, \tag{3.1}$$

$$\mathbf{E}\sup_{0\leq t\leq T} |\mathcal{T}(\varphi, t)|^2 \leq 4\mathbf{E}\mathcal{T}^2(\varphi, T) = 4\mathbf{E}\int_0^T \varphi^2(s)\,ds. \tag{3.2}$$

4. For any $N > 0$, $\varepsilon > 0$

$$\mathbf{P}\left\{\sup_{0\leq t\leq T} |\mathcal{T}(\varphi, t)| > \varepsilon\right\} \leq \frac{N}{\varepsilon^2} + \mathbf{P}\left\{\int_0^T \varphi^2(s)\,ds > N\right\}. \tag{3.3}$$

We also note the following condition for the existence of higher order moments for the process $\mathcal{T}(\varphi, t)$. Let $\mu(t) = w(t)$ be a Wiener process. If $\int_0^T \mathbf{E}\varphi^{2p}(s)\,ds < \infty$, $p > 1$, then

$$\mathbf{E}|\mathcal{T}(\varphi, t)|^{2p} \leq p(2p-1)^p \int_0^T \mathbf{E}|\varphi(s)|^{2p}\,ds < \infty. \tag{3.4}$$

A stochastic integral over a Poisson measure is defined analogously and possesses analogous properties. Let $\{\nu(t, A), \mathfrak{F}_t, t \geq 0, A \in \mathfrak{B}_0^d\}$ be a Poisson measure, $\tilde{\nu}(t, A) = \nu(t, A) - tq(A)$, where $tq(A) = \mathbf{E}\nu(t, A)$ and $\psi(s, x)$ be a random function ($(s, x) \in [0, T] \times R^d$) which is $\mathfrak{F}_s$-measurable for all $x \in R^d$ and all $s \in [0, T]$ (we say that the function $\psi(s, x)$ is adapted to the current $\{\mathfrak{F}_t, t \geq 0\}$) and $\tilde{\nu}(t, A)$ is a centered Poisson measure. The function $\psi(s, x)$ is called simple if

$$\psi(t, x) = \psi_{kl} \quad \text{for } t \in (t_k, t_{k+1}],\ x \in A_l,$$

where $A_l \in \mathfrak{B}_0^d$, $\bigcup_l A_l = R^d\setminus\{0\}$, $A_k \cap A_r = \varnothing$ for $k \neq r$ and $0 = t_0 < t_1 < \cdots < t_{n+1} = T$, and ψ_{kl} is a $\mathfrak{F}_{t_k}$-measurable random variable. The class of simple functions is denoted by $H_0^\nu = H_0^\nu(\mathfrak{F}_t)$. Set for a simple function $\psi(t, x)$

$$J(\psi) = \int_0^T \int_{R^d} \psi(s, x)\tilde{\nu}(ds, dx) \underset{\text{Def}}{=} \sum_{k,l} \psi_{kl}[\tilde{\nu}(t_{k+1}, A_l) - \tilde{\nu}(t_k, A_l)].$$

Let ψ_n be a sequence of simple functions and

$$\int_0^T \int_{R^d} |\psi(t, x) - \psi_n(t, x)|^2 q(dx)\, dt \to 0 \quad \text{as } n \to \infty. \tag{3.5}$$

Then we set

$$J(\psi) = \mathbf{P}\text{-lim } J(\psi_n).$$

This limit exists. A class of functions $\psi(t, x)$ for which there exists an approximating sequence of simple functions $\psi_n(t, x)$ satisfying relation (3.5) is denoted by $H_2(\mathfrak{F}_t, q)$.

Define

$$J(\psi, t) = J(t) = \int_0^t \int_{R^d} \psi(s, x)\tilde{\nu}(ds, dx),$$

where

$$J(\psi, t) = J(\chi_t \psi)$$

and χ_t is defined above.

The process $J(\psi, t)$ can be defined in such a way that its sample functions will belong to $D^1[0, T]$ with probability 1. In what follows we shall assume that the process $J(\psi, t)$ is defined in this manner. The integral $J(\psi, t)$ possesses the following properties:

1. it is defined for all $\psi(t, x) = \psi(t, x, \omega) \in H_2(\mathfrak{F}_t, q)$ and its sample functions belong to $D^1[0, T]$ with probability 1;
2. the process $J(\psi, t)$ is adapted to the current $\{\mathfrak{F}_t, t \in [0, T]\}$;
3. For any two constant c_1 and c_2 we have

$$J(c_1\psi_1 + c_2\psi_2, t) = c_1 J(\psi_1, t) + c_2 J(\psi_2 t);$$

4. if

$$\mathbf{E} \int_0^T \int_{R^d} \psi^2(s, x)\, dsq(dx) < \infty,$$

then $J(\psi, t)$ possesses finite moments of the second order and

$$\mathbf{E}\{J(\psi, t + h) - J(\psi, t) | \mathfrak{F}_t\} = 0,$$

$$\mathbf{E}\{[J(\psi, t + h) - J(\psi, t)]^2 | \mathfrak{F}_t\} = \mathbf{E}\left\{\int_t^{t+h} \int_{R^d} \psi^2(s, x)\, dsq(dx) \,\middle|\, \mathfrak{F}_t\right\}, \tag{3.6}$$

$$\mathbf{E} \sup_{0 \le t \le T} J^2(\psi, t) \le 4\mathbf{E}J^2(\psi, T) = 4\mathbf{E} \int_0^T \int_{R^d} \psi^2(s, x)\, dsq(dx); \tag{3.7}$$

5. for any $N \ge 0$, $\varepsilon > 0$

$$\mathbf{P}\left\{\sup_{0 \le t \le T} |J(\psi, t)| > \varepsilon\right\} \le \frac{N}{\varepsilon^2} + \mathbf{P}\left\{\int_0^T \int_{R^d} \psi^2(s, x)\, dsq(dx) > N\right\}; \tag{3.8}$$

6. if

$$\mathbf{E}\left\{\int_0^T\left[\left(\int_{R^d}|\psi(s,x)|^2q(dx)\right)^m+\int_0^T|\psi(s,x)|^{2m}q(dx)\right]ds<\infty\right\}$$

(where m is an integer and $m \geq 1$), then $\mathbf{E}|J(\psi, T)|^{2m} < \infty$.

The relations presented above can easily be extended to the case of vectors. Since in what follows we shall consider Euclidean finite-dimensional spaces R^d with a fixed coordinate system, we shall agree to identify operators with matrices, and vectors with a sequence of coordinates. Moreover, vectors will be viewed as one-column matrices. If b is a matrix, then b^* denotes the conjugate matrix. For example, if a is a vector (column-vector) then a^* is a one row matrix. If the symbol * is omitted in the notation of a vector this will always be a column-vector. Note that if $a = (a_1, \ldots, a_d)$ and $b = (b_1, \ldots, b_d)$ are two vectors then ab^* is a matrix with elements $a_i b_j$, $i, j = 1, \ldots, d$; $b^*a = \sum_{j=1}^d b_j a_j$ is the scalar product of vectors a and b. We shall also utilize the notation (a, b) for the scalar product of two vectors.

In what follows $|a|$ will denote the *norm* of a vector or of a matrix. If a is a vector then $|a| = \sqrt{(a, a)}$. If a is a matrix then $|a|$ denotes the operator norm, i.e. the smallest norm such that $|ax| \leq |a| \cdot |x|$ for all $x \in R^d$. Observe that $|a| \leq [\sum a_{ij}^2]^{1/2}$ where a_{ij} are the entries of matrix a.

Let $\varphi(t)$ and $\psi(t)$ be vector-valued functions with values in R^d. It follows from (3.1) that

$$\mathbf{E}\{\mathscr{T}(\varphi, \Delta t)\mathscr{T}^*(\psi, \Delta t)\,|\,\mathfrak{F}_t\} = \mathbf{E}\left\{\int_t^{t+\Delta t}\varphi(s)\psi^*(s)\,ds\,|\,\mathfrak{F}_t\right\}, \tag{3.9}$$

where $\mathscr{T}(\varphi, \Delta t) = \mathscr{T}(\varphi, t + \Delta t) - \mathscr{T}(\varphi, t)$.

An analogous equality is valid for stochastic integrals over a Poisson measure.

Stochastic integrals viewed as functions of the upper limit are local martingales. A *martingale* is a stochastic process $\{\xi(t), \mathfrak{F}_t, t \in [0, T]\}$ such that

a. $\mathbf{E}|\xi(T)| < \infty$,
b. $\mathbf{E}\{\xi(t)|\mathfrak{F}_s\} = \xi(s)\ \forall s, t\ (0 \leq s < t \leq T)$,
c. the sample functions of the process $\xi(t)$ belong to D.

A martingale $\{\xi(t), \mathfrak{F}_t, t \in [0, T]\}$ is called *square integrable* if $\mathbf{E}\xi^2(T) < \infty$. We also note the inequality

$$\mathbf{E}\sup_{0\leq t\leq T}|\xi(t)|^p \leq q^p\mathbf{E}|\xi(T)|^p, \qquad q = 1 - \frac{1}{p},\ p > 1$$

and, in particular,

$$\mathbf{E}\sup_{0\leq t\leq T}\xi^2(t) \leq 4\mathbf{E}\xi^2(T) \tag{3.10}$$

known as *Doob*'s *inequality*.

Given a current of σ-algebras $\{\mathfrak{F}_t, t \in [0, T]\}$ a random variable τ with values on $[0, T]$ possessing the following property: $\{\tau \le t\} \in \mathfrak{F}_t$ for any $t \in [0, T]$ is called a *random time* (random moment) on this current. A process $\{\xi(t), \mathfrak{F}_t, t \in [0, T]\}$ is called a *local square integrable martingale* if there exists a monotonically non-decreasing sequence of random moments τ_n, $n = 1, 2, \ldots$ $(\tau_n \le \tau_{n+1})$ satisfying: (a) with probability 1 there exists $n_0 = n_0(\omega)$ such that $\tau_{n_0} = T$, (b) the process $\{\xi(t \wedge \tau), \mathfrak{F}_t, t \in [0, T]\}$ is a square integrable martingale.

For brevity we shall refer, in what follows, to a local square integrable martingale as a local martingale.

A generalization of the definitions presented above to the vector case is self-evident. If for a given local martingale $\{\xi(t), \mathfrak{F}_t, t \in [0, T]\}$, $\xi(t) = (\xi_1(t), \ldots, \xi_d(t))$ there exists a random matrix $\langle \xi, \xi \rangle_t$, $t \in [0, T]$ possessing the following properties: (a) $\langle \xi, \xi \rangle_0 = 0$ and the increments $\Delta\langle \xi, \xi \rangle_t = \langle \xi, \xi \rangle_{t+\Delta t} - \langle \xi, \xi \rangle_t$, $\Delta t > 0$ are non-negative definite matrices for all t, $\Delta t > 0$; (b) entries of the matrix $\langle \xi, \xi \rangle_t$ are adapted to the current $\{\mathfrak{F}_t\}$ and are continuous with probability 1; (c) the matrix process $\xi(t)\xi^*(t) - \langle \xi, \xi \rangle_t$ is a local martingale then such a matrix is called a *characteristic* of the martingale $\xi(t)$.

The i, j-th entry of matrix $\langle \xi, \xi \rangle_t$ is completely determined by the martingales $\xi_i(t)$ and $\xi_j(t)$, is denoted by $\langle \xi_i, \xi_j \rangle_t$ and is called the *mutual characteristic of martingales* $\xi_i(t)$ and $\xi_j(t)$. For a wide class of local martingales a characteristic exists and is unique. For example, continuous martingales possess a characteristic. If one somewhat refines the notion of a characteristic replacing the continuity condition by some other broader condition then a characteristic will exist for all local martingales.

If $\xi(t)$ is a square integrable martingale, the matrix $\langle \xi, \xi \rangle_t$ possesses a finite mathematical expectation and condition (c) can be written as follows:

$$\mathbf{E}\{\Delta\xi(t)\, \Delta\xi^*(t) \mid \mathfrak{F}_t\} = \mathbf{E}\{\Delta\langle \xi, \xi \rangle_t \mid \mathfrak{F}_t\}. \tag{3.11}$$

For stochastic integrals $\mathscr{T}(\varphi, t)$ or $J(\psi, t)$ the characteristic is of the form

$$\langle \mathscr{T}, \mathscr{T} \rangle_t = \int_0^t \varphi(s)\varphi^*(s)\, ds,$$

$$\langle J, J \rangle_t = \int_0^t \int_{R^d} \psi(s, x)\psi^*(s, x)\, ds q(dx).$$

The characteristic of the sum $\mathscr{T}(\varphi, t) + J(\psi, t)$—provided the processes $w(t)$ and $v(t, A)$ are independent—is equal to the sum of characteristics of the processes $\mathscr{T}(\varphi, t)$ and $J(\psi, t)$:

$$\langle \mathscr{T} + J, \mathscr{T} + J \rangle_t = \int_0^t \varphi(s)\varphi^*(s)\, ds + \int_0^t \int_{R^d} \psi(s, y)\, \varphi^*(s, x)\, ds q(dx).$$

We introduce the following notation. The class of all martingales defined on a given probability space $(\Omega, \mathfrak{S}, \mathbf{P})$ and adapted to a current $\{\mathfrak{F}_t,$

$t \in [0, T]\}$ will be denoted by $M = M(\mathfrak{F}_t)$; the class of square integrable martingales by $M_2 = M_2(\mathfrak{F}_t)$; the class of local (square integrable) martingales by $lM_2 = lM_2(\mathfrak{F}_t)$. Subclasses of these classes consisting of processes with sample functions continuous with probability 1 will be supplemented by an upper index c (lM^c denotes the class of all continuous local martingales and so on), and subclasses of processes with continuous characteristics by an upper index r (i.e. $lM_2^r = lM_2^r(\mathfrak{F}_t)$ is the class of all local martingales with continuous characteristics adapted to a current of σ-algebras $\{\mathfrak{F}_t\}$). In some cases different probability measures will be considered on the same measurable space $\{\Omega, \mathfrak{S}\}$. In these cases in the designation of a class of martingales we shall add a notation of the measure with respect to which these are martingales (for example, $lM^c(\mathfrak{F}_t, \mathbf{P})$ denotes the class of continuous local martingales on the probability space $\{\Omega, \mathfrak{S}, \mathbf{P}\}$ adapted to the current $\{\mathfrak{F}_t\}$).

The notion of a square variation of a process is closely related to the notion of the characteristic of a local martingale. We shall restrict our attention to the case of one-dimensional processes. Let λ denote a subdivision of the interval $[0, T]$ with subdividing points $0 = t_0 < t_1 < \cdots < t_{n+1} = T$ and let $\xi(t)$ be a stochastic process. The limit of the sum

$$\sigma_\lambda^2 = \sum_{k=1}^{n} (\xi(t_{k+1}) - \xi(t_k))^2$$

in the sense of convergence in probability (as $\max(t_{k+1} - t_k) \to 0$) provided it exists, is called the *square variation of the process* $\xi(t)$ on the interval $[0, T]$ and is denoted by $[\xi, \xi]_T$. If the sample functions of the process $\xi(t)$ are continuous with probability 1 and are of bounded variation then the square variation $[\xi, \xi]_T = 0$ with probability 1. It turns out that for continuous local martingales the square variation coincides with the characteristic $[\mu, \mu]_t = \langle \mu, \mu \rangle_t$ for all $t \in [0, T]$ (where $\mu(\cdot) \in lM^c$). This equality remains valid also in the vector case provided $[\mu, \mu]$ denotes the matrix with the entries which are equal to

$$\mathbf{P}\text{-}\lim \sum_{k=1}^{n} [\mu_i(t_{k+1}) - \mu_i(t_k)][\mu_j(t_{k+1}) - \mu_j(t_k)], \qquad i, j = 1, \ldots, d,$$

where $\mu_i(t)$ are the components of the vector $\mu(t)$.

We state the following important theorem known as *Levy's theorem.*

Theorem 3.1. *A continuous local martingale* $\mu(t)$ *with characteristic* $\langle \mu, \mu \rangle_t = It$, *where* I *is the unit matrix, is a Wiener process.*

The theory of stochastic integrals is generalized to the case of integration over local martingales. In this direction results were obtained which are analogous to those presented above. Classes consisting of measurable processes $\varphi(t)$ adapted to the current $\{\mathfrak{F}_t,\ t \in [0, T]\}$ for which there exist sequences of simple functions $\varphi_n(t)$ adapted to $\{\mathfrak{F}_t\}$ such that

$$\int_0^T (\varphi(t) - \varphi_n(t))^2 \, d\langle \mu, \mu \rangle_t \to 0 \qquad \text{as } n \to \infty$$

with probability 1 replace the class $H_2\{\mu, \mathfrak{F}_t\}$ of integrable functions in this case.

The following result is basic in the theory of stochastic integration.

Theorem 3.2 (Itô's formula). *Let $f(t, x)$, $(t, x) \in [0, T] \times R^d$ be a function continuously differentiable with respect to t and twice continuously differentiable with respect to x, $\beta_k(t) \in lM^c$, $\alpha_k(t)$ be a process adapted to $\{\mathfrak{F}_t\}$ whose sample functions are with probability 1 continuous and of bounded variation on $[0, T]$, $k = 1, \ldots, d$, $\alpha(t) = (\alpha_1(t), \ldots, \alpha_d(t))$, $\beta(t) = (\beta_1(t), \ldots, \beta_d(t))$:*

$$\xi(t) = \xi(0) + \alpha(t) + \beta(t),$$

where $\xi(0)$ is an $\mathfrak{F}_0$-measurable vector. Then

$$f(t, \xi(t)) = f(0, \xi(0)) + \int_0^t \left(\frac{\partial f}{\partial t}(s, \xi(s)) \, ds + \sum_{k=1}^d \nabla^k f(s, \xi(s)) \, d\xi_k(s) \right)$$
$$+ \frac{1}{2} \int_0^t \sum_{k,j=1}^d \nabla^k \nabla^j f(s, \xi(s)) \, d\langle \beta^k, \beta^j \rangle_s . \tag{3.12}$$

Here $\quad \nabla^k f(t, x) = \dfrac{\partial f(t, x)}{\partial x_k}$ *and*

$$\int_0^t \gamma(s) \, d\xi_k(s) = \int_0^t \gamma(s) \, d\alpha_k(s) + \int^t \gamma(s) \, d\beta_k(s).$$

The first integral on the r.h.s. is the Stieltjes integral of a continuous function $\gamma(s)$ with respect to a function of bounded variation $\alpha_k(t)$, while the second is a stochastic integral over a local martingale. Formula (3.12) is basic for a special kind of stochastic differential calculus. For example, formula (3.12) yields the following rule of differentiation of a product

$$d(\xi_1 \xi_2) = \xi_1 \, d\xi_2 + \xi_2 \, d\xi_1 + d\langle \mu_1, \mu_2 \rangle_t . \tag{3.13}$$

Equality (3.13) and some other analogous to it should be understood as an equality between integrals of both sides of the equality. In the present case this means that

$$\xi_1(t)\xi_2(t) = \xi_1(0)\xi_2(0) + \int_0^t \xi_1 \, d\xi_2 + \xi_2 \, d\xi_1 + \langle \mu_1, \mu_2 \rangle_t .$$

Itô's formula may also be generalized to the case of discontinuous local martingales. For example, let

$$\xi_k(t) = \xi_k(0) + \alpha_k(t) + \beta_k(t) + \zeta_k(t), \qquad k = 1, \ldots, d,$$

where $\alpha_k(t)$ and $\beta_k(t)$ are as defined above and the process $\zeta_k(t)$ is a stochastic integral over a centered Poisson measure,

$$\zeta_k(t) = \int_0^t \int_{R^d} \gamma_k(s, u)\tilde{\nu}(ds, du), \qquad k = 1, \ldots, d,$$

while the measure $\nu(\cdot, A)$ does not depend on the martingales $\beta_k(\cdot)$, $k = 1, \ldots, d$. Vectors with components α_k, β_k, ζ_k, γ_k, ξ_k will be denoted α, β, ζ, γ, ξ respectively. Let $f(t, x)$ be a function continuously differentiable with respect to t and twice continuously differentiable with respect to x and such that the functions

$$f(t, \xi(t) + \gamma(t, u)) - f(t, \xi(t)) - (\nabla f(t, \xi(t)), \gamma(t, u)),$$
$$|f(t, \xi(t) + \gamma(t, u)) - f(\xi(t))|^2$$

are integrable over the measure $q(du)\,dt$ on $[0, T] \times R^d$. Then

$$df(t, \xi(t)) = d\eta_1 + d\eta_2, \tag{3.14}$$

where

$$d\eta_1 = \frac{\partial f}{\partial t}(t, \xi(t)) + (\nabla f(t, \xi(t)), d\alpha + d\beta) + \frac{1}{2}\sum_{k,j=1}^{d} \nabla^k \nabla^j f(t, \xi(t))\, d\langle \beta^k, \beta^j\rangle_t, \tag{3.15}$$

$$d\eta_2 = \int_{R^d} [f(t, \xi + \gamma) - f(t, \xi) - (\nabla f(t, \xi), \gamma)]q(du)\,dt + \int_{R^d} [f(t, \xi + \gamma) - f(t, \xi)]\tilde{\nu}(dt, du). \tag{3.16}$$

In particular these formulas yield the following rule of differentiation of a product

$$d(\xi_1(t)\xi_2(t)) = \xi_2\, d\xi_1 + \xi_1\, d\xi_2 + d\langle \beta_1, \beta_2\rangle_t + \int_{R^d} \gamma_1(t, u)\gamma_2(t, u)\nu(dt, du). \tag{3.17}$$

2 Stochastic Differential Equations

We shall now define a stochastic differential equation. Let $(\Omega, \mathfrak{S}, \mathbf{P})$ be a probability space, $\{\mathfrak{F}_t, t \in [0, T]\}$ be a current of σ-algebras, $\mathfrak{F}_t \subset \mathfrak{S}$, $w(t) = \{w_1(t), \ldots, w_r(t)\}$ be a Wiener process, $\nu(t, A)$ be a Poisson measure, $t \in [0, T]$, $A \in \mathfrak{B}^d$, adapted to the current $\{\mathfrak{F}_t\}$. Set

$$\mathbf{E}\nu(t, A) = tq(A), \qquad \tilde{\nu}(t, A) = \nu(t, A) - tq(A).$$

If $v(t, A)$ denotes a measure which corresponds to jumps of a process with independent increments, then it possesses the following properties

$$\int_{S_c} |x|^2 q(dx) < \infty, \qquad q(R\backslash S_c) < \infty, \tag{3.18}$$

where $S_c = \{x : |x| \le c\}$. In what follows we shall assume that measure $q(\cdot)$ possesses these properties. In that case, $\tilde{v}(t, A)$ is a random measure defined for all $A \in \mathfrak{B}^d$ whose closure does not contain point 0.

In this section the letter D will denote the space $D^d[0, T]$.

Let a function $f(t, x(\cdot))$, $(t, x(\cdot)) \in [0, T] \times D$ with values in a space Y be given. We shall call this function *non-anticipative* if for any $t \in [0, T]$ this function depends only on the values of $x(s)$ on the interval $[0, t]$. More precisely this means the following: for all $t \in [0, T]$ $\{x(s) = x'(s), \forall s \in [0, t]\}$ implies that $f(t, x(\cdot)) = f(t, x'(\cdot))$.

Assume that random non-anticipating functions $\alpha(t, x(\cdot)) = \alpha(t, x(\cdot), \omega)$, $\beta(t, x(\cdot)) = \beta(t, x(\cdot), \omega)$ and $\gamma(t, x(\cdot), y) = \gamma(t, x(\cdot), y, \omega)$, $(x(\cdot) \in D$, $y \in R^d$, $t \in [0, T])$ adapted to the current $\{\mathfrak{F}_t, t \in [0, T]\}$ be given. Let $\alpha(t, x(\cdot))$, $\gamma(t, x(\cdot), y)$ take on values in R^d and $\beta(t, x(\cdot))$ be a matrix function which maps R^r into R^d.

Set

$$\lambda(x(\cdot), t) = \int_0^t \beta(t, x(\cdot))\, dw + \int_0^t \int_{R^d} \gamma(t, x(\cdot), y)\tilde{v}(dt, dy).$$

The integrals appearing on the r.h.s. of the last equality exist for all $t \in [0, T]$ provided

$$\int_0^T |\beta(t, x(\cdot))|^2\, dt < \infty, \qquad \int_0^T \int_{R^d} |\gamma(t, x(\cdot), y)|^2\, dt q(dy) < \infty$$

with probability 1. We shall assume that these conditions are satisfied for all $x(\cdot) \in D$. In that case the function $\lambda(x(\cdot), t)$ is defined everywhere on $D \times [0, T]$ and possesses the following properties:

a. $\lambda(x(\cdot), t)$ as a function of t is adapted to the current $\{\mathfrak{F}_t\}$ for all $x(\cdot) \in D$;
b. the sample functions of the process $\lambda(x(\cdot), t)$ belong to D with probability 1
c. $\lambda(x(\cdot), \cdot)$ is a local square integrable martingale.

Denote by $\Phi = \Phi^d[\mathfrak{F}_t, [0, T]]$ the class of random processes $\varphi(t)$, $t \in [0, T]$ with values in R^d adapted to a current of σ-algebras $\{\mathfrak{F}_t\}$ whose sample functions belong to D.

Let $\varphi(\cdot) \in \Phi$. Then $\alpha(t, \varphi)$, $\beta(t, \varphi)(\gamma(t, \varphi, y))$ are $[0, t] \times \mathfrak{F}_t$, $[(0, t] \times \mathfrak{F}_t \times \mathfrak{B}^d]$-measurable functions for all $t \in [0, T]$ and stochastic integrals

$$\int_0^t \beta(s, \varphi) w(ds), \qquad \int_0^t \int_{R^d} \gamma(s, \varphi, y)\tilde{v}(ds, dy)$$

exist provided

$$\int_0^T |\beta(s, \varphi)|^2 \, ds < \infty, \qquad \int_0^T \int_{R^d} |\gamma(s, \varphi, y)|^2 q(dy) \, ds < \infty.$$

Moreover if

$$\mathbf{E} \int_0^T |\beta(s, \varphi)|^2 \, ds < \infty, \qquad \mathbf{E} \int_0^T \int_{R^d} |\gamma(s, \varphi, y)|^2 q(dy) \, ds < \infty,$$

then the stochastic integrals introduced above are square integrable martingales and

$$\mathbf{E}\left\{\left(\int_t^{t+\Delta t} \beta^*(s, \varphi) w(ds)\right)\left(\int_t^{t+\Delta t} \beta(s, \varphi) w(ds)\right) \Big| \mathfrak{F}_t\right\}$$
$$= \mathbf{E}\left\{\int_t^{t+\Delta t} \beta^*(s, \varphi)\beta(s, \varphi) \, ds \Big| \mathfrak{F}_t\right\}, \tag{3.19}$$

$$\mathbf{E}\left\{\left(\int_t^{t+\Delta t} \int_{R^d} \gamma^*(s, \varphi, y)\tilde{\nu}(ds, dy)\right)\left(\int_t^{t+\Delta t} \int_{R^d} \gamma(s, \varphi, y)\tilde{\nu}(ds, dy)\right) \Big| \mathfrak{F}_t\right\}$$
$$= \mathbf{E}\left\{\int_t^{t+\Delta t} \int_{R^d} \gamma^*(s, \varphi, y)\gamma(s, \varphi, y) q(dy) \, ds \Big| \mathfrak{F}_t\right\}. \tag{3.20}$$

Observe that if the functions $\beta(t, x(\cdot))$ and $\gamma(t, x(\cdot), y)$ are non-random, $\beta(t, x(\cdot)) = b(t, x(\cdot))$, $\gamma(t, x(\cdot), y) = c(t, x(\cdot), y)$, then $\lambda(x(\cdot), t)$ for a fixed $x(\cdot)$ is a process with independent increments and finite moments of the second order.

We shall now introduce a stochastic differential equation or more precisely a *stochastic functional-differential equation* written in the form

$$d\xi = \alpha(t, \xi(\cdot)) \, dt + \lambda(\xi(\cdot), dt), \qquad \xi(0) = \xi_0. \tag{3.21}$$

Here ξ_0 is an arbitrary $\mathfrak{F}_0$-measurable random vector in R^d.

Definition. A random process $\xi(t)$, $t \in [0, T]$, satisfying the following conditions:

a. the process $\xi(t)$ is adapted to the current $\{\mathfrak{F}_t\}$;
b. the sample functions of the process $\xi(t)$ belong to D with probability 1;
c. for all $t \in [0, T]$ with probability 1

$$\xi(t) = \xi_0 + \int_0^t \alpha(s, \xi(\cdot)) \, ds + \int_0^t \lambda(\xi(\cdot), ds) \tag{3.22}$$

is called a *solution* of equation (3.21) on the interval $[0, T]$.

The equations introduced above are termed *functional-differential* because the r.h.s. of (3.21) at a given time depends on the value of the unknown functions to be determined on the whole time interval $[0, t]$. The fact that this dependence is non-anticipative in time is quite essential.

Thus equation (3.21) is in general an *equation with a lag*, while the lag at time t is continuously distributed over the interval $[0, t]$. Evidently the case of an equation with constant lag is included in this scheme. We obtain a *stochastic differential equation without a lag* provided

$$\alpha(t, x(\cdot)) = a(t, x(t)), \qquad \beta(t, x(\cdot)) = b(t, x(t)),$$
$$\gamma(t, x(\cdot), y) = c(t, x(t), y),$$

where $a(t, x)$, $b(t, x)$ and $c(t, x, y)$ are functions of the arguments $(t, x, y) \in [0, T] \times R^d \times R^d$. Equation (3.22) in this case can be simplified:

$$\xi(t) = \xi_0 + \int_0^t a(s, \xi(s))\, ds + \int_0^t b(s, \xi(s))\, dw(s)$$
$$+ \int_0^t \int_{R^d} c(s, \xi(s), y)\tilde{v}(ds, dy)$$

or

$$d\xi(t) = a(t, \xi(t))\, dt + b(t, \xi(t))\, dw(t) + \int_{R^d} c(t, \xi(t), y)\tilde{v}(dt, dy). \tag{3.23}$$

Such an equation is called an *equation without an after-effect*: at each time t the r.h.s. of equation (3.23) depends only on the values of the function $\xi(t)$ at that time. Among equations without an after-effect the most commonly encountered are those of the diffusion type. These do not contain the discontinuous component and are of the form

$$d\xi(t) = a(t, \xi(t))\, dt + b(t, \xi(t))\, dw(t), \qquad \xi(0) = \xi_0. \tag{3.24}$$

Our immediate task is to establish a simple theorem related to the existence and uniqueness of the solution of equation (3.21).

First we shall note several properties of the operator which corresponds to the r.h.s. of equation (3.22).

For an arbitrary process $\varphi(\cdot) \in \Phi$ we define

$$\mathcal{T}(\varphi, t) = \xi_0 + \int_0^t \alpha(s, \varphi(\cdot))\, ds + \int_0^t \lambda(\varphi(\cdot), ds), \tag{3.25}$$

where ξ_0 is an $\mathfrak{F}_0$-measurable random vector in R^d. Set

$$\|x(\cdot)\|_t = \sup_{0 \le s \le t} |x(s)|.$$

We shall also assume that functions $\alpha(t, x(\cdot))$, $\beta(t, x(\cdot))$ and $\gamma(t, x(\cdot), y)$ satisfy condition

$$|\alpha(t, x(\cdot))|^2 + |\beta(t, x(\cdot))|^2 + \int_{R^d} |\gamma(t, x(\cdot), u)|^2 q(du) \le C(1 + \|x(\cdot)\|_t^2),$$

where C is a constant. Functions α, β and γ which satisfy these conditions (as well as equation (3.21)) will be called *linearly bounded*.

If α, β, γ are linearly bounded, the process $\mathscr{T}(\varphi, t)$ is then defined for any $\varphi(\cdot) \in \Phi$ and moreover $\mathscr{T}(\varphi, \cdot) \in \Phi$, i.e. the operator $\mathscr{T}$ maps Φ into itself.

Denote by Φ_2 the subspace of Φ consisting of processes $\varphi(\cdot)$ such that

$$\mathbf{E}\|\varphi\|_T^2 < \infty.$$

Lemma 3.1. *If functions α, β and γ are linearly bounded then $\mathscr{T}$ maps Φ into itself. Moreover*

$$\mathbf{E}\|\mathscr{T}(\varphi, \cdot) - \xi_0\|_T^2 \leq C_1 \int_0^T (1 + \mathbf{E}\|\varphi\|_t^2)\, dt, \tag{3.26}$$

where $C_1 = 2C(T + 4)$.

Proof. Set $\psi(t) = \mathscr{T}(\varphi, t)$. We have

$$\begin{aligned}
\|\psi(\cdot) - \xi_0\|_t^2 &\leq 2\left(\left\|\int_0^s \alpha(s, \varphi)\, ds\right\|_t^2 + \left\|\int_0^s \lambda(\varphi, ds)\right\|_t^2\right) \\
&\leq 2\left(t \cdot \int_0^t |\alpha(s, \varphi(\cdot))|^2\, ds + \left\|\int_0^t \lambda(\varphi, ds)\right\|_t^2\right. \\
&\leq 2\left(t \int_0^t C(1 + \|\varphi\|_s^2)\right)\, ds + \left\|\int_0^s \lambda(\varphi, ds)\right\|_t^2\Bigg).
\end{aligned}$$

The stochastic integral $\Lambda_t = \int_0^t \lambda(\varphi, ds)$, provided $\mathbf{E}\,\mathrm{Sp}\langle\Lambda, \Lambda\rangle_T < \infty$ is a square integrable martingale. Here $\mathrm{Sp}\, A$ denotes the *trace* of the matrix A, $\mathrm{Sp}\, A = \sum_i A_{ii}$ and $\langle\Lambda, \Lambda\rangle_t$ is the characteristic of the local martingale Λ_t. In the case under consideration

$$\begin{aligned}
\mathrm{Sp}\langle\Lambda, \Lambda\rangle_t = &\int_0^t \mathrm{Sp}\, \beta(s, \varphi(\cdot))\beta^*(s, \varphi(\cdot))\, ds \\
&+ \int_0^t \int_{R^d} |\gamma(s, \varphi(\cdot), y)|^2\, ds q(dy) \leq C \int_0^T (1 + \|\varphi(\cdot)\|_s^2)\, ds
\end{aligned}$$

and

$$\mathbf{E}\,\mathrm{Sp}\langle\Lambda, \Lambda\rangle_T \leq C \int_0^T (1 + \mathbf{E}\|\varphi(\cdot)\|_s^2)\, ds < \infty.$$

Furthermore, in view of Doob's inequality (3.10) we have

$$\mathbf{E}\|\Lambda\|_T^2 \leq 4\mathbf{E}|\Lambda(T)|^2 = 4\mathbf{E}\,\mathrm{Sp}\langle\Lambda, \Lambda\rangle_T.$$

Thus

$$\mathbf{E}\|\psi(\cdot) - \xi_0\|_t^2 \leq 2C(T + 4) \int_0^t (1 + \mathbf{E}\|\varphi(\cdot)\|_s^2)\, ds. \qquad \square$$

The inequality

$$\mathbf{E}\left\{\sup_{s\le t\le T}|\mathscr{T}(\varphi,t)-\mathscr{T}(\varphi,s)|^2\,|\,\mathfrak{F}_s\right\}$$
$$\le 2C(T+4)\int_s^T(1+\mathbf{E}\{\|\varphi\|_t^2\,|\,\mathfrak{F}_s\})\,ds \tag{3.27}$$

is verified analogously.

Lemma 3.2. *If functions* α, β, *and* γ *are non-lagging, i.e.* $\alpha(t, x(\cdot)) = \alpha_0(t, x(t))$, $\beta(t, x(\cdot)) = \beta_0(t, x(t))$ *and* $\gamma(t, x(\cdot), y) = \gamma_0(t, x(t), y)$ *and the functions* $\alpha_0(t, x)$, $\beta_0(t, x)$ *and* $\gamma_0(t, x, y)$ *satisfy the condition*

$$|\alpha_0(t,x)|^2+|\beta_0(t,x)|^2+\int|\gamma(t,x,y)|^2q(dy)\le C(1+|x|^2),$$

then

$$\mathbf{E}\|\mathscr{T}(\varphi,\cdot)-\xi_0\|_T^2\le C_1\int_0^T\mathbf{E}(1+|\varphi(t)|^2)\,dt. \tag{3.28}$$

The proof of this assertion is the same as that of lemma 3.1.

Definition. Functions $\alpha(t, x(\cdot))$, $\beta(t, x(\cdot))$, $\gamma(t, x(\cdot), y)$ are said to satisfy the *uniform Lipschitz condition* with a constant L if

$$|\alpha(t,x_1(\cdot))-\alpha(t,x_2(\cdot))|^2+|\beta(t,x_1(\cdot))-\beta(t,x_2(\cdot))|^2$$
$$+\int|\gamma(t,x_1(\cdot),y)-\gamma(t,x_2(\cdot),y)|^2q(dy)\le L^2\|x_1(\cdot)-x_2(\cdot)\|_t^2. \tag{3.29}$$

Lemma 3.3. *If* α, β, *and* γ *satisfy the uniform Lipschitz condition,* $\varphi_i \in \Phi$ $(i = 1, 2)$ *then*

$$\mathbf{E}\|\mathscr{T}(\varphi_1,\cdot)-\mathscr{T}(\varphi_2,\cdot)\|^2\le 2L^2(T+4)\int_0^T\mathbf{E}\|\varphi_1-\varphi_2\|_t^2\,dt. \tag{3.30}$$

PROOF. Set $\mathscr{T}(\varphi_i, t) = \psi_i(t)$, $i = 1, 2$. Analogously to the proof of theorem 3.1, utilizing the inequality (3.29) we obtain

$$\|\psi_1(\cdot)-\psi_2(\cdot)\|_t^2\le 2\Bigg(L^2T\int_0^t\|\varphi_1-\varphi_2\|_s^2\,ds$$
$$+\left\|\int_0^t\lambda(\varphi_1,ds)-\lambda(\varphi_2,ds)\right\|_t^2\Bigg).$$

Since $\int_0^t (\lambda(\varphi_1, ds) - \lambda(\varphi_2, ds))$ is a square integrable martingale we have

$$\mathbf{E}\left|\int_0^t \lambda(\varphi_1, ds) - \lambda(\varphi_2, ds)\right|^2$$

$$= \mathbf{E}\left|\int_0^t [\beta(s, \varphi_1) - \beta(s, \varphi_2)]\, dw(s)\right|^2 + \mathbf{E}\left|\int_0^t [\gamma(s, \varphi_1, y) - \gamma(s, \varphi_2, y)]\tilde{v}(dy)\right|$$

$$= \mathbf{E}\left\{\int_0^t |\beta(s, \varphi_1) - \beta(s, \varphi_2)|^2\, dt + \int_0^t |\gamma(s, \varphi_1, y) - \gamma(s, \varphi_2, y)|^2 q(dy)\right\}$$

$$\leq L^2 \int_0^t \mathbf{E}\|\varphi_1 - \varphi_2\|_s^2\, ds,$$

and utilizing Doob's inequality we have

$$\mathbf{E}\|\varphi_1 - \varphi_2\|_t^2 \leq 2L^2(T + 4) \int_0^t \mathbf{E}\|\varphi_1 - \varphi_2\|_s^2\, ds. \qquad \square$$

In the same manner one can prove the following lemma:

Lemma 3.4. *If α, β and γ are non-lagging functions satisfying the uniform Lipschitz condition*

$$|\alpha(t, x_1) - \alpha(t, x_2)| \leq L|x_1 - x_2|, \tag{3.31}$$

$$|\beta(t, x_1) - \beta(t, x_2)|^2 + \int_{R^d} |\gamma(t, x_1, y) - \gamma(t, x_2, y)|^2 q(dy) \leq L^2|x_1 - x_2|^2, \tag{3.32}$$

then

$$\mathbf{E}\|\mathscr{T}(\varphi_1, \cdot) - \mathscr{T}(\varphi_2, \cdot)\|_T^2 \leq 2L^2(T + 4) \int_0^T \mathbf{E}|\varphi_1(s) - \varphi_2(s)|^2\, ds. \tag{3.33}$$

Theorem 3.3. *Assume that the functions α, β and γ satisfy the uniform Lipschitz condition and let*

$$\mathscr{T}(\check{\xi}_0, t) = \xi_0 + \int_0^t \alpha(s, \check{\xi}_0)\, ds + \int_0^t \lambda(\check{\xi}_0, ds) \in \Phi_2,$$

where $\check{\xi}_0 = \check{\xi}_0(t) \equiv \xi_0$. Then equation (3.21) *possesses in Φ a unique solution $\xi(\cdot)$ which belongs to Φ_2.*

Proof. Let $\xi_0(t)$ be an arbitrary stochastic process such that $\xi_0(0) = \xi_0$ and $\xi_0(\cdot) \in \Phi_2$. In view of the conditions of the theorem such a process exists. Define by induction the sequence $\xi_n(t)$ by setting $\xi_{n+1}(t) = \mathscr{T}(\xi_n(\cdot), t)$, $n = 0, 1, \ldots$. In view of lemma 3.3 we can write

$$\mathbf{E}\|\xi_{n+1}(\cdot) - \xi_n(\cdot)\|_t^2 \leq L_1 \int_0^t \mathbf{E}\|\xi_n(\cdot) - \xi_{n-1}(\cdot)\|_s^2\, ds,$$

where $L_1 = 2L^2(T+4)$. If we set $\mathbf{E}\|\xi_1(\cdot) - \xi_0(\cdot)\|_T^2 = A$ then it follows by induction from the previous inequality that

$$\mathbf{E}\|\xi_{n+1}(\cdot) - \xi_n(\cdot)\|_t^2 \le A\frac{L_1^n t^n}{n!}.$$

Setting $\delta_n = \|\xi_{n+1}(\cdot) - \xi_n(\cdot)\|_T$, we have in view of Chebyshev's inequality

$$\mathbf{P}\left\{\delta_n > \frac{1}{n^2}\right\} \le n^4 \mathbf{E}\,\delta_n^2 \le A\frac{L_1^n t^n}{n!} n^4.$$

Hence,

$$\sum_{n=1}^{\infty} \mathbf{P}\left(\delta_n > \frac{1}{n^2}\right) < \infty.$$

By virtue of the Borel-Cantelli lemma the inequality $\delta_n > 1/n^2$ can, with probability 1, be realized only for a finite number of values of n. Thus the series $\sum_{n=1}^{\infty} \delta_n$ converges with probability 1. Therefore the series

$$\xi_0(t) + \sum_{n=0}^{\infty} (\xi_{n+1}(t) - \xi_n(t))$$

converges uniformly in $t \in [0, T]$ with probability 1. Denote the sum of this series by $\xi(t)$. The sample functions of the process $\xi(t)$ belong to D with probability 1 and the process $\xi(t)$ is adapted to $\mathfrak{F}_t$. Furthermore,

$$\mathbf{E}\|\xi(\cdot) - \xi_n(\cdot)\|_T^2 = \mathbf{E}\lim_{m\to\infty} \|\xi_{n+m}(\cdot) - \xi_n(\cdot)\|_T^2 \le \varliminf_{m\to\infty} \mathbf{E}\|\xi_{n+m}(\cdot) - \xi_n(\cdot)\|_T^2.$$

Observe that

$$\begin{aligned}\|\xi_{n+m}(\cdot) - \xi_n(\cdot)\|_T^2 &\le \left(\sum_{k=n}^{n+m-1} \|\xi_{k+1}(\cdot) - \xi_k(\cdot)\|_T\right)^2 \\ &= \left(\sum_{k=n}^{n+m-1} \frac{1}{k}\cdot k\|\xi_{k+1}(\cdot) - \xi_k(\cdot)\|_T\right)^2 \\ &\le \sum_{k=n}^{n+m-1} \frac{1}{k^2} \sum_{k=n}^{n+m-1} k^2\|\xi_{k+1}(\cdot) - \xi_k(\cdot)\|_T^2.\end{aligned}$$

Thus

$$\mathbf{E}\|\xi(\cdot) - \xi_n(\cdot)\|_T^2 \le \sum_{k=n}^{\infty} \frac{1}{k^2} \sum_{k=n}^{\infty} k^2 \frac{AL_1^k T^k}{k!} \to 0 \quad \text{as} \quad n \to \infty. \tag{3.34}$$

We now approach the limit as $n \to \infty$ in the equality $\xi_{n+1}(t) = \mathscr{T}(\xi_n, t)$. In view of lemma 3.3 and inequality (3.34) we have

$$\mathbf{E}\|\mathscr{T}(\xi, \cdot) - \mathscr{T}(\xi_n, \cdot)\|_T^2 \le C_1 \int_0^T \mathbf{E}\|\xi(\cdot) - \xi_n(\cdot)\|_s^2\, ds \to 0$$

as $n \to \infty$. Hence $\xi(t) = \mathscr{T}(\xi, t)$, $\forall t \in [0, T]$, with probability 1.

We now prove the uniqueness in Φ of a solution for equation (3.21) satisfying the conditions of theorem 3.3. Let two solutions $\xi(t)$ and $\eta(t)$ exist for equation (3.21). Set

$$\tau = \tau_N = \inf\{t : \min(|\xi(t)|, |\eta(t)|) \geq N\},$$

if the set in braces is non-empty, and set $\tau = T$ otherwise; let $\chi(t) = 1$ for $t < \tau$ and $\chi(t) = 0$ for $t \geq \tau$, and $\alpha'(t, x(\cdot)) = \chi(t)\alpha(t, x(\cdot))$, $\beta'(t, x(\cdot)) = \chi(t)\beta(t, x(\cdot))$, $\gamma'(t, x(\cdot), y) = \chi(t)\gamma(t, x(\cdot), y)$, $\xi'(t) = \chi(t)\xi(t)$, $\eta'(t) = \chi(t)\eta(t)$. If the functions α, β and γ are non-anticipative, the process $\xi'(t)$ satisfies the equation.

$$\xi'(t) = \chi(0)\xi_0 + \int_0^t \chi(s)\alpha(s, \xi(\cdot))\, ds + \int_0^t \chi(s)\lambda(\xi(\cdot), ds)$$

$$= \chi(0)\xi_0 + \int_0^t \alpha'(s, \xi'(\cdot))\, ds + \int_0^t \lambda'(\xi'(\cdot), ds). \tag{3.35}$$

The process $\eta'(t)$ also satisfies this equation while the functions α', β' and γ' satisfy the conditions of theorem 3.3 and the uniform Lipschitz condition with the same constant L; moreover $\xi'(t)$ and $\eta'(t)$ belong to Φ_2. Denote the operator corresponding to the r.h.s. of equation (3.35) by $\mathscr{T}'$,

$$\xi'(t) = \mathscr{T}'(\xi', t), \qquad \eta'(t) = \mathscr{T}'(\eta', t).$$

In view of inequality (3.34)

$$\mathbf{E}\|\xi'(\cdot) - \eta'(\cdot)\|_t^2 \leq C' \int^t \mathbf{E}\|\xi'(\cdot) - \eta'(\cdot)\|_s^2\, ds. \tag{3.36}$$

Substituting $4N^2$ into the r.h.s. of this inequality in place of $\mathbf{E}\|\xi'(\cdot) - \eta'(\cdot)\|_s^2$, we obtain

$$\mathbf{E}\|\xi'(\cdot) - \eta'(\cdot)\|_t^2 \leq 4C'N^2 t.$$

Once again, substituting the bound obtained into the r.h.s. of inequality (3.36) and continuing this process we arrive at

$$\mathbf{E}\|\xi'(\cdot) - \eta'(\cdot)\|_t^2 \leq 4N^2 \frac{(C't)^n}{n!}$$

for any n. Setting $n \to \infty$ we obtain $\mathbf{E}\|\xi'(\cdot) - \eta'(\cdot)\|_T^2 = 0$. Thus we have with probability 1 that $\chi_N(t)\sup|\xi(t) - \eta(t)| = 0$ for any $N > 0$ or $\sup_{0 \leq t \leq T} |\xi(t) - \eta(t)| = 0$ with probability 1. □

Below certain bounds on solutions for stochastic differential equations will be required.

In particular the following simple lemma will often be used.

Lemma 3.5. *If $z(t)$ is a non-negative function bounded on the interval $[0, T]$ and*

$$z(t) \le I(t) + C \int_0^t z(s)\, ds, \tag{3.37}$$

then

$$z(t) \le I(t) + C \int_0^t e^{C(t-s)} I(s)\, ds. \tag{3.38}$$

PROOF. We have

$$z(t) \le I(t) + C \int_0^t \left(I(s) + C \int_0^s z(s_1)\, ds_1 \right) ds = I(t) + C \int_0^t I(s)\, ds$$

$$+ C^2 \int_0^t \int_0^s z(s_1)\, ds_1\, ds = I(t) + C \int_0^t I(s)\, ds + C^2 \int_0^t (t-s) z(s)\, ds.$$

Once again, substituting its bound as given by inequality (3.37) into the r.h.s. of the inequality obtained in place of $z(s)$ we obtain

$$z(t) \le I(t) + C \int_0^t I(s)\, ds + C^2 \int_0^t (t-s) I(s)\, ds + \frac{C^3}{2} \int_0^t (t-s)^2 z(s)\, ds.$$

Repeating this process we arrive at

$$z(t) \le I(t) + C \int_0^t \left(1 + \sum_{k=1}^n \frac{C^k (t-s)^k}{k!} \right) I(s)\, ds$$

$$+ \frac{C^{n+2}}{(n+1)!} \int_0^t (t-s)^{n+1} z(s)\, ds.$$

Approaching the limit as $n \to \infty$ we obtain inequality (3.38). □

Lemma 3.6. *Assume that the process $\xi(t)$, $t \in [0, T]$, is a solution for equation (3.21) with linearly bounded coefficients (this is the only information available about this equation) and let $\mathbf{E}|\xi(0)|^2 < \infty$. Then*

$$\mathbf{E}\|\xi(\cdot)\|_T^2 < \infty, \tag{3.39}$$

$$\mathbf{E}\{\|\xi(\cdot) - \xi(0)\|_T^2 \mid \mathfrak{F}_0\} \le C_2 T(1 + |\xi(0)|^2), \tag{3.40}$$

$$\mathbf{E}\left\{ \sup_{s_1 \le t \le s_2} \|\xi(t) - \xi(s_1)\|^2 \mid \mathfrak{F}_0 \right\} \le C_3(1 + |\xi(0)|^2)(s_2 - s_1), \tag{3.41}$$

where $C_2 = 2C_1 e^{C_1 T}$ and $C_3 = 2C_2(1 + C_2 T)$.

PROOF. By definition the sample functions of process $\xi(t)$ are bounded with probability 1. Set $\chi_N(t) = 1$ for $\|\xi(\cdot)\|_t \le N$ and $\chi_N(t) = 0$ otherwise and let $\xi_N(t) = \chi_N(t)\xi(t)$. Then $\|\xi_N(\cdot)\|_T^2 \le N^2$ and

$$\chi_N(t)(\xi(t) - \xi(0)) = \chi_N(t)[\mathscr{T}(\xi(\cdot), t) - \xi(0)]$$
$$= \chi_N(t)[\mathscr{T}(\xi_N(\cdot), t) - \xi_N(0)],$$

since on the set $\chi_N(t) = 1$ $\mathscr{T}(\xi_N(\cdot), t) = \mathscr{T}(\xi(\cdot), t)$, $\xi_N(0) = \xi(0)$. Utilizing lemma 3.1 we obtain

$$\mathbf{E}\{\|\chi_N(\cdot)(\xi(\cdot) - \xi(0))\|_t^2 \,|\, \mathfrak{F}_0\} \le C_1 \int_0^t (1 + \mathbf{E}\{\|\xi_N(\cdot)\|_s^2 \,|\, \mathfrak{F}_0\})\, ds.$$

Setting $\quad z_N(t) = \mathbf{E}\{\|\xi_N(\cdot) - \xi_N(0)\|_t^2 \,|\, \mathfrak{F}_0\}$ we have

$$z_N(t) \le 2C_1(1 + |\xi_N(0)|^2)\, t + 2C_1 \int_0^t z_N(s)\, ds.$$

In view of lemma 3.5 one can write

$$z_N(t) \le C_2(1 + |\xi_N(0)|^2)t,$$

where $C_2 = 2C_1 e^{C_1 T}$. Utilizing Fatou's lemma and the fact that the sample functions of the process $\xi(t)$ are bounded we obtain inequality (3.40) which implies inequality (3.39). It also follows from the inequalities obtained that

$$\mathbf{E}\left\{\sup_{s_1 \le t \le s_2} |\chi_N(t)[\xi(t) - \xi(s_1)]|^2 \,|\, \mathfrak{F}_{s_1}\right\} \le C_2(1 + |\xi_N(s_1)|^2)(s_2 - s_1),$$

whence

$$\mathbf{E}\left\{\sup_{s_1 \le t \le s_2} |\chi_N(t)(\xi(t) - \xi(s_1))|^2 \,|\, \mathfrak{F}_0\right\} \le C_3(1 + |\xi_N(0)|^2)(s_2 - s_1).$$

Utilizing once again Fatou's lemma we arrive at inequality (3.40). □

In what follows we shall write $(\alpha, \beta, \gamma) \in S(C, L)$ if the functions α, β and γ are linearly bounded (by a constant C) and satisfy the uniform Lipschitz condition (with a constant L). We shall also write $(\alpha, \beta, \gamma) \in S(C, L)$ on a set A if the functions (α, β, γ) satisfy the corresponding inequalities appearing in the conditions of linear boundedness and the Lipschitz condition for the values of the arguments $(t, x(\cdot), x'(\cdot), y, \omega) \in A$.

The following lemma (to be referred to as the lemma on a local dependence of a solution for a stochastic differential equation on the coefficients of the equation) will be required below.

Lemma 3.7. *Let $\zeta_0, \alpha_i, \beta_i, \gamma_i$ $(i = 1, 2)$ be such that for $t \in [0, T]$ there exist solutions of equations*

$$d\xi_i(t) = \alpha_i(t, \xi_i(\cdot))\, dt + \lambda_i(\xi_i(\cdot), dt),\ \xi_i(0) = \zeta_0,$$

$$\lambda_i(x(\cdot), t) = \int_0^t \beta_i(s, x(\cdot))\, dw(s) + \int_0^t \int_{R^d} \gamma_i(s, x(\cdot), y)\tilde{\nu}(ds, dy)$$

and let an $N > 0$ and a random time τ (on the current of σ-algebras $\{\mathfrak{F}_t, t \in [0, T]\}$) be found such that $\alpha_1(t, x(\cdot))$, $= \alpha_2(t, x(\cdot))$, $\beta_1(t, x(\cdot)) = \beta_2(t,$

$x(\cdot))$, $\gamma_1(t, x(\cdot), y) = \gamma_2(t, x(\cdot), y)(y \in R^d, 0 \le t \le \tau, \|x(\cdot)\|_\tau \le N)$ *and let for these values of* $(t, x(\cdot), y)$ *the functions* $(\alpha_i, \beta_i, \gamma_i) \in S(C, L)$. *Then*

$$\xi_1(t) = \xi_2(t) (\text{mod } \mathbf{P})\ \forall t : t \le \tau,\ \|\xi_i(\cdot)\|_t \le N.$$

PROOF. Set $\sigma = \inf\{t : t \le \tau, \min_{i=1,2} |\xi_i(t)| > N\}$ if the set in the braces is non-void and set $\sigma = T$ otherwise. Let $\eta(t) = \xi_1(t) - \xi_2(t)$, $\chi(t) = 1$ for $t \le \sigma$ and $\chi(t) = 0$ for $t > \sigma$. Then

$$\begin{aligned}
&\mathbf{E}\chi(t)|\eta(t)|^2 \\
&\le 2\mathbf{E}\chi(t)\Bigg|\int_0^{t\wedge\sigma} [\alpha_1(s, \xi_1(\cdot)) - \alpha_2(s, \xi_1(\cdot))]\, ds \\
&+ \int_0^{t\wedge\sigma} [\beta_1(s, \xi_1(\cdot)) - \beta_2(s, \xi_1(\cdot))]\, dw(s) \\
&+ \int_0^{t\wedge\sigma} \int_{R^d} [\gamma_1(s, \xi_1(\cdot), y) - \gamma_2(s, \xi_1(\cdot), y)]\tilde{\nu}(ds, dy)\Bigg|^2 \\
&+ 2\mathbf{E}\chi(t)\Bigg|\int_0^{t\wedge\sigma} [\alpha_2(s, \xi_1(\cdot)) - \alpha_2(s, \xi_2(\cdot))]\, ds \\
&+ \int_0^{t\wedge\sigma} [\beta_2(s, \xi_1(\cdot)) - \beta_2(s, \xi_2(\cdot))]\, dw(s) \\
&+ \int_0^{t\wedge\sigma} \int_{R^d} [\gamma_2(s, \xi_1(\cdot), y) - \gamma_2(s, \xi_2(\cdot), y)]\tilde{\nu}(ds, dy)\Bigg|^2 \\
&\le 6(T+2)L^2\mathbf{E}\chi(t) \int_0^{t\wedge\sigma} \|\xi_1(\cdot) - \xi_2(s)\|_s^2\, ds \\
&\le L_1 \int_0^t \mathbf{E}\chi(s)\|\xi_1(\cdot) - \xi_2(\cdot)\|_s^2\, ds.
\end{aligned}$$

If $z(t) = \mathbf{E}\chi(t)|\eta(t)|^2$, then $z(t)$ is a bounded function satisfying the inequality

$$z(t) \le L_1 \int_0^t z(s)\, ds, \qquad z(0) = 0,$$

whence $z(t) = 0$ or $\chi(t)\|\xi_1(\cdot) - \xi_2(\cdot)\|_t = 0$. □

We say that functions $\alpha(t, x(\cdot))$, $\beta(t, x(\cdot))$, $\gamma(t, x(\cdot), y)$ satisfy a *local Lipschitz condition* if for any $N > 0$ there exists a constant L_N such that

$$\begin{aligned}
&|\alpha(t, x_1(\cdot)) - \alpha(t, x_2(\cdot))|^2 + |\beta(t, x_1(\cdot)) - \beta(t, x_2(\cdot))|^2 \\
&\qquad + \int_{R^d} |\gamma(t, x_1(\cdot), y) - \gamma(t, x_2(\cdot), y)|^2 q(dy) \le L_N \|x_1(\cdot) - x_2(\cdot)\|_t^2
\end{aligned}$$

for all $x_k(\cdot)$ satisfying $\|x_k(\cdot)\| \leq N$, $k = 1, 2$. If functions α, β and γ satisfy a local Lipschitz condition with a given system of constants L_N we shall write $(\alpha, \beta, \gamma) \in S(C, L_N)$.

Theorem 3.4. *If* $(\alpha, \beta, \gamma) \in S(C, L_N)$ *then equation* (3.21) *possesses a unique solution in* Φ.

Proof. Construct for a given $N \geq 0$ functions $\alpha_N(t, \xi(\cdot))$, $\beta_N(t, \xi(\cdot))$ and $\gamma_N(t, \xi(\cdot), y)$, which coincide with α, β and γ for $\|x(\cdot)\|_T \leq N$ and such that $(\alpha, \beta, \gamma) \in S(C, L')$ where $L' = L_N + 1$. In view of theorem 3.3 the stochastic equations

$$d\xi_N = \alpha_N(\xi_N(\cdot), t)\, dt + \lambda_N(\xi_N(\cdot), dt), \qquad \xi_N(0) = \xi_0$$

possess a unique solution satisfying

$$\mathbf{E}\|\xi_N(\cdot)\|_T^2 \leq C_1(1 + \mathbf{E}|\xi_0|^2),$$

where C_1 is a constant which depends only on C and T. On the other hand, in view of lemma 3.7 $\xi_N(t) = \xi_{N+1}(t) = \cdots$ provided $\|\xi_N(\cdot)\|_T \leq N$. Therefore all the processes $\xi_N(t)$ starting with some $N_0 = N_0(\omega)$ coincide with probability 1 and $\lim_{N\to\infty} \xi_N(t) = \xi(t)$ exists with probability 1. Clearly in equality

$$\xi_N(t) = \xi_0 + \int_0^t \alpha_N(s, \xi_N(\cdot))\, ds + \int_0^t \lambda_N(\xi_N(\cdot), ds)$$

one can approach the limit as $N \to \infty$; thus $\xi(t)$ is a solution of equation (3.21). The uniqueness of this solution is obtained analogously to the proof of theorem 3.3. □

We shall now consider *equations of the continuous type* (with $\gamma \equiv 0$) and show that if the coefficients of the equation are linearly bounded then its solutions possess moments of arbitrarily high orders. We apply Itô's formula (3.12) to the process $\xi(t)$ satisfying equation

$$d\xi = \alpha(t, \xi(\cdot))\, dt + \beta(t, \xi(\cdot))\, dw(t), \qquad \xi(0) = \xi_0, \tag{3.42}$$

and to function $f_N(x)$ which possesses the following properties: $f_N(x) = |x|^p$ $(p \geq 2)$ for $|x| \leq N$, $f_N(x)$ is twice continuously differentiable and vanishes outside some compact set, and $f_N(x) \leq f_{N+1}(x)$ $(N = 1, 2, \ldots)$. Clearly the function $f_N(x)$ and its derivatives of the first and second order possess arbitrarily high moments. In our case

$$d\alpha(s) = \alpha(s, \xi(\cdot))\, ds, \; \mu(ds) = \beta(s, \xi(\cdot))\, dw(s), \; \langle \mu^j, \mu^k \rangle_s = \int_0^s \sigma^{jk}(t, \xi(\cdot))\, dt,$$

where $\sigma = \beta\beta^*$. Substituting $t \wedge \tau_N$ in place of t in formula (3.12), where τ_N is the time of the first exit of the trajectory of the process $\xi(t)$ from the sphere of radius N we obtain

$$\begin{aligned}
&|\xi(t \wedge \tau_N)|^p \\
&= |\xi(0)|^p \\
&+ \int_0^{t\wedge\tau_N} \Big\{ p|\xi(s)|^{p-2}[(\xi(s), \alpha(s, \xi(\cdot))) + \tfrac{1}{2}\operatorname{Sp}\sigma(s, \xi(\cdot))] \\
&+ \tfrac{1}{2}p(p-2)|\xi(s)|^{p-4}(\sigma(s, \xi(\cdot))\,\xi(s), \xi(s))\Big\}\, ds \\
&+ \int_0^{t\wedge\tau_N} p|\xi(s)|^{p-2}(\xi(s), \beta(s, \xi(\cdot))\, dw).
\end{aligned} \tag{3.43}$$

Taking into account the fact that the moments of all the summands appearing in the last formula are bounded, we have

$$\begin{aligned}
\mathbf{E}\sup_{0\le t\le T} |\xi(t\wedge\tau_N)|^{2p} &\le 3\mathbf{E}\Big\{|\xi(0)|^{2p} \\
&+ T\int_0^{T\wedge\tau_N}\Big\{|\xi(s)|^{p-2}p\Big[(\xi(s), \alpha) + \frac{1}{2}\operatorname{Sp}\sigma\Big] \\
&+ \frac{p(p-2)}{2}|\xi(s)|^{p-4}(\sigma\xi, \xi)\Big\}^2 ds + p^2 \sup_{0\le t\le T}\Big|\int_0^{t\wedge\tau_N}|\xi|^{p-2}(\xi, \beta\, dw)\Big|^2\Big\}.
\end{aligned} \tag{3.44}$$

Observe that the characteristic of the martingale $\int_0^t (\xi, \beta\, dw)$ is equal to $\int_0^t (\beta\beta^*\xi, \xi)\, ds = \int_0^t (\sigma\xi, \xi)\, ds$. Applying once again Doob's inequality (3.10) we arrive at the following bound for the last integral appearing in the preceding inequality:

$$\begin{aligned}
\mathbf{E}\sup_{0\le t\le T}\Big|\int_0^{t\wedge\tau_N}|\xi|^{p-2}(\xi, \beta\, dw)\Big| &\le 4\mathbf{E}\Big|\int_0^{T\wedge\tau_N}|\xi|^{p-2}(\xi, \beta\, dw)\Big|^2 \\
&\le 4\mathbf{E}\int_0^{T\wedge\tau_N}|\xi|^{2p-4}|\beta^*\xi|^2\, ds.
\end{aligned}$$

Since the functions α and β are assumed to be linearly bounded we have

$$|(\alpha, \xi)|^2 \le C(1 + \|\xi\|_t^2)\|\xi\|_t^2, \qquad \operatorname{Sp}\sigma = |\beta|^2 \le C(1 + \|\xi\|_t^2),$$
$$|(\sigma\xi, \xi)| \le |\sigma|\cdot|\xi|^2 \le C(1 + \|\xi\|_t^2)\|\xi\|_t^2.$$

Set $\mathbf{E}\|\xi(\cdot)\|^{2p}_{t\wedge\tau_N} = z_N(t)$. Expression (3.44) implies that

$$z_N(t) \le 3\Bigg[z_N(0) + 3T\mathbf{E}\int_0^{t\wedge\tau_N} \|\xi\|_s^{2p-4}p^2(C(1+\|\xi\|_s^2)\|\xi\|_s^2$$

$$+ \tfrac{1}{4}C^2(1+\|\xi\|_s^2)^2) + \frac{p^2(p-2)^2}{4}\|\xi\|_s^{2p-8}C^2(1+\|\xi\|_s^2)^2\|\xi\|_s^4$$

$$+ 4p^2\|\xi\|_s^{2p-4}C(1+\|\xi\|_s^2)\|\xi\|_s^2\Bigg]\,ds \le 3z_n(0)$$

$$+ \mathbf{E}\int_0^t (C_1\|\xi\|^{2p}_{s\wedge\tau_N} + C_2\|\xi\|^{2p-2}_{s\wedge\tau_N} + C_3\|\xi\|^{2p-4}_{s\wedge\tau_N})\,ds,$$

where C_i ($i = 1, 2, 3$) are constants which depend only on C and p. Taking into account that $|C|^\alpha \le 1 + |C|^\beta$ for $\alpha < \beta$, we obtain

$$z_N(t) \le A + 3z_N(0) + A\int_0^t z_N(s)\,ds,$$

whence

$$z_N(t) \le [A + 3z_N(0)]e^{At}.$$

Letting $N \to \infty$ and utilizing Fatou's lemma we have

$$\mathbf{E}\|\xi(\cdot)\|_T^{2p} \le (A + 3\mathbf{E}|\xi(0)|^{2p})e^{At}, \qquad p \ge 2,\ A = A(C, p).$$

Thus the following lemma is proved.

Lemma 3.8. *For equation* (3.21) *of a continuous type with linearly bounded coefficients for any* $p > 0$ *we have*

$$\mathbf{E}\sup_{0\le t\le T}|\xi(t)|^{2p} < (A + 3\mathbf{E}|\xi(0)|^{2p})e^{AT}, \tag{3.45}$$

where A *depends only on* p, T, *and* C.

Below a bound on the quantity

$$v(t) = \mathbf{E}\sup_{t_0\le s\le t}|\xi(s) - \xi(t_0)|^p$$

for small $t - t_0$ will be required. In the case of an arbitrary stochastic differential equation satisfying the condition of linear boundedness and some other conditions which insure the existence of moments of the p-th order for the variable $|\xi(t)|$, it can be shown that

$$v(t) = O(t - t_0).$$

In the case of equations of a continuous type we show that

$$v(t) = O(t - t_0)^{p/2}.$$

First consider a simpler quantity

$$v_1(t) = \sup_{t_0 \le s \le t} \mathbf{E}|\xi(s) - \xi(t_0)|^p.$$

We shall utilize equality (3.43) with $\tau_N = \infty$ and 0 replaced by t_0, $\xi(t)$ by $\xi(t) - \xi(t_0)$. Since the variables $|\xi(t) - \xi(t_0)|$ and $\|\xi(\cdot)\|_t$ possess finite moments of arbitrary orders (this follows from the preceding lemma), all summands appearing in formula (3.43) possess finite moments and the last summand is a martingale. Therefore

$$\begin{aligned}
&\mathbf{E}|\xi(t) - \xi(t_0)|^p \\
&= \mathbf{E}\int_{t_0}^{t} \Big\{ p|\xi(s) - \xi(t_0)|^{p-2}[(\xi(s) - \xi(t_0)), \alpha(s, \xi(\cdot)) + \tfrac{1}{2}\operatorname{Sp}\sigma(s, \xi(\cdot))] \\
&+ \frac{p(p-2)}{2}|\xi(s) - \xi(t_0)|^{p-4}(\sigma(s, \xi(\cdot)), (\xi(s) - \xi(t_0)), \xi(s) - \xi(t_0))\Big\}\, ds \\
&\le \mathbf{E}\int_{t_0}^{t} \Big\{ p|\xi(s) - \xi(t_0)|^{p-1}|\alpha(s, \xi(\cdot))| \\
&+ \frac{p(p-1)}{2}|\xi(s) - \xi(t_0)|^{p-2}|\sigma(s, \xi(\cdot))|\Big\}\, ds. \qquad (3.46)
\end{aligned}$$

Clearly the l.h.s. of this inequality can be replaced by $v_1(t)$. Observe that in view of lemma 3.8

$$v_1(t) \le A, \qquad \mathbf{E}|\alpha(s, \xi(\cdot))|^p \le A, \qquad \mathbf{E}|\sigma(s, \xi(\cdot))|^{p/2} \le A,$$

where A is a constant. Utilizing Hölder's inequality we obtain

$$\begin{aligned}
v_1(t) &\le \int_{t_0}^{t} [v_1(s)]^{(p-2)/p}\left(pA^{1/p}A^{1/p} + \frac{p(p-1)}{2}A^{2/p}\right) ds \\
&\le \frac{p(p+1)}{2}A^{2/p}\int_{t_0}^{t}[v_1(s)]^{(p-2)/p}\, ds.
\end{aligned}$$

The function $v_1(t)$ is monotonically non-decreasing; hence

$$v_1(t) \le \frac{p(p+1)}{2}A^{2/p}(t - t_0)v_1(t)^{(p-2)/p},$$

this implies that

$$[v_1(t)]^{2/p} \le \frac{p(p+1)}{2}A^{2/p}(t - t_0),$$

or

$$v_1(t) \le \left[\frac{p(p+1)}{2}\right]^{p/2} A(t - t_0)^{p/2} = A_p(t - t_0)^{p/2},$$

where $A_p = A(p, T, C)$. Now a bound on $v(t)$ can easily be derived. In view of Itô's formula (3.12) we have

$$v(t) \le \mathbf{E} \sup_{t_0 \le s \le t} \eta_1(s) + \mathbf{E} \sup_{t_0 \le s \le t} \eta_2(s), \tag{3.47}$$

where

$$\eta_1(t) = \int_{t_0}^{t} \Big\{ p|\xi(s) - \xi(t_0)|^{p-2}[(\xi(s) - \xi(t_0), \alpha(s, \xi(\cdot))) + \tfrac{1}{2}\,\mathrm{Sp}\,\sigma(s, \xi(\cdot))]$$

$$+ \frac{p(p-2)}{2} |\xi(s) - \xi(t_0)|^{p-4}(\sigma(s, \xi(\cdot))(\xi(s) - \xi(t_0)), \xi(s) - \xi(t_0))\Big\}\, ds,$$

$$\eta_2(t) = \int_{t_0}^{t} p|\xi(s) - \xi(t_0)|^{p-2}(\xi(s) - \xi(t_0), \beta\, dw).$$

The variable $\sup_{t_0 \le s \le t} \eta_1(s)$ can be bounded in the same manner as the r.h.s. of formula (3.46). Thus

$$\mathbf{E} \sup_{t_0 \le s \le t} \eta_1(s) \le \frac{p(p+1)}{2} A^{2/p}(t - t_0) v_1(t)^{(p-2)/p}$$

$$= \frac{p(p+1)}{2} A^{2/p} A_p^{(p-2)/2}(t - t_0)^{p/2}.$$

The variable $\eta_2(t)$ is a square integrable martingale with the characteristic

$$\int_{t_0}^{t} p^2 |\xi(s) - \xi(t_0)|^{2p-4}(\sigma(\xi(s) - \xi(t_0)), \xi(s) - \xi(t_0))\, ds.$$

Applying once again Doob's inequality (3.10) we have

$$\mathbf{E} \sup_{t_0 \le s \le t} \eta_2(s) \le \left(\mathbf{E} \sup_{t_0 \le s \le t} |\eta_2(s)|^2\right)^{1/2} \le 2(\mathbf{E}|\eta_2(t)|^2)^{1/2},$$

$$\mathbf{E}|\eta_2(t)|^2$$

$$= \int_{t_0}^{t} p^2 \mathbf{E}|\xi(s) - \xi(t_0)|^{2p-4}(\sigma(\xi(s) - \xi(t_0), \xi(s) - \xi(t_0)))\, ds$$

$$\le p^2 \int_{t_0}^{t} \mathbf{E}|\xi(s) - \xi(t_0)|^{2p-2}|\sigma(\xi(\cdot), s)|\, ds$$

$$\le p^2 \int_{t_0}^{t} (\mathbf{E}|\xi(s) - \xi(t_0)|^{2p})^{(2p-2)/2p}(\mathbf{E}|\sigma(\xi(\cdot), s)|^p)^{1/p}\, ds$$

$$= A' \int_{t_0}^{t} (s - t_0)^{p-1}\, ds = A''(t - t_0)^p,$$

where A' and A'' are constants. Substituting the bounds obtained for the quantities $\mathbf{E} \sup_{t_0 \le s \le t} \eta_1(s)$ and $\mathbf{E} \sup_{t_0 \le s \le t} \eta_2(s)$ into inequality (3.47) we obtain

$$\mathbf{E} \sup_{t_0 \le s \le t} |\xi(s) - \xi(t_0)|^p \le B_p(t - t_0)^{p/2}, \tag{3.48}$$

where B_p is a constant. It is easy to verify that $B_p \le C_p(1 + \mathbf{E}|\xi_0|^p)$. Here C_p depends on C, p and T only.

We shall now discuss the equations without an after-effect in some detail. If for any $t > 0$, $x \in R^d$ equation (3.23) possesses a unique solution $\xi_{tx}(s)$, $s \in [t, T]$, satisfying the initial condition $\xi_{tx}(t) = x$, then the family

$$\{\xi_{tx}(s), s \in [t, T], (t, x) \in [0, T] \times R^d\}$$

is a Markov process with the transition probability

$$P(t, x, s, B) = P\{\xi_{tx}(s) \in B\}, \qquad B \in \mathscr{L}^d.$$

In particular,

$$\mathbf{E}\{f(\xi(\cdot))|\mathscr{F}_t\} = \mathbf{E}f(\xi_{tx}(\cdot))|_{x=\xi(t)},$$

where $f(x(\cdot))$ is an arbitrary non-negative Borel functional on $D^d[t, T]$.

Let $f(x)$, $x \in R^d$ be an arbitrary twice continuously differentiable function. Set

$$F(t, x) = \mathbf{E}f(\xi_{tx}(T)).$$

Theorem 3.5. *If function $f(x)$ is bounded and partial derivatives of the first and second order of this function are also bounded, functions $a(t, x)$, $b(t, x)$ and $c(t, x, y)$ are twice continuously differentiable with respect to x and moreover the first and second order derivatives of $a(t, x)$ and $b(t, x)$ with respect to x are uniformly bounded and*

$$\int_{R^d} |c(t, x, y)|^k q(dy) \le C, \qquad k = 3, 4,$$

$$\int_{R^d} \left| \frac{\partial}{\partial x_k} c(t, x, y) \right|^k q(dy) \le C, \qquad \int_{R^d} \left| \frac{\partial^2}{\partial x_k \, \partial x_j} c(t, x, y) \right|^2 q(dy) \le C,$$

then the function $F(t, x)$ is twice continuously differentiable with respect to x and once with respect to t and satisfies the integro-differential equation

$$\frac{\partial F}{\partial t} + (a(t, x), \nabla F) + \tfrac{1}{2}\,\mathrm{Sp}(\sigma(t, x), \nabla^2 F)$$

$$+ \int_{R^d} [F(t, x + c(t, x, y)) - F(t, x) - (c(t, x, y), \nabla F)]q(dy) = 0 \quad (3.49)$$

and the boundary condition

$$\lim_{t \uparrow T} F(t, x) = f(x). \quad (3.50)$$

Here $\sigma(t, x) = b(t, x)b^*(t, x)$.

For the case of *diffusion-type equations* this theorem can be somewhat strengthened.

Theorem 3.6. *Let coefficients $a(t, x)$ and $b(t, x)$ of equation*

$$d\xi = a(t, \xi(t))\,dt + b(t, \xi(t))\,dt \tag{3.51}$$

be twice continuously differentiable with respect to x, and the partial derivatives of the first order of these coefficients with respect to x be uniformly bounded, and the partial derivatives of the second order be of at most polynomial (power) order of growth as $|x| \to \infty$, the function $f(x)$ be twice continuously differentiable and its derivatives of the first and second orders as $|x| \to \infty$ increase at most as a power of $|x|$. Then the function $F(t, x)$ is twice continuously differentiable with respect to x, is differentiable with respect to t, and satisfies the following second order partial differential equation of the parabolic type

$$\frac{\partial F}{\partial t} + \sum_{k=1}^{d} a_k(x, t)\frac{\partial F}{dx_k} + \frac{1}{2}\sum_{i,k=1}^{d} \sigma_{ik}(x, t)\frac{\partial^2 F}{\partial x_i\,\partial x_k} = 0, \qquad t \in [0, T] \tag{3.52}$$

and the boundary condition (3.50).

Proofs of theorems 3.5 and 3.6 can be found in monograph [19].

Weak compactness of measures corresponding to solutions of stochastic differential equations. Let X be a metric space, $\mathfrak{A}$ be a σ-algebra of subsets of X. We say that a sequence of measures $Q_n(\cdot)$, $n = 1, 2, \ldots$, defined on $\mathfrak{A}$ *converges weakly* to a measure $Q_0(\cdot)$ provided for an arbitrary bounded and continuous function $f(x)$, $x \in X$,

$$\int_X f(x)Q_n(dx) \to \int_X f(x)Q_0(dx).$$

We shall consider only probability measures, i.e. measures such that $Q(X) = 1$. If $\{X, \mathfrak{A}, Q\}$ is viewed as a probability space, then

$$\int_X f(x)Q(dx) = \mathbf{E}f(\xi),$$

where $\xi = \xi(x) = x$ is a random element with values in X and distribution Q on $\mathfrak{A}$. Let a sequence of probability measures Q_n be defined on $\{X, \mathfrak{A}\}$. The mathematical expectation of a random variable η defined on $\{X, \mathfrak{A}, Q_n\}$ will be denoted by $\mathbf{E}_n \eta$, $n = 0, 1, 2, \ldots$. The condition for weak convergence of a sequence of measures $\{Q_n\}$ to Q_0 can be written in the form

$$\lim_{n\to\infty} \mathbf{E}_n f(\xi) = \mathbf{E}_0 f(\xi)$$

for any bounded continuous function $f(x)$. A family of measures $\{Q_u, u \in U\}$ on $\{X, \mathfrak{A}\}$ is called *weakly compact* if a weakly convergent subsequence can be extracted from an arbitrary sequence $\{Q_{u_n}\}$, $n = 1, 2, \ldots$.

Let F be a metric space of functions $\varphi(t)$ defined on $[0, T]$ with values in R^d. Cylinders in F are, by definition, the sets $A = A_{t_1, \ldots, t_n}(B)$ consisting of all

$\varphi(\cdot) \in F$ such that $\{\varphi(\cdot) : (\varphi_1(t), \ldots, \varphi_n(t)) \in B\}$, $t_k \in [0, T]$, where B is a Borel set in $(R^d)^n$. We say that A is a cylinder with basis B over coordinates $t_1, t_2, \ldots, t_n$. The smallest σ-algebra generated by cylinders will be denoted $\mathfrak{C} = \mathfrak{C}(F)$. Let $\xi(t)$, $t \in [0, T]$, be a stochastic process on the probability space $\{\Omega, \mathfrak{S}, \mathbf{P}\}$ with sample functions belonging to F with probability 1.

Then one can define on $\{F, \mathfrak{C}\}$ a measure Q which corresponds to the process $\xi(t)$ by setting

$$Q(A_{t_1, \ldots, t_n}(B)) = \mathbf{P}\{(\xi(t_1), \ldots, \xi(t_n)) \in B\}.$$

This equality uniquely defines the measure Q on $\mathfrak{C}$.

In what follows we shall utilize conditions of weak compactness of measures in $C = C^d[0, T]$ and $D = D^d[0, T]$ corresponding to families of stochastic processes. Moreover it is assumed that the space C is metrized in the usual manner:

$$\rho(x_1(\cdot), x_2(\cdot)) = \|x_1(\cdot) - x_2(\cdot)\|_T = \max_{0 \le t \le T} |x_1(t) - x_2(t)|,$$

and the metric in D is introduced by the relationship

$$\rho_D(x_1(\cdot), x_2(\cdot)) = \inf\left\{ \sup_{0 \le t \le T} |x_1(t) - x_2(\lambda(t))| + \sup_{0 \le t \le T} |t - \lambda(t)| : \lambda(\cdot) \in \Lambda\right\},$$

where Λ is the set of all continuous monotonically increasing functions such that $\lambda(0) = 0$, $\lambda(T) = T$ (i.e. $\lambda(t)$ maps $[0, T]$ into itself in a continuous and 1–1 manner; cf. [20] Chapter 6, Section 5). In this metric the space D is separable. Observe that in the spaces C and D, σ-algebras generated by the cylinders coincide with σ-algebras of the Borel sets. We denote them $\mathfrak{C}(C)$ and $\mathfrak{C}(D)$ respectively.

Let $\xi_u(t)$, $u \in U$, $t \in [0, T]$ be a family of stochastic processes with sample functions belonging to D with probability 1. The corresponding measures on $\mathfrak{C}(D)$ are denoted by Q_u.

Theorem 3.7. *In order that a family of measures Q_u, $u \in U$, be weakly compact it is necessary and sufficient that for any $\varepsilon > 0$*

$$\lim_{c \to 0} \sup_u \mathbf{P}_u\{\Delta_c[\xi_u(\cdot)] > \varepsilon\} = 0,$$

where

$$\Delta_c[x(\cdot)] = \sup_{t-c \le t' \le t \le t'' \le t+c} \{|x(t'') - x(t)| \wedge |x(t) - x(t')|\} + \sup_{0 \le t \le c} |x(t) - x(0)| + \sup_{T-c \le t \le T} |x(T) - x(t)|.$$

Assume that processes $\xi_u(t)$ are continuous and let Q_u be the corresponding measures on the space $\{C, \mathfrak{C}(C)\}$.

Theorem 3.8. *In order that a family $\{Q_u(\cdot), u \in U\}$ on $\{C, \mathfrak{C}(C)\}$ be weakly compact it is necessary and sufficient that for any $\varepsilon > 0$*

$$\lim_{c \to 0} \sup_u \mathbf{P}\left\{ \sup_{|t'-t''|<c} |\xi_u(t') - \xi_u(t'')| > \varepsilon \right\} = 0.$$

The following sufficient conditions for weak compactness of families of measures on C and D are more convenient for direct verification (cf. [20], Vol 1, Chapter 6, Sections 4 and 5).

Theorem 3.9. *Let for some $p > 0$, an $l > 0$ and $H > 0$ exist such that*

$$\mathbf{E}|\xi_u(t_2) - \xi_u(t_1)|^p |\xi_u(t_3) - \xi_u(t_2)|^p \leq H(t_3 - t_1)^{1+l} \tag{3.53}$$

for all t_1, t_2 and t_3 $(0 < t_1 < t_2 < t_3 \leq T)$ and $u \in U$. Then the family of measures $\{Q_u, u \in U\}$ on $\{D, \mathfrak{C}(D)\}$ corresponding to processes $\xi_u(t)$ with sample functions in D is weakly compact. If $\xi_u(t)$ are continuous processes and if there exist $p > 0$, $l > 0$ and $H > 0$ such that

$$\mathbf{E}|\xi_u(t_2) - \xi_u(t_1)|^p \leq H|t_2 - t_1|^{1+l}, \tag{3.54}$$

then the family of measures $\{Q_u, u \in U\}$ on $\{C, \mathfrak{C}(C)\}$ is also weakly compact.

It is useful to improve the last conditions in order to avoid excessive requirements concerning the existence of moments for the processes under consideration.

Assume that

$$\lim_{N \to \infty} \sup_{u \in U} \mathbf{P}\left\{\sup_t |\xi_u(t)| > N\right\} = 0. \tag{3.55}$$

Let $\tau_N = \inf\{t : \sup|\xi_u(t)| > N\}$, $(\inf \varnothing = T)$ and $\xi_u^N(t) = \xi_u(t)$ for $t < \tau_N$, $\xi_u^N(t) = 0$ for $t \geq \tau_N$. Then

$$\mathbf{P}\{\Delta_c(\xi_u(\cdot)) > \varepsilon\} \leq \mathbf{P}\{\tau_N < T\} + \mathbf{P}\{\Delta_c(\xi_u^N(\cdot)) > \varepsilon\}.$$

This inequality implies the following assertion.

Lemma 3.9. *If a family of stochastic processes $\xi_u(t)$ with sample functions in $D(C)$ satisfies condition (3.55) and for some $p > 0$, $l > 0$ the expression*

$$\mathbf{E}|\xi_u^N(t_2) - \xi_u^N(t_1)|^p |\xi_u^N(t_3) - \xi_u^N(t_3)|^p \leq H^N(t_3 - t_1)^{1+l},$$

$$\forall (t_1, t_2, t_3)\ (0 < t_1 < t_2 < t_3 \leq T),\ u \in U,$$

$$(\mathbf{E}|\xi_u^N(t_2) - \xi_u^N(t_1)|^p \leq H^N(t_2 - t_1)^{1+l})$$

is valid then the family of measures $\{Q_u, u \in U\}$ corresponding to the processes $\xi_u(t)$ in D (C respectively) is weakly compact in $D(C)$.

We apply these results to stochastic differential equations.

Theorem 3.10. *Let a family of stochastic processes* $\{\xi_u(t), u \in U\}$ *satisfying the equations*

$$d\xi_u = \alpha_u(t, \xi_u(\cdot))\,dt + \lambda_u(\xi_u(\cdot), dt),$$
$$\xi_u(0) = \xi_u$$

be given with coefficients which are linearly bounded uniformly in $u \in U$, *i.e.*

$$|\alpha_u(t, x(\cdot)|^2 + |\beta_u(t, x(\cdot))|^2 + \int |\gamma_u(t, x(\cdot), y)|^2 q(dy) \le L(1 + \|x\|_t^2)$$

and let $\sup_u \mathbf{E}|\xi_u(0)|^2 < \infty$. *Then the family of measures* $\{Q_u, u \in U\}$ *in* $\{D, \mathfrak{C}(D)\}$ *corresponding to stochastic processes* $\xi_u(t)$ *is weakly compact in* D. *If* $\gamma_u \equiv 0$, *then the measures* $\{Q_u, u \in U\}$ *on* $\{C, \mathfrak{C}(C)\}$ *are weakly compact in* C.

PROOF. Introduce processes $\xi_u^N(t) = \chi_N(t)\,\xi_u(t)$, where $\chi_N(t) = 1$ for $\|\xi_u(\cdot)\|_t \le N$ and $\chi_N(t) = 0$ otherwise. We utilize the bounds derived in lemma 3.6. We have

$$\begin{aligned}
&\mathbf{E}|\xi_u^N(t_3) - \xi_u^N(t_2)|^2 |\xi_u^N(t_2) - \xi_u^N(t_1)|^2 \\
&\quad = \mathbf{E}\{|\xi_u^N(t_2) - \xi_u^N(t_1)|^2 \mathbf{E}\{|\xi_u^N(t_3) - \xi_u^N(t_2)|^2 \,|\, \mathfrak{F}_{t_2}\}\} \\
&\quad \le \mathbf{E}\{|\xi_u^N(t_2) - \xi_u^N(t_1)|^2 C_2(1 + |\xi_u^N(t_2)|^2)(t_3 - t_2)\} \\
&\quad \le C_3(1 + N^2)^2(t_3 - t_1)^2.
\end{aligned}$$

Moreover in view of lemma 3.6.

$$\mathbf{P}\left\{\sup_{0 \le t \le T} |\xi_u(t)|^2 > N\right\} \le \frac{1}{N^2}\mathbf{E}\{\|\xi_u(\cdot)\|_T^2\} \le \frac{C_3}{N^2}(\mathbf{E}|\xi_u(0)|^2 + 1) \to 0$$

uniformly in u as $N \to \infty$. By virtue of lemma 3.9, the family of measures $\{Q_u, u \in U\}$ is weakly compact.

If $\gamma = 0$, then processes $\xi_u(t)$ possess moments of all orders. In view of the bound (3.48)

$$\mathbf{E}|\xi_u(t + h) - \xi_u(t)|^4 = O(h^2),$$

thus the weak compactness of the family $\{Q_u, u \in U\}$ in C follows in this case as well. □

The weak convergence of distributions of a sequence of random variables in general does not imply convergence of a sequence to a limit, say in probability. Nevertheless, the following theorem can be proved in this direction (cf. [41] p. 12).

Theorem 3.11. *Let a sequence of stochastic processes* $\xi(t)$, $n = 1, 2, \ldots$, $t \in [0, T]$, *with values in* R^d *be given and let the following conditions be satisfied*:

a. *the distribution of variables* $\xi_n(t)$ *is stochastically continuous from the right (left) on* $[0, T]$;

b. *for all* $\varepsilon > 0$

$$\lim_{h \downarrow 0} \overline{\lim_{n \to \infty}} \sup_{|t-s|<h} \mathbf{P}\{|\xi_n(t) - \xi_n(s)| > \varepsilon\} = 0;$$

c.

$$\lim_{N \to \infty} \overline{\lim_{n \to \infty}} \sup_{t \in [0, T]} \mathbf{P}\{|\xi_n(t)| > N\} = 0.$$

Then one can find a subsequence n_k *and processes* $\tilde{\xi}_{n_k}(t)$ *with values in* R^d *defined on some generally different probability space, with finite-dimensional (marginal) distributions of these processes coinciding with finite-dimensional (marginal) distributions of processes* $\xi_{n_k}(t)$ *such that as* $k \to \infty$ $\tilde{\xi}_{n_k}(t)$ *converge in probability for each* $t \in [0, T]$ *to a stochastically continuous process* $\xi(t)$.

Densities of measures.

We now present some results on densities of measures which correspond to solutions of stochastic differential equations. These results are connected with an interesting theorem due to Girsanov. This theorem allows us by applying a measure substitution on a probability space to obtain very general theorems dealing with the existence of stochastic differential equations which are necessary for the solution of control problems.

Let $\mathfrak{F}_t$, $t \in [0, T]$, be a current of σ-algebras of sets on a space Ω; Q and P be two measures defined on $\mathfrak{F}_T$. Denote the contractions of these measures to σ-algebra $\mathfrak{F}_t$ by Q_t and P_t respectively. If Q is absolutely continuous with respect to P, so is Q_t with respect to P_t. Set

$$\rho_t = \rho_t(\omega) = \frac{dQ_t}{dP_t}.$$

The stochastic process ρ_t is non-negative, adapted to the current $\{\mathfrak{F}_t, t \in [0, T]\}$ and is a non-negative martingale on $\{\Omega, \mathfrak{F}_T, P\}$. The last assertion is easily verified: for any $A \in \mathfrak{F}_s$, $s < t$

$$\int_A \rho_s \, dP = Q(A) = \int_A \rho_t \, dP.$$

Also observe that the set of random variables $\rho_t(\omega)$, $t \in [0, T]$, in view of known theorems on martingales (cf. [20] Vol 1, pp. 61–62) is uniformly integrable, $\rho_t = \mathbf{E}\{\rho_T | \mathfrak{F}_t\}$ and the measure P is absolutely continuous with respect to Q if and only if $\rho_T(\omega) > 0$ (mod P). In this case $dP/dQ = [\rho_T(\omega)]^{-1}$. Now let an arbitrary non-negative martingale $\{\rho_t, \mathfrak{F}_t\}$, $t \in [0, T]$, with $\mathbf{E}\rho_T = \int \rho_T \, dP = 1$ be given. Introduce in $\{\Omega, \mathfrak{S}\}$ a new measure $Q(A) = \int_A \rho_T \, dP$. Then $Q(\cdot)$ is a probability measure on Ω and for any $A \in \mathfrak{F}_t$ the expression

$$Q(A) = \int_A \rho_T \, dP = \int_A \mathbf{E}\{\rho_T | \mathfrak{F}_t\} \, dP = \int_A \rho_t \, dP$$

is valid, i.e. the contraction of measure Q to $\mathfrak{F}_t$ is given by the formula

$$Q_t(A) = \int_A \rho_t \, dP.$$

We now present a formula which expresses the conditional mathematical expectation calculated with respect to measure Q via the conditional mathematical expectation corresponding to probability measure P. In the case when different measures are considered on the space of elementary events $\{\Omega, \mathfrak{S}\}$ we shall adjoin a subscript to the symbol $\mathbf{E}$. This subscript indicates with respect to what measure the calculation of mathematical expectation is carried out.

Let η be an arbitrary $\mathfrak{F}_t$-measurable non-negative random variable and $N_t = \{\omega : \rho_t(\omega) = 0\}$. For any $A \in \mathfrak{F}_s$, $s < t$, in view of the definition of a conditional mathematical expectation we have

$$\int_A \mathbf{E}_Q\{\eta \mid \mathfrak{F}_s\} \, dQ = \int_A \eta \, dQ = \int_A \eta\rho_t \, d\mathbf{P}$$

$$= \int_A \mathbf{E}_\mathbf{P}\{\eta\rho_t \mid \mathfrak{F}_s\} \, d\mathbf{P} = \int_A \frac{\chi(N_s)}{\rho_s} \mathbf{E}_\mathbf{P}\{\eta\rho_t \mid \mathfrak{F}_s\} \, dQ,$$

where $\chi(N_s) = \chi(N_s, \omega)$ is the indicator of the set N_s and $\chi(N_s)/\rho_s = 0$ provided $\rho_s = 0$. Thus

$$\mathbf{E}_Q\{\eta \mid \mathfrak{F}_s\} = \frac{\chi(N_s)}{\rho_s} \mathbf{E}_\mathbf{P}\{\eta\rho_t \mid \mathfrak{F}_s\} \pmod{Q}. \tag{3.56}$$

If we set that $\mathbf{E}_Q\{\eta \mid \mathfrak{F}_s\} = 0$ everywhere on N_s (this does not contradict the definition of a conditional mathematical expectation since $Q(N_s) = 0$), equality (3.56) will then hold $\mathbf{P}$-almost everywhere.

In the case when $\{\mathfrak{F}'_t, t \in [0, T]\}$ is a current of σ-algebras and $\mathfrak{F}'_t \subset \mathfrak{F}_t$, $\forall t \in [0, T]$, formula (3.56) should be modified. Note that for any $A \in \mathfrak{F}'_s$

$$Q(A) = \int_A \rho_T \, d\mathbf{P} = \int_A \mathbf{E}_\mathbf{P}\{\rho_T \mid \mathfrak{F}'_s\} \, d\mathbf{P},$$

and moreover

$$\mathbf{E}_\mathbf{P}\{\rho_T \mid \mathfrak{F}'_s\} = \mathbf{E}_\mathbf{P}\{\mathbf{E}_\mathbf{P}\{\rho_T \mid \mathfrak{F}_s\} \mid \mathfrak{F}'_s\} = \mathbf{E}_\mathbf{P}\{\rho_s \mid \mathfrak{F}'_s\}.$$

This implies that for all $A \in \mathfrak{F}'_s$

$$\int_A \mathbf{E}_Q(\eta \mid \mathfrak{F}'_s) \, dQ = \int_A \chi'_s \frac{\mathbf{E}_\mathbf{P}(\eta\rho_t \mid \mathfrak{F}'_s)}{\mathbf{E}_\mathbf{P}(\rho_s \mid \mathfrak{F}'_s)} \, dQ,$$

where χ'_s is the indicator function of the set $[\omega : \mathbf{E}_\mathbf{P}\{\rho_s \mid \mathfrak{F}'_s\} = 0]$ and $\chi'_s/\mathbf{E}_\mathbf{P}\{\rho_s \mid \mathfrak{F}'_s\} = 0$ if the numerator is 0.

Consequently for an arbitrary non-negative $\mathfrak{F}_t$-measurable random variable η and a current $\{\mathfrak{F}'_t, t \in [0, T]\}$, $\mathfrak{F}'_t \subset \mathfrak{F}_t$,

$$\mathbf{E}_Q\{\eta \mid \mathfrak{F}'_s\} = \chi'_s \frac{\mathbf{E}_\mathbf{P}\{\eta\rho_t \mid \mathfrak{F}'_s\}}{\mathbf{E}_\mathbf{P}\{\rho_s \mid \mathfrak{F}'_s\}}. \tag{3.57}$$

We now describe the structure of an arbitrary positive continuous martingale. The case of the continuous local martingales is the simplest.

Theorem 3.12. *A positive process $\{\zeta(t), \mathfrak{F}_t, t \geq 0\}$ is a continuous local martingale if and only if*

$$\zeta(t) = \exp\{\mu(t) - \tfrac{1}{2}\langle\mu, \mu\rangle_t\}, \tag{3.58}$$

where $\{\mu(t), \mathfrak{F}_t, t \geq 0\}$ is a continuous local martingale.

Proof. Let $\zeta(t) \in lM^c$ and $\zeta(t) > 0$, $\forall t > 0$. Set

$$\tau_n = \inf\left\{t : \left(\zeta(t) \leq \frac{1}{n}\right) \vee (\zeta(t) \geq n)\right\}, \qquad \zeta_n(t) = \zeta(t \wedge \tau_n),$$

$$f(x) = \ln x \quad \text{for} \quad x \in \left[\frac{1}{n}, \infty\right)$$

and assume that $f(x)$ is defined for $x < 1/n$ in such a manner that it is twice differentiable. Utilizing Itô's formula for $f(x)$ we obtain

$$\ln \zeta_n(t) - \ln \zeta_n(0) = \int_0^{t \wedge \tau_n} \frac{1}{\zeta}\, d\zeta - \frac{1}{2}\int_0^{t \wedge \tau_n} \frac{1}{\zeta^2}\, d\langle\zeta, \zeta\rangle_s.$$

Set $\mu(t) = \int_0^t \frac{1}{\zeta(s)}\, d\zeta(s) + \ln \zeta(0)$. Then $\mu(t) \in lM^c$,

$$\langle\mu, \mu\rangle_t = \int_0^1 \frac{1}{\zeta^2(s)}\, d\langle\zeta, \zeta\rangle_s \text{ and } \zeta_n(t) = \exp\{\mu(t \wedge \tau_n) - \tfrac{1}{2}\langle\mu, \mu\rangle_{t \wedge \tau_n}\}.$$

Approaching the limit as $n \to \infty$ we arrive at (3.58).

Assume now that $\mu(t) \in lM^c$ and construct by means of formula (3.58) the process $\zeta(t)$. Utilizing Itô's formula once again and setting this time $f(x) = e^x$, $\xi(t) = \mu(t) - \frac{1}{2}\langle\mu, \mu\rangle_t$, we have

$$\zeta(t) = \zeta(0) + \int_0^t \zeta(s)\left[d\mu(s) - \frac{1}{2}\, d\langle\mu, \mu\rangle_s\right]$$
$$+ \frac{1}{2}\int_0^t \zeta(s)\, d\langle\mu, \mu\rangle_s = \zeta(0) + \int_0^t \zeta(s)\, d\mu(s).$$

Thus $\zeta(t) \in lM^c$ and $\zeta(t) > 0$, $\forall t > 0$. □

Observe that the process $\zeta(t)$ defined by formula (3.58) satisfies equation

$$d\zeta = \zeta \, d\mu. \tag{3.59}$$

It is easy to verify that for $s < t$

$$\mathbf{E}\{\zeta(t) \mid \mathfrak{F}_s\} \le \zeta(s).$$

A process $\{\zeta(t), \mathfrak{F}_t, t \ge 0\}$ satisfying this inequality is called a *supermartingale*. The preceding inequality implies that a local martingale $\zeta(t)$ is a martingale if and only if

$$\mathbf{E}\zeta(t) = \mathbf{E}\zeta(0) \qquad \forall t > 0.$$

Indeed if $\xi(t)$ is a martingale, the equality $\mathbf{E}\zeta(t) = \mathbf{E}\zeta(0)$, $\forall t > 0$, is self-evident. However, if this condition is fulfilled, then

$$\mathbf{E}(\zeta(s) - \mathbf{E}\{\zeta(t) \mid \mathfrak{F}_s\}) = \mathbf{E}\zeta(s) - \mathbf{E}\zeta(t) = 0.$$

Since $\zeta(t)$ is a supermartingale it follows that $\zeta(s) = \mathbf{E}\{\zeta(t) \mid \mathfrak{F}_s\}$ (mod **P**).

Return now to formula (3.58). Assume that $\mu(0) = 0$. In view of the remarks above, in order that $\zeta(t)$ be a martingale it is necessary and sufficient that condition $\mathbf{E}\zeta(t) = 1$ for all $t > 0$ be satisfied. In some cases it is of interest to express the last condition in a more convenient form for verification. The best result in this direction is probably the following (cf. [20], Vol III, Chapter 3, Section 1, Theorem 12).

Theorem 3.13 (Novikov). *If* $\mathbf{E}\exp\{\frac{1}{2}\langle\mu, \mu\rangle_T\} < \infty$, *then* $\zeta(t) = \exp\{\mu(t) - \frac{1}{2}\langle\mu, \mu\rangle_t\}$, $t \in [0, T]$, *is a martingale.*

Let $\{w(t), \mathfrak{F}_t, t \in [0, T]\}$ be a d-dimensional Wiener process defined on the probability space $\{\Omega, \mathfrak{S}, \mathbf{P}\}$ and $\psi(t)$, $t \in [0, T]$, be a stochastic process with values in R^d adapted to the current $\{\mathfrak{F}_t\}$ such that

$$\int_0^T |\psi(t)|^2 \, dt < \infty \quad \text{with probability 1.}$$

Set

$$\eta(t) = w(t) - \int_0^t \psi(s) \, ds,$$

$$\rho_t = \exp\left\{\int_0^t (\psi(s), dw(s)) - \frac{1}{2}\int_0^t |\psi(t)|^2 \, dt\right\}.$$

The preceding remarks are applicable to the process $\{\rho_t, \mathfrak{F}_t\}$. The role of the local martingale $\mu(t)$ is taken here by the process $\int_0^t (\psi(s), dw(s))$ and $\langle\mu, \mu\rangle_t = \int_0^t |\psi(s)|^2 \, ds$. We are also assuming that $\mathbf{E}\rho_T = 1$.

Introduce on $\mathfrak{S}$ a new measure Q by setting

$$Q(A) = \int_A \rho_T(w) \, d\mathbf{P}.$$

Theorem 3.14 (Girsanov). *Under the preceding assumptions the process* $\{\eta(t), \mathfrak{F}_t, t \in [0, T]\}$ *is a Wiener process on the probability space* $\{\Omega, \mathfrak{S}, Q\}$.

PROOF. First we verify that $\{\eta(t), \mathfrak{F}_t, Q\}$ is a local martingale. Let $\Theta \in R^d$, $\eta_\Theta(t) = (\Theta, \eta(t))$. Equation (3.59) implies that $d\rho_t = \rho_t \, d\mu$. Utilizing the formula for stochastic differentials of a product of two processes (3.13) we obtain

$$\begin{aligned} d(\eta_\Theta(t)\rho_t) &= \rho_t \, d\eta_\Theta(t) + \eta_\Theta(t) \, d\rho_t + \rho_t(\Theta, \psi) \, dt \\ &= \rho_t(\Theta, dw) + \eta_\Theta \rho_t(\psi, dw) = \rho_t(\Theta + \psi\eta_\Theta, dw). \end{aligned}$$

Thus $\eta_\Theta(t)\rho_t$ is a local martingale (with respect to the measure Q on Ω) for any $\Theta \in R^d$; consequently $\eta(t)\rho_t \in lM^c(\mathfrak{F}_t, \mathbf{P})$.

Let

$$\tau_n = \inf\{t : |\eta(t)| \geq n\} \qquad \left(\tau_n = T, \quad \text{if} \quad \sup_{0 \leq t \leq T} |\eta(t)| < n\right).$$

Utilizing formula (3.56) we have $(0 \leq s < t \leq T)$

$$\begin{aligned} \tilde{\mathbf{E}}\{\eta_\Theta(t \wedge \tau_n) \,|\, \mathfrak{F}_{s \wedge \tau_n}\} &= \frac{1}{\rho(s \wedge \tau_n)} \mathbf{E}\{\eta_\Theta(t \wedge \tau_n)\rho(t \wedge \tau_n) \,|\, \mathfrak{F}_{s \wedge \tau_n}\} \\ &= \frac{1}{\rho(s \wedge \tau_n)} \eta(s \wedge \tau_n)\rho(s \wedge \tau_n) = \eta(s \wedge \tau_n). \end{aligned}$$

Thus $\eta(t) \in lM^c\{\mathfrak{F}_t, Q\}$. Furthermore let the square variation of the process $\eta(t)$ coincide with the square variation of the process $w(t)$, i.e. $\langle \eta, \eta \rangle_t = \mathscr{I} t$. In view of Levy's theorem the process $\{\eta(t), \mathfrak{F}_t, Q\}$ is a Wiener process. □

We now present the following important result due to Clark (see e.g. [35]).

Let $w(t)$ be a Wiener process in R^d, $\mathfrak{F}_t$ be the completion of the smallest σ-algebra $\sigma\{w(s), s \leq t\}$, $\{\eta(t), \mathfrak{F}_t, t \in [0, T]\}$ be a square integrable martingale. Then there exists a process $\beta(t)$ adapted to the current $\mathfrak{F}_t$ with values in R^d such that

$$\mathbf{E}\int_0^T |\beta(t)|^2 \, dt < \infty \quad \text{and} \quad \eta(t) = \mathbf{E}\eta(0) + \int_0^t (\beta(s), dw(s)).$$

Clearly this result is generalized to the case of local square integrable martingales: if $\eta(0) = 0$ and $\eta(t)$ is a local, continuous from the right and square integrable martingale, then there exists an $\mathfrak{F}_t$-adapted process $\beta(t)$ such that

$$\int_0^T |\beta(s)|^2 \, ds < \infty \quad (\text{mod } \mathbf{P}) \quad \text{and} \quad \eta(t) = \int_0^t (\beta(s), dw(s)).$$

3 Controlled Stochastic Differential Equations

A *controlled stochastic differential equation* is an equation of the form

$$d\xi = \alpha(t, \xi(\cdot), \eta(\cdot), u)\, dt + \lambda(\xi(\cdot), \eta(\cdot), dt),$$
$$\xi(0) = \xi_0, \tag{3.60}$$

where

$$\lambda(x(\cdot), u(\cdot), t) = \int_0^t \beta(s, x(\cdot), u(s))\, dw(s) + \int_0^t \int_{R^d} \gamma(s, x(\cdot), u(s), y)\tilde{v}(ds, dy),$$
$$\alpha(t, x(\cdot), u),\ \beta(t, x(\cdot), u),\ \gamma(t, x(\cdot), u, y),$$
$$(t, x(\cdot), u) \in [0, \infty) \times D^d \times U,\ y \in R^d,$$

are random functions satisfying the conditions stipulated below. The equations under consideration have the same meaning as the corresponding equations in the preceding Section. The new element introduced here is the dependence of functions α, β and γ on a parameter u taking on values in a certain space U. In equation (3.60) a stochastic process $\eta(t)$ with values in U is substituted for the values of the second parameter. Within certain limits (which depend on a specific problem) processes $\eta(t)$ introduced in equation (3.60) can be chosen arbitrarily and can therefore "control" the solution of equation (3.60). In view of this, the function $\eta(t)$ is called a *control or a strategy*, and the class of all controls $\eta(t)$ which can be introduced into (3.60) in a given problem is called the set of *admissible controls* or *admissible strategies*. This set is denoted by $\mathfrak{U}$. A solution of equation (3.60) corresponding to a given control $\eta = \eta(t)$, $t \geq 0$, is denoted by $\xi^{(\eta)}(t)$ or $\xi^{(\eta)}(\cdot)$. We recall here some restrictions to which admissible controls are subjected.

Firstly, in order to use the machinery of stochastic differential equations, it is necessary that for a given control $\eta(t)$ equation (3.60) will possess a solution. In a number of cases it is required that this solution be unique in a certain sense. In practical problems it is often assumed that a continuous observation of realizations of a solution of equation (3.60) is arranged, and it is required that the control $\eta(t)$ at each instant of time be a functional on the observed realization of the solution of equation (3.60) on the time interval $[0, t]$, $\eta(t) = g(t, \xi(\cdot))$, where $g(t, x(\cdot))$ is a non-anticipative function. Such control is often called a "*feedback*" *under a complete observation of the controlled object*. It is assumed in a number of problems that an incomplete realization of the solution of equation (3.60) is observed, namely some function $\xi(t)$ of it, or that some of its components are generally masked by noise. It is required to obtain a control of the form $\eta(t)h(t, \xi(t))$ where $h(t, y(\cdot))$ is also a non-anticipative function. Such controls are called *feedback controls with incomplete data*.

If we introduce a stochastic process adapted to a given current of σ-algebras $\{\mathfrak{F}_t\}$ given in advance as a control into expression (3.60), then this control can be viewed as a control without feedback. We shall call such a control a *generalized control.* A solution of equation (3.60) in this case is simpler to obtain as compared to the case when $\eta(t)$ is given in the form $\eta(t) = g(t, \xi(\cdot))$.

Secondly, feedback controls $\eta(t) = g(t, \xi(\cdot))$—provided they exist—are contained in the class of all generalized controls. Therefore in a number of cases it is preferable to consider generalized controls. Moreover feedback controls are a particular case of generalized controls without feedback. It should be noted that in practice it is difficult to realize a generalized control.

In order for it to be possible to consider equation (3.60) and for the stochastic integral $\int_0^t \lambda(\xi(\cdot), \eta(s), ds)$ to make sense, one must require that an admissible control $\eta(t)$ be such that at least the σ-algebra $\mathfrak{F}_t^{(\eta)} = \sigma\{\eta(s), s \in [0, t]\}$ will not depend on the σ-algebra $\mathfrak{F}^{(w,\, v)}[t, \infty) = \sigma\{w(s) - w(t), v(s, A) - v(t, A), s \geq t, A \in \mathfrak{B}_0\}$. The class of all stochastic processes $\{\eta(t), t \geq 0\}$ whose sample functions are Borel with probability 1 with values in U satisfying the last condition be denoted by $\mathfrak{U}$ and we shall always assume that the generalized controls $\eta(t)$ belong to $\mathfrak{U}$.

The purpose of a control is to minimize the mean value, (i.e. the mathematical expectation) of a functional $F(\xi^{(\eta)}(\cdot), \eta(\cdot))$ which depends on the chosen control $\eta(t)$, $t \geq 0$, and on the resulting trajectory of the controlled object $\xi^{(\eta)}(t)$ (a solution of equation (3.60)). The function $F(x(\cdot), u(\cdot))$ characterizes the cost of the loss or of the errors associated with the functioning of the controlled object and the choice of a given control $u(\cdot)$. It is called the *loss function* and its mean value

$$\bar{F}(\eta) = \mathbf{E}F(\xi^{(\eta)}(\cdot), \eta(\cdot))$$

is called the *cost of control.* Unless stated otherwise we shall always assume that the function $F(x(\cdot), u(\cdot))$ is defined on $D \times B$. Here $D = D_{[0,\, T]}(R^d)$ is the space of all functions on $[0, T]$ with values in R^d continuous from the right which possess left-hand limits at each point $t \in [0, T)$ ($(0, T]$ respectively), while $B = B_{[0,\, T]}(U)$ is the space of all Borel functions on $[0, T]$ with values in U.

Set

$$Z = \inf_{\eta \in \mathfrak{U}} \bar{F}(\eta) = \inf_{\eta \in \mathfrak{U}} \mathbf{E}F(\xi(\cdot), \eta(\cdot)).$$

The quantity Z is called the *optimal cost* (or *the cost*) of a control in a given class of admissible controls $\mathfrak{U}$. If the cost of a control $\bar{F}(\eta)$ for some $\eta = \eta_0(\cdot)$ takes on its smallest value on $\mathfrak{U}$ then η_0 is called *optimal* in $\mathfrak{U}$. A control η' ($\eta' \in \mathfrak{U}$) such that $\bar{F}(\eta') - Z < \varepsilon$ is called *ε-optimal.*

The basic problems of the theory of optimal control are as follows: to determine the optimal cost, to find out whether the optimal controls exist; to describe these optimal controls and the methods for their determination; to check whether there exist ε-optimal controls in a given class of "simple" controls and to obtain simple methods for constructing ε-optimal controls.

Denote by $\mathfrak{U} = \mathfrak{U}(D)$ the σ-algebra of Borel sets in D. This σ-algebra coincides (cf. [20], Chapter 6, Section 5) with the σ-algebra generated by the cylinders in D. Denote by $\mathfrak{U}_t$ the σ-algebra in D generated by the cylinders in D with bases over $[0, t]$, i.e. by the sets of the form

$$\{x(\cdot): (x(s_1), x(s_2), \ldots, x(s_n)) \in B^{(n)}\},$$

where $B^{(n)}$ is an arbitrary Borel set in R^{d_n}, n is an arbitrary integer, and $s_1, \ldots, s_n$ are arbitrary numbers belonging to $[0, t]$. The family $\{\mathfrak{U}_t, t \geq 0\}$ forms a current of σ-algebras.

Introduce on space $[0, T] \times D$ a σ-algebra $\mathfrak{G} = \mathfrak{G}(D)$ defined in the following manner: $\mathfrak{G}(D)$ is the minimal σ-algebra containing the sets of the form $(t, s] \times A$, where $A \in \mathfrak{A}_t$ and $0 \leq t < s \leq T$.

Furthermore it is assumed that the following conditions are satisfied: (a) U is a metric compact space; (b) functions $\alpha = a(t, x(\cdot), u)$, $\beta = b(t, x(\cdot), u)$, $\gamma = c(t, x(\cdot), u, y)$ are non-random, defined for $(t, x(\cdot), u) \in [0, T] \times D \times U$, $y \in R^d$ and moreover $a(t, x(\cdot), u)$ and $c(t, x(\cdot), u, y)$ are vector-valued functions with values in R^d, $b(t, x(\cdot), u)$ is an operator function which maps R^{d_1} into R^d. For a fixed u and y these functions are $\mathcal{G}$-measurable and continuous jointly in the variables $(t, x(\cdot), u)$; (c) functions $a(t, x(\cdot), u)$, $b(t, x(\cdot), u)$ and $c(t, x(\cdot), u, y)$ are linearly bounded uniformly in u, i.e. there exists a constant C which does not depend on u and such that

$$|a(t, x, u)| + |b(t, x, u)| + |c(t, x, u, y)| \leq C(1 + \|x(\cdot)\|_t), \forall t \in [0, T].$$

Many of the results obtained in the present Section can be easily extended to the cases when the functions a, b and c are themselves random. We shall not dwell on this point however.

Recall that $w(t)$, $t \geq 0$, denotes a Wiener process with values in R^{d_1}, $v(t, A)$ is a Poisson measure on $\mathfrak{B}_0^d$ where $\mathfrak{B}_0^d$ is the class of all Borel sets in R^d whose closures do not contain point 0 and $\tilde{v}(t, A) = v(t, A) - tq(A)$, $tq(A) = \mathbf{E}v(t, A)$ and the processes $v(t, A)$ and $w(t)$ are independent.

When dealing with equation (3.60) it is assumed that processes $w(t)$, $v(t, A)$, $\eta(t)$, $t \in [0, T]$, $A \in \mathfrak{B}_0^d$, are defined on a probability space which is not fixed a priori and which can be constructed arbitrarily. Moreover in the case of generalized controls, different probability spaces may correspond to different processes $\eta(t)$.

It was already mentioned that in the course of solving a controlled stochastic equation (3.60) two distinct situations are encountered depending on whether generalized controls or feedback controls are considered. In the first case a definite random control process $\eta(t)$, $t \geq 0$, is preassigned and a process $\xi(t)$ which satisfies equation (3.60) is sought. In the second case the process $\eta(t)$ is unknown beforehand. It is required to obtain this process simultaneously with the process $\xi(t)$ if it is known that the processes are related by a relation of the form $\eta(t) = u(t, \xi(\cdot))$, where $u(t, x(\cdot))$ is a $G(u)$-measurable function.

In the first case the problem of solving equation (3.60) is actually identical to the problem of solving an uncontrolled equation of the form

$$d\xi = \alpha(t, \xi(\cdot))\, dt + \tilde{\lambda}(\xi(\cdot), dt), \tag{3.61}$$

where

$$\tilde{\lambda}(x(\cdot), t) = \int_0^t \beta(s, x(\cdot))\, dw(s) + \int_0^t \int_{R^d} \gamma(s, x(\cdot), y)\tilde{v}(ds, dy),$$

$$\alpha(t, x(\cdot)) = a(t, x(\cdot), \eta(t)), \qquad \beta(t, x(\cdot)) = b(t, x(\cdot), \eta(t)),$$

$$\gamma(t, x(\cdot), y) = c(t, x(\cdot), \eta(t), y).$$

Thus for example, if the functions a, b, and c satisfy a local Lipschitz condition then, in view of theorem 3.4 equation (3.60) possesses a unique solution and possesses finite moments of the second order for any process $\eta(\cdot) \in \bar{\mathfrak{U}}$.

Naturally one can also use other theorems of existence and uniqueness of the solution for equation (3.61).

In the second case the situation is substantially more complex if one does not require that the function $u = u(t, x(\cdot))$ satisfy Lipschitz's condition with respect to $x(\cdot)$; in such cases it is necessary to solve an equation of the form

$$d\xi = a[t, x(\cdot), u(t, \xi(\cdot))]\, dt + b[t, \xi(\cdot), u(t, \xi(\cdot))]\, dw$$
$$+ \int_{R^d} c[t, \xi(\cdot), u(t, \xi(\cdot), y)]\tilde{v}(ds, dy); \tag{3.62}$$

unfortunately for an arbitrary dependence between $u(t, x(\cdot))$ and $x(\cdot)$ there are no existence theorems for solutions of such an equation.

Observe, however that the existence of a solution for equation

$$d\xi = b(t, \xi(\cdot))\, dw, \qquad \xi(0) = \xi_0, \qquad t \in [0, T], \tag{3.63}$$

implies the existence of a solution for equation

$$d\xi = a[t, \xi(\cdot), u(t, \xi(\cdot))]\, dt + b(t, \xi(\cdot))\, dw, \qquad t \in [0, T], \tag{3.64}$$

for an arbitrary control $u = u(t, x(\cdot))$ such that the function $a[t, x(\cdot), u(t, x(\cdot))]$ is $\mathcal{G}$-measurable and $\mathbf{E}_{\mathbf{P}}\, e^{\zeta(T)} = 1$. Here

$$\zeta(T) = \int_0^T (b^{-1}(t, x(\cdot))a[t, x(\cdot), u(t, x(\cdot))])^*\, dw$$
$$- \frac{1}{2}\int_0^T |b^{-1}(t, x(\cdot))a[t, x(\cdot), u(t, x(\cdot))]|^2\, dt,$$

and $\mathbf{P}$ is the measure in C induced by a solution of equation (3.63). This equation is discussed in Section 6 in more detail.

In what follows we shall limit ourselves to problems of optimal control for solutions of stochastic differential equations based on complete data and on a fixed finite interval of time $[0, T]$. Here C and D denote the spaces $C_{[0,T]}(R^d)$, $D_{[0,T]}(R^d)$.

First we shall consider the problem of existence of piecewise constant ε-optimal controls.

The theorem on continuous dependence of a solution of a stochastic differential equation on a control will be required. We shall agree to write $(a, b, c) \in S(C, \cdot)$ if the coefficients (a, b, c) of equation

$$d\xi = a(t, \xi(\cdot), \eta(t))\, dt + b(t, \xi(\cdot), \eta(t))\, dw + \int c(t, \xi(\cdot), \eta(\cdot), y)\tilde{\nu}(dt, dy) \tag{3.65}$$

are linearly bounded uniformly in u; $(a, b, c) \in S(\cdot, L_N)$ if they satisfy the local Lipschitz condition (in $x(\cdot)$) with a family of constants L_N which do not depend on u; $(a, b, c) \in S(\cdot, L)$ if (a, b, c) satisfies the uniform Lipschitz condition in $x(\cdot)$ where the constant L does not depend on u; also we set $S(C, L_N) = S(C, \cdot) \cap S(\cdot, L_N)$.

Let a sequence of stochastic processes $\xi_n(t)$, $n = 0, 1, \ldots, t \in [0, T]$, which are solutions of the equations

$$d\xi_n = a(t, \xi_n(\cdot), \eta_n(t))\, dt + \lambda(\xi_n(\cdot), \eta_n(t), dt), \qquad \xi_n(0) = \xi_0 \tag{3.66}$$

be given.

Theorem 3.15. *Assume that $(a, b, c) \in S(C, L)$ and the processes $\eta_n(t) \to \eta_0(t)$ almost for all $t \in [0, T]$ (with respect to the Lebesgue measure) with probability 1. Then*

$$\mathbf{E} \sup_{t \in [0,T]} |\xi_n(t) - \xi_0(t)|^2 \to 0 \quad \text{as } n \to \infty. \tag{3.67}$$

Proof. Set

$$z_n(t) = \mathbf{E}\|\xi_n(\cdot) - \xi_0(\cdot)\|_t^2. \tag{3.68}$$

It is known from the above that the function $z_n(t)$ is bounded on $[0, T]$. Set

$$\mathcal{T}(\xi, \eta, t) = \xi_0 + \int_0^t a(s, \xi(\cdot), \eta(s))\, ds + \int_0^t b(s, \xi(\cdot), \eta(s))\, dw(s) + \int_0^t \int_{R^d} c(s, \xi(\cdot), \eta(s), y)\tilde{\nu}(ds, dy).$$

Then

$$z_n(t) = \mathbf{E}\|\mathcal{T}(\xi_n, \eta_n, \cdot) - \mathcal{T}(\xi_0, \eta_0, \cdot)\|_t^2 \le 2(\mathbf{E}\|\mathcal{T}(\xi_n, \eta_n, \cdot) - \mathcal{T}(\xi_0, \eta_n, \cdot)\|_t^2 + \mathbf{E}\|\mathcal{T}(\xi_0, \eta_n, \cdot) - \mathcal{T}(\xi_0, \eta_0, \cdot)\|_t^2).$$

Lemma 3.3 implies that

$$\mathbf{E}\|\mathscr{T}(\xi_n, \eta_n, \cdot) - \mathscr{T}(\xi_0, \eta_n, \cdot)\|_t^2 \le L_1 \int_0^t \mathbf{E}\|\xi_n(\cdot) - \xi_0(\cdot)\|_s^2 \, ds,$$

where $L_1 = 2L^2(T + 4)$. Furthermore,

$$\begin{aligned}\mathbf{E}\|\mathscr{T}(\xi_0, \eta_n, \cdot) - \mathscr{T}(\xi_0, \eta_0, \cdot)\|_t^2 \le 3\mathbf{E}\Bigg\{\Bigg(\int_0^t |a(s, \xi_0(\cdot), \eta_n(s))\\ - a(s, \xi_0(\cdot), \eta_0(s))| \, ds\Bigg)^2 + \Bigg\|\int_0^t [b(s, \xi_0(\cdot), \eta_n(s))\\ - b(s, \xi_0(\cdot), \eta_0(s))] \, dw(s)\Bigg\|_t^2 + \Bigg\|\int_0^t \int_{R^d} [c(s, \xi_0(\cdot), \eta_n(s), y)\\ - c(s, \xi_0(\cdot), \eta_n(s), y)]\tilde{\nu}(ds, dy)\Bigg\|_t^2\Bigg\}.\end{aligned}$$

Utilizing Doob's inequality we obtain that the r.h.s. of the last inequality does not exceed

$$\begin{aligned}I(t) = \mathbf{E}\Bigg\{T \int_0^t |a(s, \xi_0(\cdot), \eta_n(s)) - a(s, \xi_0(\cdot), \eta_0(s))|^2 \, ds\\ + 4 \int_0^t |b(s, \xi_0(\cdot), \eta_n(s)) - b(s, \xi_0(\cdot), \eta_0(s))|^2 \, ds\\ + 4 \int_0^t \int_{R^d} |c(s, \xi_0(\cdot), \eta_n(s), y) - c(s, \xi_0(\cdot), \eta_0(s), y)|^2 q(dy) \, ds\Bigg\}.\end{aligned}$$

Thus,

$$z_n(t) \le I(t) + L_1 \int_0^t z_n(s) \, ds,$$

whence utilizing lemma 3.5 we obtain

$$z_n(t) \le I(t) + \int_0^t I(s) e^{L_1(t-s)} \, ds \le L_2 I(t),$$

where $L_2 = L_2(T, L)$. The expression for $I(t)$ can be represented in the form $I(t) = \int_\Omega \int_0^t \Phi_n(\omega, s) \, ds \, dP$ and the function $\Phi_n(\omega, s)$ admits the bound $\Phi_n(\omega, s) \le C'(1 + \|\xi_0(\cdot)\|^2)$ where C' is a constant independent of n and thus the function $\Phi_n(\omega, s)$ is uniformly integrable in (ω, s). However, with probability 1, $\Phi_n(\omega, s) \to 0$ for almost all s, i.e., $\Phi_n(\omega, s) \to 0$ $(P \times l)$ almost everywhere (l is the Lebesgue measure on the line). Therefore $I(t) \to 0$ and hence also $z_n(t) \to 0$. □

The theorem just proved generalizes to the case of an equation in the class $S(C, L_N)$.

Theorem 3.16. *If $(a, b, c) \in S(C, L_N)$ and $\eta_N(t) \to \eta_0(t)$ almost for all $t \in [0, T]$ with probability 1, then*

$$\mathbf{P}\{\|\xi_n(\cdot) - \xi_0(\cdot)\|_T > \varepsilon\} \to 0 \qquad \text{as } n \to \infty \quad \forall \varepsilon > 0.$$

Proof. Choose some $N > 0$. Construct functions $a_N(t, x(\cdot), u)$, $b_N(t, x(\cdot), u)$ and $c_N(t, x(\cdot), u, y)$ in such a manner that they will coincide with $a(t, x(\cdot), u)$, $b(t, x(\cdot), u)$ and $c(t, x(\cdot), u, y)$ respectively on the sphere $\|x(\cdot)\|_T \le N$ and will be linearly bounded by a constant C and satisfy a uniform Lipschitz condition with some constant $L' = L'(N)$. One may for example set $a_N(t, x(\cdot)) = a(t, x(\cdot))g_N(t, x(\cdot))$ where $g_N(t, x(\cdot)) = 1$ for $\|x(\cdot)\|_t \le N$, $g_N(t, x(\cdot)) = N + 1 - \|x(\cdot)\|_t$ for $\|x(\cdot)\|_t \in [N, N + 1]$, $g_N(t, x(\cdot)) = 0$ for $\|x(\cdot)\|_t \ge N + 1$.

It is easy to verify that $a_N(t, x(\cdot), u)$ satisfies the uniform Lipschitz condition with a constant L' which depends only on N, L_N and L_{N+1}. Functions b_N and c_N can be defined analogously.

Consider now processes $\xi_n^N(t)$, $n = 0, 1, \ldots$, which are solutions of the equations

$$d\xi_n^N = a_N(t, \xi_n^N(\cdot), \eta_n(t))\, dt + \lambda_N(\xi_n^N(\cdot), \eta_n(t), dt),$$

where

$$\lambda_N(x(\cdot), u, dt) = b_N(t, x(\cdot), u)\, dw + \int_{R^d} c_N(t, x(\cdot), u, y)\tilde{v}(dt, dy).$$

These equations possess a unique solution and theorem 3.15 is applicable to them. In view of lemma 3.7 $\xi_n^V(t) = \xi_n(t)$ for $t \le \tau$, where $\tau = \inf\{t : |\xi_n(t)| \ge N\}$ ($\inf \varnothing = T$). Therefore

$$\mathbf{P}\{\|\xi_n(\cdot) - \xi_0(\cdot)\|_T > \varepsilon\} \le \mathbf{P}\{\|\xi_n(\cdot)\|_T \ge N\} + \mathbf{P}\{\|\xi_0(\cdot)\|_T \ge N\} + \mathbf{P}\{\|\xi_n^N(\cdot) - \xi_0^N(\cdot)\|_T > \varepsilon\}.$$

Lemma 3.6 implies that

$$\mathbf{P}\{\|\xi_n(\cdot)\|_T \ge N\} \le \frac{C_1(1 + \mathbf{E}|\xi_0|^2)}{N^2}, \quad n = 0, 1, \ldots,$$

where C_1 does not depend on C and T; in view of theorem 3.15

$$\mathbf{P}\{\|\xi_n^N(\cdot) - \xi_0^N(\cdot)\|_T > \varepsilon\} \le \frac{1}{\varepsilon^2}\mathbf{E}\|\xi_n^N(\cdot) - \xi_0^N(\cdot)\|_T^2 \to 0,$$

$\forall \varepsilon > 0$, as $n \to \infty$ and N is fixed. Choosing first an N sufficiently large so that $\mathbf{P}\{\|\xi_n(\cdot)\|_T \ge N\} \le \varepsilon/3$ for all n, and then n_0 such that for $n \ge n_0$ the inequality $\varepsilon^{-2}\mathbf{E}\|\xi_n^N(\cdot) - \xi_0^N(\cdot)\|_T^2 < \varepsilon/3$ is fulfilled, we obtain

$$\mathbf{P}\{\|\xi_n(\cdot) - \xi_0(\cdot)\|_T > \varepsilon\} < \varepsilon, \qquad \forall n \ge n_0.$$

□

Step controls. Let the current $\{\mathfrak{F}_t, t \in [0, T]\}$ of σ-algebras be defined on a certain probability space and the processes $w(t)$, $v(t, A)$ be adapted to this current. A control $\eta(t)$, $t \in [0, T]$ is called a *step control* if there exist $t_1, t_1, \ldots, t_n$* such that $\eta(t) = \eta_k$ for $t \in (t_k, t_{k+1}]$, where η_k is an $\mathfrak{F}_{t_k}$-measurable random variable with values in U. The class of all step controls with given $t_1, t_2, \ldots, t_n$ will be denoted by $\mathfrak{U}_\delta$ or $\mathfrak{U}(t_1, t_2, \ldots, t_n)$. Here δ denotes the subdivision of the interval $[0, T]$ by means of the given points $t_1, \ldots, t_n$. Let $\mathfrak{U}_0 = \bigcup_{(t_1, \ldots, t_n)} \mathfrak{U}(t_1, \ldots, t_n)$ be the class of all step controls.

Let $\mathfrak{U}$ and $\mathfrak{U}_1$ be two classes of admissible controls, $\mathfrak{U}_1 \subset \mathfrak{U}$. We say that the class $\mathfrak{U}_1$ is *dense* in $\mathfrak{U}$ (for a given controlled object) if

$$\inf_{\eta \in \mathfrak{U}_1} \mathbf{E}F(\xi(\cdot), \eta(\cdot)) = \inf_{\eta \in \mathfrak{U}} \mathbf{E}F(\xi(\cdot), \eta(\cdot)).$$

Define $\mathfrak{G}(\mathfrak{F}_t)$ to be the minimal σ-algebra in $[0, T] \times \Omega$ containing the sets of the form $(t, s] \times A$, $[0, T] \times \Omega$, where $A \in \mathfrak{F}_t$. Denote by $\tilde{\mathfrak{A}}$ the class of all $\mathfrak{G}(\mathfrak{F}_t)$ measurable functions taking on values in U.

Lemma 3.10. *Let* $(a, b, c) \in S(C, L_N)$, *the loss function* $F(x(\cdot), u(\cdot))$ *be bounded and continuous with respect to the metric*

$$\rho[(x_1(\cdot), u_1(\cdot)), (x_2(\cdot), u_2(\cdot))] = \sup_{0 \le t \le T} |x_1(t) - x_2(t)| + \int_0^T |u_1(t) - u_2(t)|\, dt \quad (3.69)$$

$(x(\cdot) \in D, u(\cdot) \in \tilde{\mathfrak{A}})$. *Then the class of controls* $\mathfrak{A}_0$ *for equation* (3.65) *is dense in* $\tilde{\mathfrak{A}}$.

Proof. Let $\eta(t) = u(t, \omega)$, $t \in [0, T]$, be an arbitrary control in $\mathfrak{U}$. As it is known there exists a sequence of functions $u_n(t, \omega)$ of the form $u_n(t, \omega) = \Sigma c_k \chi_{\Delta_k}(t) \chi_{\Lambda_k}(\omega)$ where $\Delta_k = (a_k, b_k]$, $\Lambda_k \in \mathfrak{F}_{a_k}$, $c_k \in U$ such that $u_n(t, \omega) \to u(t, \omega)$ almost everywhere in measure $l \times \mathbf{P}$ (l is the Lebesgue measure on $[0, T]$).

By means of controls $u(t, \omega)$ and $u_n(t, \omega)$ we construct solutions of equation (3.65). Denote them $\xi(t)$ and $\xi_n(t)$ respectively. In view of theorem 3.16 $\|\xi(\cdot) - \xi_n(\cdot)\|_T \to 0$ in probability. Therefore $\rho[(\xi(\cdot), u(\cdot)), (\xi_n(\cdot), u_n(\cdot))] \to 0$ in probability and

$$\mathbf{E}F(\xi_n(\cdot), u_n(\cdot)) \to \mathbf{E}F(\xi(\cdot), u(\cdot)).$$

This implies that

$$\inf_{\eta(\cdot) \in \mathfrak{U}_0} \mathbf{E}F(\xi(\cdot), \eta(\cdot)) = \inf_{\eta(\cdot) \in \tilde{\mathfrak{U}}} \mathbf{E}F(\xi(\cdot), \eta(\cdot)). \qquad \square$$

Remark. Let $\{u_1, u_2, \ldots, u_N\}$ be an arbitrary sequence of points in $\mathfrak{U}$. Denote by $\mathfrak{U}(t_1, \ldots, t_n, u_1, \ldots, u_N)$ the subset of $\mathfrak{U}(t_1, \ldots, t_n)$ consisting of step controls taking on values belonging solely to $\{u_1, u_2, \ldots, u_N\}$.

* With $0 = t_0 < t_1 < \cdots < t_n < t_{n+1} = T$.

Since $\mathfrak{U}$ is a compact set, for any given $\varepsilon > 0$ an ε-net can be constructed in $\mathfrak{U}$. Let this be the sequence $\{u_1, \ldots, u_N\}$. Clearly for any function $u(t) = \mathfrak{U}_0(t_1, \ldots, t_n)$ a sequence $u_k(t)$, $k = 1, 2, \ldots, u_k(\cdot) \in \mathfrak{U}(t_1, \ldots, t_n, u_1, \ldots, u_N)$ can be found which converges to $u(\cdot)$ uniformly in (t, ω). In view of lemma 3.10 (provided the conditions are satisfied) for any $\varepsilon > 0$ and an arbitrary countable everywhere dense sequence of points $(u_1, u_2, \ldots, u_N, \ldots)$ in U there exists an ε-optimal control belonging to some $\mathfrak{U}_0(t_1, \ldots, t_n, u_1, \ldots, u_N)$.

Usually, controls discussed in theorem 3.15 are generalized ones. We shall now present a general method for constructing ε-optimal feedback controls.

We introduce the following approximations to solutions for stochastic differential equations.

Set for an arbitrary control $\eta(\cdot) \in \mathfrak{U}(t_1, t_2, \ldots, t_{n-1})$

$$\xi_{k+1} = \xi_k + \int_{t_k}^{t_{k+1}} a(s, \check{\xi}(\cdot), \eta_k)\, ds + \int_{t_k}^{t_{k+1}} \lambda(\check{\xi}(\cdot), \eta_k, ds),$$

$$k = \overline{0, n-1},\ \xi(0) = \xi_0, \tag{3.70}$$

where $\check{\xi}(t) = \xi_k$ for $t \in [t_k, t_{k+1})$. The process $\check{\xi}(t)$ can be viewed as a piecewise constant approximation to the solution of equation (3.67) corresponding to control $\eta(\cdot)$. However for processes with discontinuities such an approximation is not fully satisfactory, it is therefore replaced by the processes

$$\bar{\xi}(t) = \xi_0 + \int_0^t a(s, \check{\xi}(\cdot), \eta(s))\, ds + \int_0^t \lambda(\check{\xi}(\cdot), \eta(s), ds), \tag{3.71}$$

or

$$\bar{\xi}(t) = \xi_k + \int_{t_k}^t a(s, \check{\xi}(\cdot), \eta(s))\, ds + \int_{t_k}^t \lambda(\check{\xi}(\cdot), \eta(s), ds) \quad \text{as } t > t_k.$$

The process $\bar{\xi}(t)$ is called *a finite-difference approximation* of the solution of equation (3.67) corresponding to the control $\eta(\cdot)$.

The sequence $\{\xi_k, \eta_k, k = 0, \ldots, n\}$ can be viewed as a discrete controlled object with controlling sequence $\{\eta_0, \eta_1, \ldots, \eta_N)$. To determine the conditional probabilities which define this object observe that functions $a(s, \check{\xi}(\cdot), u)$ for $s \in [t_k, t_{k+1}]$ depend only on $s, \xi_0, \xi_1, \ldots, \xi_k$ and $a(s, \check{\xi}(\cdot), u) = a_k(s, \xi_0, \xi_1, \ldots, \xi_k, u)$. Analogously $b(s, \check{\xi}(\cdot), u) = b_k(s, \xi_0, \ldots, \xi_k, u)$, $c(s, \check{\xi}(\cdot), u, y) = c_k(s, \xi_0, \ldots, \xi_k, u, y)$. Set

$$\alpha_k = \int_{t_k}^{t_{k+1}} a_k(s, \xi_0, \ldots, \xi_k, \eta_k)\, ds,$$

$$\beta_k = \int_{t_k}^{t_{k+1}} b_k(s, \xi_0, \ldots, \xi_k, \eta_k)\, dw(s),$$

$$\gamma_k = \int_{t_k}^{t_{k+1}} \int_{R^d} c_k(s, \xi_0, \ldots, \xi_k, \eta_k, y)\tilde{v}(ds\, dy).$$

Then $\xi_{k+1} = \xi_k + \alpha_k + \beta_k + \gamma_k$. Moreover the conditional distribution of the variables β_k and γ_k given $\xi_0, \xi_1, \ldots, \xi_k, \eta_0, \ldots, \eta_k$ coincides with the corresponding conditional distributions given $\xi_0, \ldots, \xi_k, \eta_k$ and is Gaussian for variable β_k and is infinitely divisible (without a Gaussian component) with mean 0 and finite variance for γ_k. Moreover, under the stated conditions α_k is a conditional constant while β_k and γ_k are conditionally independent. Thus

$$\mathbf{P}\{\xi_{k+1} \in A \mid \xi_0 = x_0, \ldots, \xi_k = x_k, \eta_0 = u_0, \ldots, \eta_k = u_k\}$$
$$= \mathbf{P}\{\alpha + \beta + \gamma + x_k \in A\},$$

where α is a constant, β and γ are independent, β is Gaussian and γ is infinitely divisible without a Gaussian component,

$$\alpha = \int_{t_k}^{t_{k+1}} a_k(s, x_0, \ldots, x_k, u_k)\, ds,$$

$$\beta = \int_{t_k}^{t_{k+1}} b_k(s, x_0, \ldots, x_k, u_k)\, dw(s),$$

$$\gamma = \int_{t_k}^{t_{k+1}} \int_{R^d} c_k(s, x_0, \ldots, x_k, u_k, y)\tilde{\nu}(ds, dy).$$

We shall derive a bound on the precision of the finite-difference approximation to a solution of equation (3.65), uniform in $\eta(\cdot)$. For this purpose some additional assumptions on the coefficients of the equation will be needed. Let $K(B)$ be a fixed measure on Borel subsets of the interval $[0, T]$, $K[0, T] = K < \infty$.

Introduce in D the class of semi-norms $|||x(\cdot)|||_t$ by setting

$$|||x(\cdot)|||_t^2 = \int_0^t |x(s)|^2 K(ds).$$

Observe that

$$|||x(\cdot)|||_t \le K^{1/2} \|x(\cdot)\|_t.$$

We say that a measurable function $\alpha(t, x(\cdot))$, $(t, x(\cdot)) \in [0, T] \times D$ with values in R^d satisfies a *uniform Lipschitz condition* with respect to a semi-norm if

$$|\alpha(t, x(\cdot)) - \alpha(t, y(\cdot))| \le L |||x(\cdot) - y(\cdot)|||_t, \qquad \forall t \in [0, T],$$

for some L. The class of functions $\alpha(t, x(\cdot))$ satisfying this condition with a given constant L is denoted by $\tilde{S}(\cdot, L)$ and let

$$\tilde{S}(C, L) = S(C, \cdot) \cap \tilde{S}(\cdot, L).$$

Set

$$\mathcal{T}_t(\xi(\cdot)) = \xi(0) + \int_0^t \alpha(t, \xi(t))\, dt + \int_0^t \lambda(\xi(\cdot), dt),$$

where $\xi(\cdot)$ is a random process adapted to the current of σ-algebras $\{\mathfrak{F}_t, t \in [0, T]\}$,

$$\lambda(x(\cdot), dt) = \beta(t, \xi(\cdot))\, dw + \int_{R^d} \gamma(t, x(\cdot), y)\tilde{v}(dt, dy),$$

$$\alpha, \beta, \gamma \in \tilde{S}(C, L) \quad \text{and} \quad \int_0^T \mathbf{E}|\xi(t)|^2 < \infty.$$

Analogously to lemma 3.3 one can verify the validity of the inequality

$$\mathbf{E}\|\mathscr{T}_t(\xi) - \mathscr{T}_t(\eta)\|_T \leq L_1 \int_0^T \mathbf{E}|||\xi_1(\cdot) - \xi_2(\cdot)|||_t^2\, dt, \tag{3.72}$$

where $L_1 = 2L^2(T + 4)$.

We now return to finite-difference approximations for the solution of a stochastic differential equation. Let $a, b, c \in \tilde{S}(C, L)$. Consider an arbitrary subdivision $\delta = (t_0, t_1, \ldots, t_n)$ of the interval $[0, T]$ $(t_0 = 0, t_n = T)$ and choose a control $\eta(\cdot) \in \mathfrak{U}_\delta$. Based on a given δ and $\eta(\cdot)$ we construct a controlled sequence $\{\xi_k, \eta_k, k = 0, 1, \ldots, n\}$ and random processes $\breve{\xi}_\delta(t)$, $\bar{\xi}_\delta(t)$ in the way mentioned above and defining $\breve{\xi}_\delta(t)$ and $\bar{\xi}_\delta(t)$ according to (3.71). We now bound the quantity $\|\xi_\delta(\cdot)\|_T$. Utilizing lemma 3.1 we obtain

$$\mathbf{E}\|\bar{\xi}_\delta(\cdot) - \xi_0\|_T^2 \leq C_1 \int_0^t (1 + \mathbf{E}\|\breve{\xi}_\delta(\cdot)\|_s^2)\, ds.$$

Since $\|\breve{\xi}_\delta(\cdot)\|_t \leq \|\bar{\xi}_\delta(\cdot)\|_t$, it follows that

$$\mathbf{E}\|\bar{\xi}_\delta(\cdot)\|^2 < 2(\mathbf{E}|\xi_0|^2 + C_1 T) + 2C_1 \int_0^t \mathbf{E}\|\bar{\xi}_\delta(\cdot)\|_s^2\, ds.$$

In view of lemma 3.5 we have

$$\mathbf{E}\|\bar{\xi}(\cdot)\|_T^2 \leq C_2(1 + \mathbf{E}|\xi_0|^2), \tag{3.73}$$

where C_2 is a constant which depends on C and T only. Set

$$z(t) = \mathbf{E}\|\xi(\cdot) - \bar{\xi}_\delta(\cdot)\|_t^2,$$

where $\xi(t)$ is a solution of equation (3.67) which corresponds to the control $\eta(\cdot)$. Inequality (3.72) yields that

$$z(t) \leq L_1 \int_0^t \mathbf{E}|||\xi(\cdot) - \breve{\xi}_\delta(\cdot)|||_s^2\, ds.$$

Let $t \in [t_k, t_{k+1})$. Using lemma 3.1 and bound (3.73) we obtain

$$\mathbf{E}|\bar{\xi}_\delta(t) - \breve{\xi}_\delta(t)|^2 \leq C_1 \int_{t_k}^{t_{k+1}} (1 + \mathbf{E}\|\breve{\xi}_\delta(\cdot)\|_s^2)\, ds$$

$$\leq C_1 \int_{t_k}^t (1 + \mathbf{E}\|\bar{\xi}_\delta(\cdot)\|_s^2)\, ds \leq C_3(1 + \mathbf{E}|\xi_0|^2)\, \Delta t_k,$$

where $\Delta t_k = t_{k+1} - t_k$. Therefore

$$\int_0^t \mathbf{E}|||\xi(\cdot) - \check{\xi}_\delta(\cdot)|||_s^2\, ds \le 2\int_0^t \mathbf{E}|||\xi(\cdot) - \bar{\xi}_\delta(\cdot)|||_s^2\, ds$$

$$+ 2\int_0^t ds \int_0^s \mathbf{E}|\bar{\xi}_\delta(s') - \check{\xi}_\delta(s')|^2 K(ds')$$

$$\le 2K\int_0^t \mathbf{E}\|\xi(\cdot) - \bar{\xi}_\delta(\cdot)\|_s^2\, ds + 2KTC_3|\delta|(1 + \mathbf{E}|\xi_0|^2),$$

where $|\delta| = \max_k \Delta t_k$. Thus

$$z(t) \le 2KL_1 \int_0^t z(s)\, ds + C_4|\delta|(1 + \mathbf{E}|\xi_0|^2), \qquad C_4 = 2KTC_3,$$

whence in view of lemma 3.5 we finally arrive at $z(t) \le C_4(1 + \mathbf{E}|\xi_0|^2)e^{2KL_1T}|\delta|$. We have thus obtained the following result.

Theorem 3.17. *If $(a, b, c) \in \tilde{S}(C, L)$ then uniformly in all $\eta \in \mathfrak{U}_\delta$*

$$\mathbf{E}\|\xi(\cdot) - \bar{\xi}_\delta(\cdot)\|_T^2 \le C_4 L_2(1 + \mathbf{E}|\xi_0|^2)|\delta|, \tag{3.74}$$

where C_4 depends only on C, K, T and L_2 depends on L, K and T only.

In the case when the process $\xi(t)$ is discontinuous, the difference $\check{\xi}_\delta(t) - \bar{\xi}_\delta(t)$ as $|\delta| \to 0$ can be large with a positive probability.

However, if in the stochastic differential equation the discontinuous term is absent, a bound analogous to (3.74) is valid also for approximating the process $\xi(t)$ by means of a process $\check{\xi}_\delta(t)$. In this case the Lipschitz condition on the seminorm can be replaced by a Lipschitz condition on the norm $\|\cdot\|_t$.

Indeed, assume that $c \equiv 0$ and $(a, b) \in S(C, L)$. Let $z(t)$ be as defined above. In view of lemma 3.3

$$z(t) \le L_1 \int_0^t \mathbf{E}\|\xi(\cdot) - \check{\xi}_\delta(\cdot)\|_s^2\, ds \le L_1 \int_0^t 2z(s)\, ds + A,$$

where

$$A = 2L_1 \int_0^t \mathbf{E}\|\check{\xi}_\delta(\cdot) - \bar{\xi}_\delta(\cdot)\|_s^2\, ds.$$

We have

$$\mathbf{E}\|\bar{\xi}_\delta(\cdot) - \check{\xi}(\cdot)\|_T^2 \le \mathbf{E}\max_k \max_{t_k \le t_1 \le t_{k+1}} \left|\int_{t_k}^t a(s, \check{\xi}_\delta(\cdot), \eta(s))\, ds\right.$$

$$\left. + \int_{t_k}^t \lambda(\check{\xi}_\delta(\cdot), \eta(s), ds)\right|^2 \le 2(A_1 + A_2).$$

Here

$$A_1 = \mathbf{E} \max_k \max_{t \in [t_k, t_{k+1}]} \Delta t_k \left| \int_{t_k}^t C^2(1 + \|\check{\xi}_\delta(\cdot)\|_s^2\, ds) \right|,$$

$$A_2 = \mathbf{E} \max_k \max_{t \in [t_k, t_{k+1}]} \left| \int_{t_k}^t b(s, \check{\xi}_\delta(\cdot), \eta(s))\, dw(s) \right|^2.$$

Utilizing inequality (3.73) we obtain

$$A_1 \le \max_k \Delta t_k\, \mathbf{E} \int_{t_k}^{t_{k+1}} C_3(1 + \|\check{\xi}_\delta(\cdot)\|_T^2)\, ds$$

$$\le C_3 \max(\Delta t_k)^2 (1 + \mathbf{E}|\xi_0|^2) \to 0 \quad \text{as} \quad |\delta| \to 0.$$

Furthermore

$$A_2 = \mathbf{E}\left(\max_k \max_{t \in [t_k, t_{k+1}]} \left| \int_{t_k}^t b(s, \check{\xi}_\delta(\cdot), \eta(s))\, dw(s) \right|^4 \right)^{1/2}$$

$$\le \left[\mathbf{E} \max_k \max_{t \in [t_k, t_{k+1}]} \left| \int_{t_k}^t b(s, \check{\xi}_\delta(\cdot), \eta(s))\, dw(s) \right|^4 \right]^{1/2}$$

$$\le \left[\mathbf{E} \sum_{k=1}^{n-1} \max_{t \in [t_k, t_{k+1}]} \left| \int_{t_k}^t b(s, \check{\xi}_\delta(\cdot), \eta(s))\, dw(s) \right|^4 \right]^{1/2}.$$

Since the stochastic integral is a martingale, using inequalities (3.4) and (3.10) we have

$$\mathbf{E} \max_{t \in [t_k, t_{k+1}]} \left| \int_{t_k}^t b(s, \check{\xi}_\delta(\cdot), \eta(s))\, dw(s) \right|^4 \le \left(\frac{4}{3}\right)^4 \mathbf{E} \left| \int_{t_k}^{t_{k+1}} b\, dw(s) \right|^4$$

$$\le \left(\frac{4}{3}\right)^4 2 \cdot 3^2\, \Delta t_k \int_{t_k}^{t_{k+1}} \mathbf{E}|bb^*|^2\, dt \le C\, \Delta t_k \int_{t_k}^{t_{k+1}} \mathbf{E}(1 + \|\check{\xi}_\delta\|_s^4)\, ds.$$

We show that the quantity $\mathbf{E}|\check{\xi}_\delta(t)|^4$ is uniformly bounded. Then $A_2 \le C' \max_k \Delta t_k$ and $A_2 \to 0$. Since

$$\xi_{k+1} = \xi(0) + \int_0^{t_{k+1}} a(s, \check{\xi}_\delta(\cdot), \eta(s))\, ds + \int_0^{t_{k+1}} b(s, \check{\xi}_\delta(\cdot), \eta(s))\, dw(s),$$

we have

$$\mathbf{E}\|\check{\xi}_\delta(t)\|_{t_{k+1}}^4 \le 3^3 \mathbf{E}\left\{ |\xi(0)|^4 + \left\| \int_0^t a(s, \check{\xi}_\delta(\cdot), \eta(s))\, ds \right\|_{t_{k+1}}^4 \right.$$

$$\left. + \left\| \int_0^t b(s, \check{\xi}_\delta(\cdot), \eta(s))\, dw(s) \right\|_{t_{k+1}}^4 \right\} \le 3^3 \left\{ \mathbf{E}|\xi(0)|^4 \right.$$

$$\left. + t_{k+1}^3 \int_0^{t_{k+1}} \mathbf{E}|a(s, \check{\xi}_\delta(\cdot), \eta(s))|^4\, ds + \mathbf{E} \left\| \int_0^t b\, dw \right\|_{t_{k+1}}^4 \right\}.$$

Observe that in view of the bounds on the stochastic integrals (3.10) we obtain

$$\mathbf{E}\left\|\int_0^t b\, dw\right\|_{t_{k+1}}^4 \leq \left(\frac{4}{3}\right)^4 \cdot 2 \cdot 3^2 \cdot t_{k+1} \int_0^{t_{k+1}} \mathbf{E}(bb^*)^2\, dt.$$

Thus

$$\mathbf{E}\|\check{\xi}_\delta(\cdot)\|_{t_{k+1}}^4 \leq 27\mathbf{E}|\xi(0)|^4 + C'T + C' \int_0^{t_{k+1}} \mathbf{E}\|\check{\xi}_\delta(\cdot)\|_s^4\, ds,$$

where C' is a constant which depends on C and T only. Note that $\|\check{\xi}_\delta(\cdot)\|_t^4 = \|\check{\xi}_\delta(\cdot)\|_{t_k}^4$ for $t \in [t_k, t_{k+1})$. Therefore we have for all $t \in [0, T]$

$$\mathbf{E}\|\check{\xi}_\delta(\cdot)\|_t^4 \leq 27\mathbf{E}|\xi(0)|^4 + C'T + C' \int_0^t \mathbf{E}\|\check{\xi}_\delta(\cdot)\|_s^4\, ds.$$

Whence utilizing inequality (3.22) we obtain the following bound on $\check{\xi}_\delta(t)$ uniformly in δ:

$$\mathbf{E}\|\check{\xi}_\delta(\cdot)\|^4 \leq C''(\mathbf{E}|\xi(0)|^4 + C'T). \tag{3.75}$$

Thus it is verified that $\mathbf{E}\|\bar{\xi}_\delta(\cdot) - \check{\xi}_\delta(\cdot)\|_T^2 \to 0$ as $\delta \to 0$ uniformly in $\eta(\cdot)$. Moreover

$$\mathbf{E}\|\bar{\xi}_\delta(\cdot) - \check{\xi}_\delta(\cdot)\|_T^2 \leq C'''|\delta|[\mathbf{E}|\xi(0)|^2 + (\mathbf{E}|\xi(0)|^4)^{1/2}].$$

Hence Theorem 3.17 admits in the continuous case the following refinement.

Lemma 3.11. *If* $(a, b) \in S(C, L)$ $(c \equiv 0)$ *and* $\mathbf{E}|\xi(0)|^4 < \infty$ *then*

$$\mathbf{E}\|\xi(\cdot)) - \check{\xi}_\delta(\cdot)\|_T^2 = O(\mathbf{E}|\xi(0)|^4)^{1/2}|\delta|. \tag{3.76}$$

We shall now discuss some corollaries of the bounds obtained. First we shall consider the continuous case.

Lemma 3.12. *Let the function* $F(x(\cdot), u(\cdot))$, $(x(\cdot), u(\cdot)) \in C \times \mathfrak{U}_0$ *satisfy the condition*

$$|F(x(\cdot), u(\cdot)) - F(y(\cdot), u(\cdot))| \leq \varphi(\|x(\cdot) - y(\cdot)\|_T), \tag{3.77}$$

where $\varphi(t)$ *is a bounded function, positive for* $t > 0$, *and* $\varphi(t) \to 0$ *as* $t \downarrow 0$. *Assume that* $(a, b) \in S(C, L)$ and $c \equiv 0$. *Then*

$$\lim_{|\delta| \to 0} [\mathbf{E}F(\check{\xi}_\delta(\cdot), \eta_\delta(\cdot)) - \mathbf{E}F(\xi_\delta(\cdot), \eta_\delta(\cdot))] = 0 \tag{3.78}$$

uniformly in all $\eta_\delta(\cdot)$ *where* $\eta_\delta(\cdot)$ *is an arbitrary step control in* $\mathfrak{U}_\delta$, $\xi_\delta(\cdot)$ *is a solution of* (3.67) *corresponding to the control* $\eta_\delta(\cdot)$.

The proof of this assertion follows from the fact that in view of (3.76) and (3.77), $F(\check{\xi}_\delta(\cdot), \eta_\delta(\cdot)) - F(\xi_\delta(\cdot), \eta_\delta(\cdot)) \to 0$ in probability as $|\delta| \to 0$.

Introduce the metric $\rho_D(x(\cdot), y(\cdot))$ on the space D:

$$\rho_D(x(\cdot), y(\cdot)) = \inf_{\lambda \in \Lambda} \{\sup |x(t) - y(\lambda(t))| + \sup |t - \lambda(t)|\},$$

where Λ is the set of all continuous one-to-one mappings $\lambda(t)$ of the interval $[0, T]$ onto itself such that $\lambda(0) = 0$ and $\lambda(T) = T$.

Lemma 3.13. *Relation* (3.78) *is satisfied if* $(a, b, c) \in \tilde{S}(C, L)$ *and the control cost* $F(x(\cdot), u(\cdot))$ *satisfies the condition*

$$|F(x(\cdot), u(\cdot)) - F(y(\cdot), u(\cdot))| \leq \varphi[\rho_D(x(\cdot), y(\cdot))] \tag{3.79}$$

for all step controls, where $\varphi(t)$ *is a bounded positive function,* $t \geq 0$ *and* $\varphi(t) \to 0$ *as* $t \downarrow 0$.

PROOF. Since

$$|\mathbf{E}F(\check{\xi}_\delta(\cdot), \eta_\delta(\cdot)) - \mathbf{E}F(\xi_\delta(\cdot), \eta_\delta(\cdot))| \leq \mathbf{E}F(\check{\xi}_\delta(\cdot), \eta_\delta(\cdot)) - \mathbf{E}F(\bar{\xi}_\delta(\cdot), \eta_\delta(\cdot))| + |\mathbf{E}F(\bar{\xi}_\delta(\cdot), \eta_\delta(\cdot)) - \mathbf{E}F(\xi_\delta(\cdot), \eta_\delta(\cdot))|,$$

in view of lemma 3.10 it remains to show that

$$|\mathbf{E}F(\check{\xi}_\delta(\cdot), \eta_\delta(\cdot)) - \mathbf{E}F(\bar{\xi}_\delta(\cdot), \eta_\delta(\cdot))| \to 0$$

as $|\delta| \to 0$. We utilize the following bound on the distance in D between the function $x(t)$ and its finite-difference approximation $\check{x}(t)$ ($\check{x}(t) = x(t_k)$ for $t \in [t_k, t_{k+1})$, $k = 0, 1, \ldots, n-1$):

$$\rho_D(x(\cdot), \check{x}(\cdot)) \leq |\delta| + 4\,\Delta_{2|\delta|}(x(\cdot)), \tag{3.80}$$

where $\Delta_c(x(\cdot))$ is defined in Theorem 3.7. Inequality (3.80) is contained in lemma 2 of Section 5 in Chapter 6, vol I of [20]. Thus

$$|\mathbf{E}F(\check{\xi}_\delta(\cdot), \eta_\delta(\cdot)) - \mathbf{E}F(\bar{\xi}_\delta(\cdot), \eta_\delta(\cdot))| \leq \mathbf{E}\varphi(|\delta| + 4\,\Delta_{2|\delta|}(\bar{\xi}_\delta(\cdot))).$$

The definition of the quantity $\Delta_c(x(\cdot))$ yields that

$$|\Delta_c(\xi_\delta(\cdot)) - \Delta_c(\bar{\xi}_\delta(\cdot))| \leq 2 \sup_{0 \leq t \leq T} |\xi_\delta(t) - \bar{\xi}_\delta(t)|.$$

Let $\varphi(t) \leq K$ and $\varphi(t) < \varepsilon/2$ for $t < \varepsilon' = \varepsilon'(\varepsilon)$. Then for $|\delta| < \varepsilon'/2$ we have

$$\mathbf{E}\varphi[|\delta| + 4\,\Delta_{2|\delta|}(\bar{\xi}_\delta(\cdot))] \leq K\mathbf{P}(|\delta| + 4\,\Delta_{2|\delta|}(\bar{\xi}_\delta(\cdot)) > \varepsilon) + \frac{\varepsilon}{2}$$

$$\leq K\mathbf{P}(\bar{B}_N) + K\mathbf{P}\left(B_N \cap \left\{\Delta_{2|\delta|}(\bar{\xi}_\delta(\cdot)) > \frac{\varepsilon'}{8}\right\}\right) + \frac{\varepsilon}{2},$$

where B_N denotes the event $\{\sup_{0 \leq t \leq T} |\xi_\delta(t)| \leq N\}$ and $\bar{B}_N = \Omega \backslash B_N$. As it follows from (3.79)

$$\mathbf{P}(\bar{B}_N) \leq \frac{C'(1 + |x|^2)}{N^2},$$

where C' does not depend on δ. Furthermore, in view of lemma 3.10

$$\mathbf{P}\left(B_N \cap \left\{\Delta_{2|\delta|}(\bar{\xi}_\delta(\cdot)) > \frac{\varepsilon'}{8}\right\}\right) \leq \mathbf{P}\left(B_N \cap \left\{\Delta_{2|\delta|}(\xi_\delta(\cdot)) > \frac{\varepsilon}{16}\right\}\right)$$

$$+ \mathbf{P}\left\{\sup_{0 \leq t \leq T} |\xi_\delta(t) - \bar{\xi}_\delta(t)| > \frac{\varepsilon'}{32}\right\}$$

$$\leq K'(N, \varepsilon)|\delta| + \mathbf{P}\left(B_N \cap \left\{\Delta_{2|\delta|}(\xi_\delta(\cdot)) > \frac{\varepsilon}{16}\right\}\right).$$

To bound the last summand we shall utilize the following assertion.

Lemma 3.14. *If $\xi(t)$ is a separable stochastically continuous process, $t \in [0, T]$, and for all $\varepsilon > 0$, $t_i \in [0, T]$, $i = 1, 2, 3$, $t_1 < t_2 < t_3$ and some $r > 0$*

$$\mathbf{P}\{B_N \cap (|\xi(t_2) - \xi(t_1)| \wedge |\xi(t_3) - \xi(t_2)| > \varepsilon)\} < \frac{K_N |t_3 - t_1|^{1+r}}{\varepsilon^p},$$

then

$$\mathbf{P}\{B_N \cap \Delta_c(\xi(\cdot)) > \varepsilon\} \leq \frac{K'K_N}{\varepsilon^p} c^r,$$

where K' depends on r and p only.

Lemma 3.14 is a corollary of theorem 1 in Section 4 of Chapter 3 Vol I of [20].

We shall apply this lemma to the process $\xi_\delta(t)$. We have

$$\mathbf{P}\{B_N \cap (|\xi_\delta(t_2) - \xi_\delta(t_1)| \wedge |\xi_\delta(t_3) - \xi_\delta(t_2)| > \varepsilon)\}$$

$$\leq \frac{1}{\varepsilon^4} \mathbf{E}\chi(B_N)|\xi_\delta(t_2) - \xi_\delta(t_1)|^2 |\xi_\delta(t_3) - \xi_\delta(t_2)|^2.$$

Let B_N be the event $\{\sup_{0 \leq t \leq t_2} |\xi(t)| > N\}$. Then utilizing lemma 3.6 we obtain

$$\mathbf{E}\chi(B_N)|\xi_\delta(t_2) - \xi_\delta(t_1)|^2 |\xi_\delta(t_3) - \xi_\delta(t_2)|^2$$

$$\leq \mathbf{E}\{\chi(B_N)|\xi_\delta(t_2) - \xi_\delta(t_1)|^2 \mathbf{E}\{|\xi_\delta(t_3) - \xi_\delta(t_2)|^2 | \mathfrak{F}_{t_2}\}\}$$

$$\leq \mathbf{E}\chi(B_N) L(1 + |\xi_\delta(t_2)|^2)|\xi_\delta(t_3) - \xi_\delta(t_2)|^2 (t_2 - t_1).$$

In turn, the expression on the r.h.s. of the inequality does not exceed

$$L(1 + N^2)\mathbf{E}|\xi_\delta(t_3) - \xi_\delta(t_2)|^2 (t_2 - t_1)$$

$$\leq L(1 + N^2)(1 + \mathbf{E}|\xi_\delta(0)|^2)(t_3 - t_1)^2.$$

Thus lemma 3.14 is applicable with $r = 1$ and $p = 4$ and the probability $\mathbf{P}(B_N \cap \{\Delta_{2|\delta|}(\xi_\delta(\cdot)) > \varepsilon/16\})$ for a given ε and N becomes arbitrarily small for $|\delta|$ sufficiently small. Consequently, choosing first an arbitrary $\varepsilon > 0$

one can find an $N = N(\varepsilon)$ large enough and then a $\delta_0 > 0$, $\delta_0 = \delta_0(N, \varepsilon)$, such that for $|\delta| < \delta_0$

$$\mathbf{E}\varphi[|\delta| + 4\,\Delta_{2|\delta|}(\bar{\xi}_\delta(\cdot))] < \varepsilon.$$

Relation (3.78) and lemma 3.13 are thus verified. □

For a given subdivision δ, consider a controlled sequence $\{\xi_n, \eta_n, k = 0, 1, \ldots, n\}$ constructed in a manner indicated above. We also introduce the loss function

$$F_\delta(x_0, \ldots, x_n, u_0, \ldots, u_n) = F(\check{x}_\delta(\cdot), \check{u}_\delta(\cdot)),$$

where $\check{x}_\delta(t) = x_k$, $\check{u}_\delta(t) = u_k$ for $t \in [t_k, t_{k+1})$, $k = 0, \ldots, n$. We now determine a control $(\eta_0^*, \eta_1^*, \ldots, \eta_n^*)$ by means of sequence $\{\xi_k, \eta_k, k = 0, \ldots, n\}$ which minimizes the cost function, i.e. a control $\{\eta_k^*\}$ such that

$$\begin{aligned}&\mathbf{E}F_\delta(\xi_0, \xi_1^*, \ldots, \xi_n^*, \eta_0^*, \ldots, \eta_n^*)\\ &\quad = \inf_{\bar{\eta} = (\eta_0, \ldots, \eta_n)} \mathbf{E}F_\delta(\xi_0, \ldots, \xi_1^\eta, \ldots, \xi_n^\eta, \ldots, \eta_0, \ldots, \eta_n).\end{aligned}$$

Here $\{\xi_k^*\}$ denotes a sequence of variables $\{\xi_k\}$ constructed by means of control $\{\eta_k^*\}$, and $\{\xi_k^\eta\}$ by means of control $\{\eta_k\}$. It is known from the above that such an optimal sequence exists and moreover it can be assumed that

$$\eta_k^* = g_\delta(\xi_0, \xi_1^*, \ldots, \xi_k^*, \eta_0^*, \eta_1^*, \ldots, \eta_{k-1}^*),$$

where $g_\delta(x_0, \ldots, x_k, u_0, \ldots, u_{k-1})$ are non-random functions. These functions can be obtained in the manner described in Section 3 of Chapter 1.

Now set $\eta_\delta^*(t) = \eta_k^*$ for $t \in [t_k, t_{k+1})$. We show that a piecewise-constant control $\eta_\delta^*(\cdot)$ belonging to the class of feedback controls will be—for $|\delta|$ sufficiently small—an ε-optimal control in the class of all generalized controls for a solution of equation (3.65) under the loss function $F(x(\cdot), u(\cdot))$.

Theorem 3.18. *If* $(a, b, c) \in \tilde{S}(C, L)$ $(S(C, L))$ *and the loss function is continuous in metric* (3.69) *and satisfies condition* (3.79) (*correspondingly* (3.77)) *then*

$$Z = \lim_{|\delta| \to 0} \mathbf{E}F_\delta(\xi_0, \xi_1^*, \ldots, \xi_n^*, \eta_0^*, \ldots, \eta_n^*). \tag{3.81}$$

Proof. Since $F_\delta(\xi_0, \xi_1^*, \ldots, \xi_n^*, \eta_0^*, \ldots, \eta_n^*) = F(\check{\xi}_\delta^*(\cdot), \eta_\delta^*(\cdot))$, where $\eta_\delta^*(t) = \eta_k^*$ for $t \in [t_k, t_{k+1})$, $k = 0, 1, \ldots, n$, and $\check{\xi}_\delta^*(t)$ is the step approximation to a solution of equation (3.65) constructed by means of control $\eta_\delta^*(t)$ introduced above, it follows from lemma 3.13 (lemma 3.12) that

$$\begin{aligned}Z &= \inf_{\eta(\cdot) \in \mathfrak{U}} \mathbf{E}F(\xi^{(\eta)}(\cdot), \eta(\cdot))\\ &\le \lim_{|\delta| \to 0} \mathbf{E}F_\delta(\xi_0, \xi_1^*, \ldots, \xi_n^*, \eta_0^*, \ldots, \eta_n^*).\end{aligned} \tag{3.82}$$

However, in view of theorem 3.17 it follows that for any $\varepsilon > 0$ one can find a step control $\eta_\delta(t)$ such that $\mathbf{E}F(\xi_\delta(\cdot), \eta_\delta(\cdot)) < Z + \varepsilon$, where $\xi_\delta(t)$ is a solution of equation (3.65) which corresponds to the control $\eta_\delta(t)$.

Consider all the possible subdivisions $\delta' = (0 = t'_0 < t'_1 \cdots < t'_m = T)$ of the interval $[0, T]$ which are refinements of the subdivision δ; let $\eta_{\delta'}(t) = \eta_\delta(t)$. Then $\xi_{\delta'}(t) = \xi_\delta(t)$, $\forall t \in [0, T]$. In view of lemma 3.13 (lemma 3.12)

$$\lim_{|\delta'| \to 0} \mathbf{E}F(\check{\xi}_{\delta'}, \eta_{\delta'}(\cdot)) = \mathbf{E}F(\xi_\delta(\cdot), \eta_\delta(\cdot)).$$

Therefore for $|\delta'|$ sufficiently small $\mathbf{E}F(\check{\xi}_{\delta'}(\cdot), \eta_{\delta'}(\cdot)) < Z + \varepsilon$; in that case, however

$$\mathbf{E}F_{\delta'}(\xi_0, \xi_1^*, \ldots, \xi_m^*, \eta_0^*, \ldots, \eta_m^*) < Z + \varepsilon,$$

so that

$$\overline{\lim_{|\delta'| \to 0}}\, \mathbf{E}F_{\delta'}(\xi_0, \xi_1^*, \ldots, \xi_m^*, \eta_0^*, \ldots, \eta_m^*) \leq Z.$$

This fact, together with inequality (3.82) imply the assertion of the theorem. □

Corollary. *Under the conditions of the theorem, for $|\delta|$ sufficiently small, the optimal control $\{\eta_k^*, k = 0, \ldots, n\}$ by means of sequence $\{\xi_k, \eta_k, k = 0, \ldots, n\}$ is an ε-optimal control for the solution of the stochastic differential equation* (3.65).

Remark. Consider the preceding construction for a fixed value of $x(0) = x$. Then we have a family of optimization problems where x serves as a parameter.

The preceding bounds on convergence are of uniform nature in x in any finite sphere $|x| \leq N$. In particular the convergence in formula (3.82) is uniform in x ($\xi_0 = x$ is non-random) for $|x| \leq N$, $\forall N > 0$.

4 Evolutional Loss Functions

Consider the problem of optimal controlling by means of a solution of the stochastic differential equation

$$d\xi = a(t, \xi(\cdot), \eta(\cdot))\, dt + \lambda(\xi(\cdot), \eta(\cdot), dt), \qquad \xi(0) = \xi_0, \tag{3.83}$$

on a fixed time interval $[0, T]$ in the case when the loss function $F(x(\cdot), u(\cdot))$ is a special form. Namely, we shall assume that $F(x(\cdot), u(\cdot)) = F_0(x(\cdot), u(\cdot))$, where $F_t(x(\cdot), u(\cdot))$, $t \in [0, T]$, is a family of functionals of the form

$$F_t(x(\cdot), u(\cdot)) = h(x(T), u(T)) \exp\left\{\int_t^T g(s, x(s), u(s))\, ds\right\} + \int_t^T \exp\left\{\int_t^s g(\theta, x(\theta), u(\theta))\, d\theta\right\} f(s, x(s), u(s))\, ds, \tag{3.84}$$

and $h(x, u), f(t, x, u), g(t, x, u)$ are given real functions. In what follows we shall assume that the following condition is satisfied: functions $h(x, u), f(t, x, u), g(t, x, u), (t, x, u) \in [0, T] \times R^d \times U$, are bounded and continuous jointly in their arguments.

Observe that in particular for $g(t, x, u) \equiv 0$

$$F_t(x(\cdot), u(\cdot)) = h(x(T), u(T)) + \int_t^T f(s, x(s), u(s))\, ds;$$

if also $h \equiv 0$, then $F_t(x(\cdot), u(\cdot))$ is an additive functional on the trajectory of the system and the selected strategy. For $g \equiv f \equiv 0$

$$F_t(x(\cdot), u(\cdot)) = h(x(T), u(T)),$$

and the loss function depends only on $\xi(T)$ and $u(T)$ and the time of the completion of the control. Thus the class of loss functions described by formula (3.84) is quite extensive. For problems of optimal control with the loss function (3.84) one can obtain a number of deeper results.

Functionals of the form (3.84) will be called *evolutional(ary) functionals*.

Observe that these functionals viewed as functions in t satisfy the differential equation

$$-\frac{dF_t}{dt} = g(t, x(t), u(t))F_t + f(t, x(t), u(t)), \qquad 0 \le t \le T,$$

and the boundary condition $F_T = h(x(T), u(T))$. In order to indicate the dependence of functions F_t on T below we shall also use the notation $F_{[t,T]}(x(\cdot), u(\cdot))$ or $F_{[t,T]}$ in addition to the common notation $F_t = F_t(x(\cdot), u(\cdot))$. The quantities $F_{[t,T]}$ are functionals of the values of functions $x(\cdot), u(\cdot)$ on the time interval $[t, T]$ and satisfy the following relation:

$$F_{[t_1,T]} = F^a_{[t_1,t_2]} + G[t_1, t_2]F_{[t_2,T]},$$

where

$$G[t_1, t_2] = G_{[t_1,t_2]}(x(\cdot), u(\cdot)) = \exp\left\{\int_{t_1}^{t_2} g(s, x(s), u(s))\, ds\right\},$$

$$F^a_{[t_1,t_2]} = \int_{t_1}^{t_2} G[t_1, s] f(s, x(s), u(s))\, ds, \qquad F^a_T = F^a_{[t,T]}.$$

We shall apply the general method of constructing ε-optimal controls to the case under consideration. Observe that the evolutional functionals introduced above satisfy the continuity condition stipulated in theorem 3.18. Therefore if all the coefficients of the equation (3.83) satisfy the conditions of this theorem then the stepwise optimal controls corresponding to a subdivision $\delta = \{0 = t_0 < t_1 < \cdots < t_n = T\}$ of the interval $[0, T]$ as $|\delta| \to 0$ will be ε-optimal controls for solutions of equation (3.83). We shall now describe their structure more precisely.

For a given δ the controlled sequence $\{\xi_k, k = 0, 1, \ldots, n\}$ introduced above is determined by recurrence relations

$$\xi_{k+1} = \xi_k + \int_{t_k}^{t_{k+1}} a(s, \xi_0, \ldots, \xi_k, u_k)\, ds + \int_{t_k}^{t_{k+1}} b(s, \xi_0, \ldots, \xi_k, u_k)\, dw$$
$$+ \int_{t_k}^{t_{k+1}} \int_{R^d} c(s, \xi_0, \ldots, \xi_k, u_k, y)\tilde{v}(ds, dy), \tag{3.85}$$

where

$$a(s, \xi_1, \ldots, \xi_k, u_k) = a(s, \check{\xi}(\cdot)) \quad \text{for } s \in [t_k, t_{k+1});$$

functions $b(s, \xi_1, \ldots, \xi_k, u)$, and $c(s, \xi_1, \ldots, \xi_k, u, y)$ are defined analogously.

The optimal control $\eta^* = \{\eta_1^*, \eta_2^*, \ldots, \eta_n^*\}$ for this chain—in accordance with the concepts of dynamic programming—can be obtained using induction on a decreasing index starting with η_n^*. Moreover in the class of all generalized controls there exists an optimal feedback control of the form $\eta_k^* = g_k(\xi_0, \xi_1, \ldots, \xi_k)$ where $g_k(x_0, \ldots, x_k)$ are Borel functions of their arguments. Formulas to determine functions $g_k(x_0, \ldots, x_k)$ and the optimal cost of control can be obtained in the following manner:

Let $\bar{\eta} = \{\eta_0, \eta_1, \ldots, \eta_n\}$ be an arbitrary generalized control for the sequence $\{\xi_k, k = 0, 1, \ldots, n\}$, $\bar{F}(\bar{\eta}) = \mathbf{E}F(\check{\xi}(\cdot), \check{\eta}(\cdot))$, $\check{\xi}(t) = \xi_k$, $\check{\eta}(t) = \eta_k$ for $t \in [t_k, t_{k+1})$,

$$\mathfrak{F}_k = \sigma\{\eta_0, \ldots, \eta_{k-1}, w(s), v(s, A) : s \in [0, t_k), A \in \mathfrak{B}_0^d\}.$$

Set $Z_n(x) = \min_u h(x, u)$. Then

$$\begin{aligned} \bar{F}(\bar{\eta}) &= \mathbf{E}\mathbf{E}\{F(\check{\xi}(\cdot), \check{\eta}(\cdot)) \mid \mathfrak{F}_n\} \\ &= \mathbf{E}[F_0^a(\check{\xi}(\cdot), \check{\eta}(\cdot)) + G[0, T]\, \mathbf{E}\{h(\xi_n, \eta_n) \mid \mathfrak{F}_n\}] \\ &\geq \mathbf{E}[F_0^a(\check{\xi}(\cdot), \check{\eta}(\cdot)) + G[0, T]\, Z_n(\xi_n)]. \end{aligned}$$

If $h(x, \eta_n) = Z_n(x)$ for all x the sign $\geq$ in the above relation should be replaced by $=$. Let $u = g_n(x)$ be a Borel function which yields for each x the value of u at which $h(x, u)$ attains its minimum. This function is the solution of equation

$$h(x, g_n(x)) = \min_u h(x, u) = Z_n(x).$$

We replace control $\bar{\eta}$ by $\bar{\eta}^{(n)}$ by setting $\bar{\eta}^{(n)} = \{\eta_0, \eta_1, \ldots, \eta_{n-1}, \eta_n^*\}$ where $\eta_n^* = g_n(\xi_n)$. Control $\bar{\eta}^{(n)}$ is at least as good as $\bar{\eta}$, i.e. $\bar{F}(\bar{\eta}) \geq \bar{F}(\bar{\eta}^{(n)})$. Furthermore, we have

$$\bar{F}(\check{\eta}^{(n)}) = \mathbf{E}(F^a_{[0, t_{n-1}]}(\check{\xi}, \check{\eta}) + G[0, t_{n-1}]\mathbf{E}\{F_{t_{n-1}}(\check{\xi}(\cdot), \check{\eta}(\cdot)) \mid \mathfrak{r}_{n-1}\}),$$
$$\mathbf{E}\{F_{t_{n-1}}(\check{\xi}(\cdot), \check{\eta}(\cdot)) \mid \mathfrak{F}_{n-1}\} = Z^*_{n-1}(\check{\xi}(\cdot), \eta_{n-1}),$$

where

$$Z^*_{n-1}(\check{x}(\cdot), u) = F^a_{t_{n-1}}(x_{n-1}, u) + G_{[t_{n-1}, T]}(x_{n-1}, u) \times \mathbf{E}Z_n[\xi_n(x(\cdot), u)],$$

$\check{x}(t) = x_k$, $\check{u}(t) = u_k$ for $t \in [t_k, t_{k+1})$, $\xi_n(x(\cdot), u)$ are determined in accordance with formula (3.85):

$$\xi_n = x_{n-1} + \int_{t_{n-1}}^{t_n} a(s, x_0, \ldots, x_{n-1}, u)\, ds + \int_{t_{n-1}}^{t_n} \lambda(\check{x}(\cdot), u, ds),$$

and the substitution of x and u in place of $x(\cdot)$ and $u(\cdot)$ in the functionals signifies that the values of these functionals on the functions $x(\cdot) = x =$ const, $u(\cdot) = u =$ const are considered. Thus

$$\bar{F}(\bar{\eta}^{(n)}) = \mathbf{E}[F^a_{[0, t_{n-1}]}(\check{\xi}(\cdot), \check{\eta}(\cdot)) + G[0, t_{n-1}]Z^*_{n-1}(\check{\xi}(\cdot), \eta_{n-1})].$$

With a minor modification we are back in the initial situation; only now the interval $[0, t_{n-1}]$ replaces $[0, T]$ and it is required to minimize the value of the functional $Z^*_{n-1}(x(\cdot), u)$ choosing an appropriate value of u.

In accordance with that proven above, if we set

$$\eta^*_{n-1} = g_{n-1}(\check{\xi}(\cdot)), \bar{\eta}^{(n-1)} = \{\eta_0, \eta_1, \ldots, \eta^*_{n-1}, \eta^*_n\},$$

where $g_{n-1}(\check{x}(\cdot)) = g_{n-1}(x_0, x_1, \ldots, x_{n-1})$ is a Borel function of its arguments satisfying the equation

$$Z^*_{n-1}(\check{x}(\cdot), g_{n-1}(\check{x}(\cdot))) = \min_u Z^*_{n-1}(\check{x}(\cdot), u),$$

then for any η_{n-1}, $\bar{F}(\bar{\eta}) \geq \bar{F}(\bar{\eta}^{(n-1)})$. The sign $\geq$ may be replaced by $=$ provided $\eta_{n-1} = \eta^*_{n-1}$, $\eta_n = \eta^*_n$. Set $Z_{n-1}(\check{x}(\cdot)) = \min Z^*_{n-1}(\check{x}(\cdot), u)$. Continuing the preceding constructions we arrive at a sequence of functionals $Z_k(\check{x}(\cdot))$, $Z^*_k(x(\cdot), u)$ which depend on values of $x(\cdot)$ on the time interval $[0, t_k]$ such that

$$\left.\begin{aligned} &Z_k(\check{x}(\cdot)) = \min_u Z^*_k(\check{x}(\cdot), u),\ k = n, n-1, \ldots, 0, \\ &Z^*_n(\check{x}(\cdot), u) = h(x, u), \\ &Z^*_k(x(\cdot), u) = F^a_{[t_k, t_{k+1}]}(x(\cdot), u(\cdot)) \\ &\qquad + G[t_k, t_{k+1}]\mathbf{E}Z_{k+1}(\xi_{k+1}(\check{x}(\cdot), u)), \end{aligned}\right\} \tag{3.86}$$

where $Z^*_k(x(\cdot), u)$ and $Z_k(x(\cdot))$ depend on the values of the function $x(\cdot)$ on the time interval $[0, t_k]$ and are such that if $g_k(x(\cdot))$ is a Borel function satisfying the equation

$$Z^*_k(\check{x}(\cdot), g_k(\check{x}(\cdot))) = Z_k(\check{x}(\cdot)), \qquad k = 0, 1, \ldots, n \tag{3.87}$$

and which therefore depends only on the values $x_0, x_1, \ldots, x_k$, then for any control $\bar{\eta} = \{\eta_1, \eta_2, \ldots, \eta_n\}$ and any $j \leq n$ the control $\bar{\eta}^{(j)} = \{\eta_1, \ldots, \eta_{j-1}, \eta^*_j, \ldots, \eta^*_n\}$ ($\eta^*_k = g_k(\check{\xi}(\cdot))$) will be at least as good as the control $\bar{\eta}$: $\bar{F}(\bar{\eta}) \geq \bar{F}(\bar{\eta}^{(j)})$. In particular the control $\bar{\eta}^{(0)} = \{\eta^*_0, \ldots, \eta^*_n\}$ is an optimal control for the chain $\{\xi_k, k = 0, 1, \ldots, n\}$ and the optimal cost of the control given that $\xi_0 = x$ is equal to $Z_0(x)$.

Thus, the problem of the structure of an optimal control for chain $\{\xi_k, k = 0, \ldots, n\}$ is constructively solved. At the same time a method for constructing ε-optimal controls for solutions of equation (3.83) was obtained. Moreover, it turns out that under the conditions of theorem 3.16 for any $\varepsilon > 0$ there exists an ε-optimal feedback control of the form

$$\eta(t) = g_k(\xi_0, \xi_1, \ldots, \xi_k) \quad \text{for } t \in [t_k, t_{k+1}).$$

Functions $Z_k(x(\cdot))$ have the following meaning: they present the value of the optimal cost of the control for the chain $\{\xi_k, \xi_{k+1}, \ldots, \xi_n\}$ such that the variables ξ_j, $j = k + 1, \ldots, n$, are determined by relations (3.85) under the condition that the values $\xi_0 = x_0, \ldots, \xi_k = x_k$ of the terms of the controlled sequence up to time k are given. Moreover the optimal control and the cost of control are characterized by the relations established above.

We now introduce the function representing the optimal cost of control for the solution of equation (3.83) on the time interval $[t, T]$

$$Z(t, x(\cdot)) = \inf_{\eta(\cdot) \in \mathfrak{U}_0[t, T]} \mathbf{E}F_t(\xi_t^{(\eta)}(\cdot), \eta(\cdot))$$

under the additional assumptions: $\xi_t^{(\eta)}(s) = x(s)$, $s \in [0, t]$. Here $\mathfrak{U}_0[t, T]$ is the set of step controls on $[t, T]$. However first of all it is necessary to define the function $Z(t, x(\cdot))$ more precisely, since the set $\mathfrak{U}_0[t, T]$ is uncountable and a lower bound of an uncountable set of measurable functions may be a non-measurable function. To avoid this difficulty in what follows, we shall interpret the l.u.b. of a certain set of measurable functions as the essential l.u.b. of this set and will not resort to a new notation. Thus

$$\inf_{\eta(\cdot) \in \mathfrak{U}_0[t, T]} \mathbf{E}F_t(\xi_t^{(\eta)}(\cdot), \eta(\cdot)) = \operatorname{ess} \inf_{\eta(\cdot) \in \mathfrak{U}_0[t, T]} \mathbf{E}F_t(\xi_t^{(\eta)}(\cdot), \eta(\cdot)).$$

The following theorem is often called the *optimality principle* or *Bellman's principle.*

Theorem 3.19. *Let $(a, b, c) \in \tilde{\mathfrak{S}}(C, L)$. Then for all $s \in [t, T]$,*

$$Z(t, x(\cdot)) = \inf_{\eta(\cdot) \in \mathfrak{U}_0[t, s]} \mathbf{E}\{F^a_{[t, s]}(\xi_t^{(\eta)}(\cdot), \eta(\cdot)) + G[t, s]Z(s, \xi_t^{(\eta)}(\cdot))\}. \tag{3.88}$$

PROOF. Let t, $x(\cdot)$ and $\varepsilon > 0$ be fixed. For any $\varepsilon > 0$ a δ and a $\eta(\cdot) \in \mathfrak{U}_\delta$ can be found such that

$$\begin{aligned} Z(t, x) + \varepsilon &> \mathbf{E}F_t(\xi_t^{(\eta)}(\cdot), \eta(\cdot)) \\ &= \mathbf{E}[F^a_{[t, s]}(\xi_t^{(\eta)}(\cdot), \eta(\cdot)) + G[t, s]F_s(\xi_{tx}^{(\eta)}(\cdot), \eta(\cdot))], \end{aligned}$$

the expression on the r.h.s. is at least

$$\mathbf{E}[F^a_{[t, s]}(\xi^{(\eta)}(\cdot), \eta(\cdot)) + G[t, s]Z(s, \xi_t^{(\eta)}(\cdot))].$$

This implies that

$$Z(t, x) \geq \inf_{\eta \in \mathfrak{U}_0[t, T]} \mathbf{E}[F^a_{[t, s]}(\xi^{(\eta)}_t(\cdot), \eta(\cdot)) + G[t, s]Z(s, \xi^{(\eta)}_t(\cdot))].$$

We now show that the reverse inequality is also valid. Since $\mathbf{E}F_{[s, T]}(\xi^{(\eta^*)}_s(\cdot), \eta^*(\cdot))$ converges to $Z(t, x(\cdot))$ as $|\delta| \to 0$ uniformly in $x(\cdot)$ on the sphere $\|x(\cdot)\|_t \leq N$ (here η^* is an optimal control for the approximating sequence $\{\xi^{(\eta)}_k, k = 0, 1, \ldots, n\}$ corresponding to the given subdivision δ) and $\sup_{0 \leq s \leq T} |\xi^{(\eta)}_t(s)|$ is a stochastically bounded random variable uniform in the above sphere for all $N > 0$, a δ can be found such that

$$\mathbf{P}\{\mathbf{E}\{F_{[s, T]}(\xi^{(\eta^*)}_s(\cdot), \eta^*(\cdot)) \,|\, \mathfrak{F}_s\} < Z(s, \xi^{(\eta^*)}_t(s)) + \varepsilon\} > 1 - \varepsilon.$$

We have

$$\begin{aligned} Z(t, x(\cdot)) &= \inf_{\eta \in \mathfrak{U}_0[t, T]} \mathbf{E}F_{[t, T]}(\xi^{(\eta)}_t(\cdot), \eta(\cdot)) \\ &= \inf_{\eta \in \mathfrak{U}_0[t, T]} \mathbf{E}[F^a_{[t, s]}(\xi^{(\eta)}_t(\cdot), \eta(\cdot)) \\ &\quad + G[t, s]\mathbf{E}\{F_{[s, T]}(\xi^{(\eta)}_t(\cdot), \eta(\cdot)) \,|\, \mathfrak{F}_s\}. \end{aligned}$$

Each one of the controls $\eta(\cdot) \in \mathfrak{U}_0[t, T]$ can be viewed as consisting of two components: $\eta(\cdot) = [\eta_1(\cdot), \eta_2(\cdot)]$, where $\eta_1(\cdot) \in \mathfrak{U}_0[t, s]$ and $\eta_2(\cdot) \in \mathfrak{U}_0[s, T]$. Therefore

$$\begin{aligned} &Z(t, x(\cdot)) \\ &\leq \inf_{\eta \in \mathfrak{U}_0[t, s]} \mathbf{E}(F^a_{[t, s]}(\xi^{(\eta)}_t(\cdot), \eta(\cdot)) + G[t, s][Z(s, \xi^{(\eta^*)}_t(s)) + \varepsilon + C_1\varepsilon]) \\ &\leq C_2\varepsilon + \inf_{\eta \in \mathfrak{U}_0[t, s]} \mathbf{E}(F^a_{[t, s]}(\xi^{(\eta)}_t(\cdot), \eta(\cdot)) + G[t, s]Z(s, \xi^{(\eta^*)}_t(s))). \end{aligned}$$

Here

$$C_1 = \sup_{x(\cdot), u(\cdot)} F_{[s, T]}(x(\cdot), u(\cdot)), \; C_2 = C_1 + \sup_{x(\cdot), u(\cdot)} G[0, T].$$

Since $\varepsilon > 0$ is arbitrarily small the last inequality implies

$$Z(t, x(\cdot)) \leq \inf_{\eta(\cdot) \in \mathfrak{U}_0[t, T]} \mathbf{E}\{F^a_{[t, s]}(\xi^{(\eta)}_{tx}(\cdot), \eta(\cdot)) + G[t, s]Z(s, \xi^{(\eta)}_t(s))\},$$

which proves the assertion of the theorem. □

The results just obtained are simplified for the case of equations without an aftereffect. Assume that functionals on $x(\cdot)$, $a(t, x(\cdot), u)$, $b(t, x(\cdot), u)$, $c(t, x(\cdot), u, y)$ for any t, u, y, $(t, u) \in [0, T] \times U$, $y \in R^d$ depend only on the values of the function $x(\cdot)$ at time t, i.e.

$$a(t, x(\cdot), u) = a(t, x(t), u), \qquad b(t, x(\cdot), u) = b(t, x(t), u),$$
$$c(t, x(\cdot), u, y) = c(t, x(t), u, y),$$

where $a(t, x, u)$, $b(t, x, u)$, $c(t, x, u, y)$ are non-random functions of arguments $(t, x, u) \in [0, T] \times R^d \times U$. Equation (3.83) then becomes a non-lagging stochastic differential equation:

$$d\xi(t) = a(t, \xi(t), \eta(t))\, dt + b(t, \xi(t), \eta(t)) + \int_{R^d} c(t, \xi(t), \eta(t), y)\tilde{\nu}(dt, dy), \tag{3.89}$$

or more briefly

$$d\xi = a(t, \xi(t), \eta(t))\, dt + \lambda(\xi(t), \eta(t), dt),$$

where

$$\lambda(\xi(t), \eta(t), t) = \int_0^t b(s, \xi(s), \eta(s))\, dw(s) + \int_0^t c(s, \xi(s), \eta(s), y)\tilde{\nu}(ds, dy).$$

If the function $\eta(t) = u(t)$ is non-random and equation (3.89) possesses a unique solution, it then defines a Markov process. We say that this equation or the corresponding controlled object is an equation (or a system) without an aftereffect. If $c \equiv 0$ the solution of the equation is called a *diffusion solution.* The absence of an aftereffect simplifies the structure of the optimal control and in the case of an evolutional loss functional allows us to ascertain more efficient methods for the determination of optimal controls. In this case the preceding relations become as follows. The approximating control sequence $\{\xi_k, k = 0, 1, \ldots, n\}$ is defined by the recurrence relations

$$\xi_{k+1} = \xi_k + \int_{t_k}^{t_{k+1}} a(s, \xi_k, u_k)\, ds + \int_{t_k}^{t_{k+1}} \left[b(s, \xi_k, u_k)\, dw(s) + \int_{R^d} c(s, \xi_k, u_k, y)\tilde{\nu}(ds, dy)\right]$$

and is a Markov chain for fixed $u_0, \ldots, u_n$. Since $\xi_{k+1}(\check{x}(\cdot), u) = \xi_{k+1}(x_k, u)$, it follows that $Z_k(x_k, u) = \mathbf{E}Z_{k+1}(\xi_{k+1}(\check{x}(\cdot), u))$ depends on x_k and u only:

$$Z_k(x_k, u) = \mathbf{E}Z_{k+1}\left(x_k + \int_{t_k}^{t_{k+1}} a(s, x_k, u)\, ds + \int_{t_k}^{t_{k+1}} b(s, x_k, u)\, dw + \int_{t_k}^{t_{k+1}} \int_{R^d} c(s, x_k, u, y)\tilde{\nu}(ds, dy)\right).$$

The function $g_k(x(\cdot))$ also depends only on x_k and satisfies equation

$$Z_k(x) = \min_u Z_k^*(x, u) = Z_k^*(x, g_k(x)).$$

Thus an optimal control for the chain $\{\xi_0, \ldots, \xi_n\}$ at each step is of the form $\eta_k = g_k(\xi_k)$, i.e. is Markovian. Observe, however that the control for the solution of equation (3.89) of the form

$$\eta(t) = g_k(\xi_k(t_k)), \qquad t \in [t_k, t_{k+1}),$$

which for $|\delta|$ sufficiently small is ε-optimal ($\varepsilon > 0$), is not a Markov control.

Denote by $\xi_{sx}^{(\eta)}(t)$ the solution of equation (3.89) for $t \geq s$ under the initial condition $\xi(s) = x$ which corresponds to the control $\eta(t)$, $t \geq s$. We introduce the optimal cost of control

$$Z(t, x) = \inf_{\eta(\cdot) \in \mathfrak{U}_0[t, T]} \mathbf{E}F_t(\xi_{tx}^{(\eta)}(\cdot), \eta(\cdot)).$$

In the case under consideration this function coincides with function $Z(t, x(\cdot))$ introduced above. The optimality principle now becomes as given in

Theorem 3.20. *Let* $(a, b, c) \in \tilde{\mathfrak{S}}(C, L)$. *Then*

$$Z(t, x) \leq \mathbf{E}\{F_{[t, s]}^a(\xi_{tx}^{(\eta)}(\cdot), \eta(\cdot)) + G[t, s]Z(s, \xi_{tx}^{(\eta)}(s))\}, \tag{3.90}$$

for all $t \leq s \leq T$, *with the equality sign valid if and only if the control is optimal.*

The analytic properties of function $Z(t, x)$ are of importance. Unfortunately, such properties of $Z(t, x)$ as differentiability is usually difficult to prove. We shall show that this function is continuous jointly in the variable.

Let $t < s$. We bound the difference

$$d = \mathbf{E}F_t(\xi_{tx}(\cdot), \eta(\cdot)) - \mathbf{E}F_s(\xi_{sy}(\cdot), \eta(\cdot)),$$

where $\xi_{tx}(\cdot) = \xi_{tx}^{(\eta)}(\cdot)$, $\eta(t) \in \mathfrak{U}_\delta$, $\delta = \{t = t_0, t_1, \ldots, t_n = T\}$ and the control $\eta(t)$ is of the form $\eta(t) = \varphi_k(\xi_{tx}(t_k))$ for $t \in [t_k, t_{k+1})$. We may assume without loss of generality that $s = t_j$ for some j. We thus have

$$\begin{aligned} d = {} & \mathbf{E}F_{[t, s]}^a(\xi_{tx}(\cdot), \eta(\cdot)) + \mathbf{E}(G[t, s] - 1)F_s(\xi_{tx}(\cdot), \eta(\cdot)) \\ & + \mathbf{E}[F_s(\xi_{tx}(\cdot), \eta(\cdot)) - F_s(\xi_{sy}(\cdot), \eta(\cdot))]. \end{aligned}$$

Since the functions f, g and h are bounded, the first two summands on the r.h.s. are bounded from above by $L(s - t)$, where L is a constant. Furthermore, let

$$d_1 = \mathbf{E}[F_s(\xi_{tx}(\cdot), \eta(\cdot)) - F_s(\xi_{sy}(\cdot), \eta(\cdot))].$$

Then

$$\begin{aligned} d_1 &= \mathbf{E}\mathbf{E}\{F_s(\xi_{tx}(\cdot), \eta(\cdot)) - F_s(\xi_{sy}(\cdot), \eta(\cdot)) \mid \mathfrak{F}_s\} \\ &= \mathbf{E}\mathbf{E}\left([F_s(\xi_{sz}(\cdot), \eta(\cdot)) - F_s(\xi_{sy}(\cdot), \eta(\cdot))] \Big|_{z = \xi_{tx}(s)} \right). \end{aligned}$$

The functional $F_{[t,s]}(x(\cdot), u(\cdot))$ is continuous in $x(\cdot)$ uniformly in $u(\cdot)$. Therefore

$$|F_{[t,s]}(x(\cdot), u(\cdot)) - F_{[t,s]}(x'(\cdot), u(\cdot))| \le \lambda\Big(\sup_{s \le \theta \le t} |x(\theta) - x'(\theta)|\Big),$$

where $\lambda(r)$, $r \in [0, \infty)$, is a bounded function, continuous at the point $r = 0$, $\lambda(0) = 0$. It follows from lemma 3.6 that

$$\begin{aligned}\mathbf{E} \sup_{s \le \theta \le t} |\xi_{s\xi_{tx}(s)}(\theta) - \xi_{sy}(\theta)|^2 &\le L\mathbf{E}|\xi_{tx}(s) - y|^2 \\ &\le 2L(|x - y|^2 + \mathbf{E}|\xi_{tx}(s) - x|^2) \\ &\le 2L[|x - y|^2 + L_1(s - t)(1 + |x|^2)].\end{aligned}$$

Thus $\sup_{s \le \theta \le t} |\xi_{s\xi_{tx}(s)}(\theta) - \xi_{sy}(\theta)|^2 \to 0$ in probability as $t \downarrow s$ and $x \to y$ and the convergence is uniform in $\eta(\cdot)$. Since

$$d_1 \le \mathbf{E}\left[\lambda\Big(\sup_{s \le \theta \le t} |\xi_{sz}(\theta) - \xi_{sy}(\theta)|\Big)\Big|_{z = \xi_{tx}(s)}\right],$$

we have $d_1 \to 0$ and simultaneously $d \to 0$ as $x \to y$, $s \downarrow t$ uniformly with respect to $\eta(\cdot) \in \mathfrak{U}_0[t, T]$. Furthermore,

$$\begin{aligned}&Z(t, x) - Z(s, y) \\ &= \inf_{\eta(\cdot) \in \mathfrak{U}_0[t, T]} \mathbf{E}F_t(\xi_{tx}(\cdot), \eta(\cdot)) - \inf_{\eta(\cdot) \in \mathfrak{U}_0[s, T]} \mathbf{E}F_s(\xi_{sy}(\cdot), \eta(\cdot)) \\ &\le \sup_{\eta(\cdot) \in \mathfrak{U}_0[t, T]} \mathbf{E}\big(F_t(\xi_{tx}(\cdot), \eta(\cdot)) - F_t(\xi_{sy}(\cdot), \eta(\cdot))\big) \le \sup_{\eta(\cdot)} d.\end{aligned}$$

Analogously we bound the difference $Z(s, y) - Z(t, x)$ and obtain that $Z(t, x) - Z(s, y) \to 0$ as $t \to s$ and $x \to y$. Hence the following assertion is verified.

Lemma 3.15. *If $(a, b, c) \in \tilde{S}(C, L)$ then the function $Z(t, x)$ is continuous in (t, x).*

If some stronger smoothness properties are imposed on $Z(t, x)$ one can show by means of formal calculations that this function is a solution of a non-linear equation of the parabolic type. This equation is often called *Bellman's equation.* The converse statement is of substantial importance. If there exists a sufficiently smooth solution of the Bellman equation then it is the optimal cost of control, and using this solution it is easy to determine an optimal control.

Formally Bellman's equation may be obtained in the following manner. For $t < s$ we have in view of equality (3.90)

$$\begin{aligned}\frac{1}{s-t}&[Z(t, x) - Z(s, x)]\\ &= \inf_{\eta \in \mathfrak{U}_0[t, s]} \mathbf{E}\left[\frac{1}{s-t} F^a_{[t, s]}(\xi_{tx}(\cdot), \eta(\cdot))\right.\\ &\quad + \frac{1}{s-t}(G[t, s] - 1)Z(s, \xi_{tx}(s))\\ &\quad \left.+ \frac{1}{s-t}(Z(s, \xi_{tx}(s)) - Z(s, x))\right]. \end{aligned} \tag{3.91}$$

Moreover for $\eta \in \mathfrak{U}_0$

$$\lim_{s \downarrow t} \frac{1}{s-t} F^a_{[t, s]}(\xi_{tx}(\cdot), \eta(\cdot)) = f(t, x, \eta(t)),$$

$$\lim_{s \downarrow t} \frac{1}{s-t}(G[t, s] - 1) = g(t, x, \eta(t)).$$

Furthermore, utilizing generalized Itô's formula (3.15), (3.16) and assuming that the function $Z(t, x)$ possesses all the properties required for applicability of this formula, we have

$$\begin{aligned}Z(s, \xi_{tx}(s)) - Z(s, x) &= \int_t^s LZ(\theta, \xi_{tx}(\theta), \eta(\theta))\, d\theta\\ &\quad + \int_t^s \Lambda Z(\theta, \xi_{tx}(\theta), \eta(\theta), d\theta),\end{aligned} \tag{3.92}$$

where

$$LZ(t, x, u) = L^a Z(t, x, u) + L^w Z(t, x, u) + L^\pi Z(t, x, u),$$

$$L^a Z(t, x, u) = \sum_1^d \frac{\partial Z(t, x)}{\partial x_k} a_k(t, x, u) = (\nabla Z(t, x), a(t, x, u)),$$

$$\begin{aligned}L^w Z(t, x, u) &= \frac{1}{2} \sum_{k, r=1}^d \frac{\partial^2 Z(t, x)}{\partial x_k\, \partial x_r} \sum_j b_{kj}(t, x, u)\, b_{rj}(t, x, u)\\ &= \frac{1}{2} \operatorname{Sp}(\nabla^2 Z(t, x), bb^*),\end{aligned}$$

$$\begin{aligned}L^\pi Z(t, x, u) = \int_{R^d} &[Z(t, x + c(t, x, u)y) - Z(t, x)\\ &- (\nabla Z(t, x), c(t, x, u, y))]q(dy),\end{aligned}$$

and the second integral on the r.h.s. of equality (3.92) is a local martingale:

$$\Lambda Z(t, x, u, dt) = (\nabla Z(t, x), b(t, x, u)\, dw)$$
$$+ \int_{R^d} [Z(t, x + c(t, x, u, y)) - Z(t, x)]\tilde{v}(dt, dy).$$

Therefore

$$\lim_{s \downarrow t} \frac{1}{s - t} \mathbf{E}(Z(s, \xi_{tx}(s)) - Z(s, x)) = LZ(t, x, \eta(t)).$$

Interchanging in (3.91) the order of passage to the limit and of the operation of taking the l.u.b., we arrive at the following equation

$$-\frac{\partial Z(t, x)}{\partial t} = \inf_{u \in U} \{L_u Z(t, x) + g(t, x, u)Z(t, x) + f(t, x, u)\}, \quad (3.93)$$

where $t \in [0, T)$, $L_u Z(t, x) = LZ(t, x, u)$. The following initial condition should be adjoined to this equation

$$Z(T, x) = \min_{u \in U} h(x, u). \quad (3.94)$$

We now show that if the Bellman equation possesses a solution it coincides with the cost of the control.

Consider a stochastic differential equation (3.89) with arbitrary linearly bounded non-anticipative coefficients $a(t, x, u)$, $b(t, x, u)$ and $c(t, x, u, y)$. Assume that there exists a function $Z(t, x)$, $(t, x) \in [0, T] \times R^d$ continuously differentiable with respect to t and twice continuously differentiable with respect to x which satisfy the boundedness conditions

$$|Z(t, x)| \le C(1 + |x|^2), \qquad |\nabla Z(t, x)| \le C(1 + |x|), \; |\nabla^2 Z(t, x)| \le C$$

and the generalized Bellman's equation (3.93) as well as the initial condition (3.94).

Denote by $\mathfrak{U}$ the class of all generalized controls for which equation (3.89) possesses a solution.

Theorem 3.21. *Under the above stipulated conditions the function $Z(t, x)$ coincides with the optimal cost of control in the class of admissible controls $\mathfrak{U}$. Moreover there exists an optimal Markovian control $\eta(t) = \varphi(t, \xi(t))$ provided there exists a Borel function $\varphi(t, x)$, $(t, x) \in [0, T] \times R^d$ satisfying the equation*

$$\inf_{u \in U} [L_u Z(t, x) + g(t, x, u)Z(t, x) + f(t, x, u)]$$
$$= L_{\varphi(t, x)} Z(t, x) + g(t, x, \varphi(t, x))Z(t, x) + f(t, x, \varphi(t, x)). \quad (3.95)$$

PROOF. Let $\eta(t)$ be an arbitrary admissible control in $\mathfrak{U}$. Using this control we construct a solution of equation (3.89). Denote $\zeta(s) = Z(s, \xi_{tx}(s))\, G_{[t,s]}(\xi_{tx}(\cdot), \eta(\cdot)), s \geq t$. This process possesses a stochastic differential. To obtain this differential we utilize the generalized Itô formula. Since the second multiplier in the expression for $\xi(s)$ is absolutely continuous in s, the classical rule for the derivative of a product of two functions is valid. Therefore we have

$$d\zeta(s) = G[t, s]\left\{\frac{\partial Z}{\partial t} + L_\eta Z(s, \xi_{tx}(s)) + \Lambda_\eta Z(s, \xi_{tx}(s))\right\}$$
$$+ Z(s, \xi_{tx}(s))G[t, s]g(s, \xi_{tx}(s), \eta(s)).$$

The boundedness of $G[t, s]$ and the properties of $Z(t, x)$ imply that processes $G[t, s]L_\eta Z$, $G[t, s]\Lambda_\eta Z$, $G[t, s]\partial Z/\partial t$ possess finite moments. Whence,

$$\mathbf{E}(\zeta(T) - \zeta(t)) = \mathbf{E}\int_t^T G[t, s]\left\{\frac{\partial Z}{\partial t} + L_\eta Z + gZ\right\} ds.$$

Equation (3.93) yields that $-((\partial Z/\partial t) + L_\eta Z + gZ) \leq f$. Moreover if $\eta(s) = \varphi(s, \xi(s))$ where the function $\varphi(t, x)$ is determined by equation (3.95) then in the last relation the inequality sign $\leq$ should be replaced by the equality $=$. Noting that $\zeta(T) = G[t, T]\, Z(T, \xi_{tx}(T))$, $\zeta(t) = Z(t, x)$ we obtain

$$\begin{aligned} &Z(t, x)\\ &= -\mathbf{E}\int_t^T G[t, s]\left(\frac{\partial Z}{\partial t} + L_\eta Z + gZ\right) ds + \mathbf{E}G[t, T]\, Z(T, \xi_{tx}(T))\\ &\leq \mathbf{E}\int_t^T G[t, s]f\, ds + \mathbf{E}G[t, T]Z(T, \xi_{tx}(T)) \leq \mathbf{E}F_{[t,T]}(\xi_{tx}(\cdot), \eta(\cdot)), \end{aligned}$$

or $Z(t, x) \leq \mathbf{E}F_{[t,T]}(\xi_{tx}(\cdot), \eta(\cdot))$. The equality sign holds provided $\eta(s) = \varphi(s, \xi_{tx}(s))$, $s \in [0, T]$. □

Utilizing approximate solutions for the Bellman equation one can obtain bounds on the optimal cost of control.

Let a solution of equation (3.89) possess moments of order $2p$, and the function $Z_\varepsilon(t, x)$, $(t, x) \in [0, T] \times R^d$ be twice continuously differentiable with respect to x, once with respect to t and its growth at infinity be at most of the order of $|x|^p$, the growth of $\nabla Z_\varepsilon(t, x)$ at ∞ be at most as that of $|x|^{p-1}$ and the growth of $\nabla^2 Z_\varepsilon(t, x)$ be at most as that of $|x|^{p-2}$.

Assume that $Z_\varepsilon(t, x)$ satisfies the differential inequality

$$\frac{\partial Z_\varepsilon(t, x)}{\partial t} + L_u Z_\varepsilon(t, x) + g(t, x, u)Z_\varepsilon(t, x) + f(t, x, u) + \varepsilon \geq 0 \quad (3.96)$$

for all $u \in U$, $t \in [0, T)$ and

$$Z(T, x) \leq h(x) + \varepsilon, \qquad h(x) = \min_u h(x, u). \quad (3.97)$$

We show that using function $Z_\varepsilon(t, x)$ one can obtain a bound on the optimal cost of control and construct an ε_1-optimal step control where ε_1 is arbitrarily close to ε ($\varepsilon_1 > \varepsilon$). Evidently, it would be necessary to impose stronger analytic conditions on the coefficients of the stochastic differential equation. In particular, the following lemma is needed.

Lemma 3.16. *Assume that*

a. *the function $Z_\varepsilon(t, x)$ satisfies relation (3.96), is bounded, possesses bounded first order derivatives with respect to t and x_k ($k = 1, \ldots, m$) and of the second order with respect to the space variables; moreover the second order derivatives with respect to x_k ($k = 1, \ldots, m$) are uniformly bounded and possess a concave modulus of continuity.*
b. *functions $f(t, x, u)$ and $g(t, x, u)$ are uniformly continuous in (t, x) and possess a concave modulus of continuity.*
c. *functions $(\partial/\partial x_k)a(t, x, u)$, $(\partial/\partial x_k)b(t, x, u)$ ($k = 1, \ldots, m$) and $(\partial/\partial x_k)c(t, x, u, y)$ are uniformly bounded, $(a, b, c) \in S(C, L)$.*

Then there exists a function $\kappa(t)$, $t \in [0, T]$, $\kappa(t) \to 0$ as $t \to 0$ such that

$$\begin{aligned} &Z_\varepsilon(t, x) \\ &\le \inf \mathbf{E}\{F^a_{[t_1, t_2]}(\xi_{t_1, x}(\cdot), u) + G_{[t_1, t_2]}(\xi_{t_1, x}(\cdot), u)\, Z_\varepsilon(t_2, \xi_{t_1, x}(t_2))\} \\ &+ [\varepsilon + \kappa(t_2 - t_1)](t_2 - t_1),\ \forall t_1 < t_2\ (t_i \in [0, T]), \end{aligned} \tag{3.98}$$

where $\xi_{tx}(s)$ are solutions of equation (3.89) under the control $u(t) = u =$ const.

PROOF. Inequality (3.96) implies that

$$\begin{aligned} 0 &\le \int_{t_1}^{t_2} \inf_{u \in U} \left[\frac{\partial Z_\varepsilon(t, x)}{\partial t} + L_u Z_\varepsilon(t, x) + g(t, x, u) Z_\varepsilon(t, x) + f(t, x, u)\right] dt \\ &+ \varepsilon(t_2 - t_1) \\ &\le \inf_{u \in U} \int_{t_1}^{t_2} \left[\frac{\partial Z_\varepsilon(t, x)}{\partial t} + L_u Z_\varepsilon(t, x) + g(t, x, u) Z_\varepsilon(t, x) + f(t, x, u)\right] dt \\ &+ \varepsilon(t_2 - t_1). \end{aligned}$$

We have

$$\begin{aligned} &-\varepsilon(t_2 - t_1) \\ &\le \inf_{u \in U} \mathbf{E} \int_{t_1}^{t_2} \left[\frac{\partial Z_\varepsilon(t, \xi_{t_1, x}(t))}{\partial t} + L_u Z_\varepsilon(t, \xi_{t_1, x}(t)) \right. \\ &\left. + g(t, \xi_{t_1, x}(t), u)\, Z_\varepsilon(t, \xi_{t_1, x}(t)) + f(t, \xi_{t_1, x}(t), u)\right] dt + \delta_1(t_1, t_2, x), \end{aligned}$$

and for $\delta_1(t_1, t_2, x)$ the bound

$$\delta_1(t_1, t_2, x)$$

$$\leq \sup_{u \in U} \left| \mathbf{E}\left\{\int_{t_1}^{t_2} [L_u Z_\varepsilon(t, \xi_{t_1, x}(t)) - L_u Z_\varepsilon(t, x)]\, dt + \int_{t_1}^{t_2} g(t, x, u) \times \right.\right.$$

$$(Z_\varepsilon(t, \xi_{t_1, x}(t)) - Z_\varepsilon(t, x)) + (g(t, \xi_{t_1, x}(t), u) - g(t, x, u))Z_\varepsilon(t, \xi_{t_1, x}(t))$$

$$\left.\left. + (f(t, \xi_{t_1, x}(t), u) - f(t, u, u)) + \left(\frac{\partial Z_\varepsilon}{\partial t}(t, \xi_{t_1, x}(t)) - \frac{\partial Z_\varepsilon(t, x)}{\partial t}\right)\right\}\right| dt$$

is valid. It follows from the properties of solutions of stochastic differential equations and the conditions of the lemma that $\delta_1(t_1, t_2, x) = o(t_2 - t_1)$ uniformly in x.

Utilizing the generalized Itô formula we obtain

$$\mathbf{E} \int_{t_1}^{t_2} \left[\frac{\partial Z_\varepsilon}{\partial t}(t, \xi_{t_1, x}(t)) + L_u Z_\varepsilon(t, \xi_{t_1, x}(t))\right] dt$$

$$= \mathbf{E} Z_\varepsilon(t_2, \xi_{t_1, x}(t_2)) - Z_\varepsilon(t_1, x).$$

Thus

$$Z_\varepsilon(t_1, x) - \varepsilon(t_2 - t_1)$$

$$\leq \inf_{u \in U} \mathbf{E}\left\{Z_\varepsilon(t_2, \xi_{t_1, x}(t_2)) + \int_{t_1}^{t_2} g(t, \xi_{t_1, x}(t), u) Z_\varepsilon(t, \xi_{t_1, x}(t))\, dt\right.$$

$$\left. + \int_{t_1}^{t_2} f(t, \xi_{t_1, x}(t), u)\, dt\right\} + \delta_1(t_1, t_2, x)$$

$$\leq \inf_{u \in U} \mathbf{E}\{F^a_{[t_1, t_2]}(\xi_{t_1, x}(\cdot), u) + G_{[t_1, t_2]}(\xi_{t_1, x}(\cdot), u)\, Z_\varepsilon(t_2, \xi_{t_1, x}(t_2))\}$$

$$+ \delta_2(t_1, t_2, x),$$

where

$$\delta_2(t_1, t_2, x) = \delta_1(t_1, t_2, x) + \delta_1^*(t_1, t_2, x),$$

$$\delta_1^*(t_1, t_2, x)$$

$$\leq \sup_{u \in U} \mathbf{E}\left[Z_\varepsilon(t_2, \xi_{t_1, x}(t_2)) + \int_{t_1}^{t_2} g(t, \xi_{t_1, x}(t), u)\right.$$

$$\left. \times Z_\varepsilon(t_2, \xi_{t_1, x}(t))\, dt - G_{[t_1, t_2]}(\xi_{t_1, x}(\cdot), u) Z_\varepsilon(t_2, \xi_{t_1, x}(t_2))\right]$$

$$+ \sup_{u \in U} \mathbf{E}\left[\int_{t_1}^{t_2} f(t, \xi_{t_1, x}(t), u)\, dt - F^a_{[t_1, t_2]}(\xi_{t_1, x}(\cdot), u)\right] = \delta_1' + \delta_1''.$$

Moreover,

$$\delta_1' \le \sup_{u \in U} \mathbf{E} \int_{t_1}^{t_2} g(t, \xi_{t_1, x}(t), u)[Z_\varepsilon(t, \xi_{t_1, x}(t)) - Z_\varepsilon(t_2, \xi_{t_1, x}(t_2))]\, dt$$

$$+ \sup_{u \in U} \mathbf{E} Z_\varepsilon(t_2, \xi_{t_1, x}(t_2)) \left[1 + \int_{t_1}^{t_2} g(t, \xi_{t_1, x}(t), u)\, dt - G_{[t_1, t_2]}(\xi_{t_1, x}(\cdot), u) \right].$$

In view of the uniform boundedness of the quantity $(\partial/\partial t)Z_\varepsilon(t, x)$, the quantity $(\partial/\partial x_k)Z_\varepsilon(t, x)$, (the first summand on the r.h.s. of the last inequality) is of order $O(t_2 - t_1)^2 + (t_2 - t_1)O(\mathbf{E}|\xi_{t_1, x}(t_2) - x|)$ where O is uniform in both x and u.

It is known from the theory of stochastic differential equations that the bound

$$\mathbf{E}|\xi_{t_1, x}(t_2) - x| \le K_1(t_2 - t_1)^{1/2},$$

where K_1 does not depend on x and u provided the coefficients of equation (3.89) satisfy a uniform Lipschitz condition. Thus the summand under consideration is of order $o(t_2 - t_1)$ (also uniformly in x). The second summand appearing in the bound for δ_1' is of order $O(t_2 - t_1)^2$ uniformly in x since the quantities $Z_\varepsilon(t, x)$ and $g(t, x, u)$ are bounded. Consequently,

$$\delta_1' = o(t_2 - t_1).$$

Furthermore,

$$\delta_1'' = \sup_{u \in U} \mathbf{E} \int_{t_1}^{t_2} f(t, \xi_{t_1, x}(t), u)(1 - G_{[t_1, t_2]}(\xi_{t_1, x}(\cdot), u))\, dt = o(t_2 - t_1)$$

uniformly in x due to the uniform boundedness of the functions $f(t, x, u)$ and $g(t, x, u)$.

Thus we finally obtain inequality

$$Z_\varepsilon(t, x) \le \inf_{u \in U} \mathbf{E}\{F^a_{[t_1, t_2]}(\xi_{t_1, x}(\cdot), u) + G_{[t_1, t_2]}(\xi_{t_1, x}(\cdot), u)$$
$$\times Z_\varepsilon(t_2, \xi_{t_1, x}(t_2))\} + \varepsilon(t_2 - t_1) + o(t_2 - t_1).$$

which proves the lemma. □

Now let an arbitrary subdivision $\delta = \{0 = t_0 < t_1 < \cdots < t_n = T\}$ of the interval $[0, T]$ be given. In accordance with the procedure described above we construct a sequence of functions $Z_k^*(x, u)$ and $Z_k(x)$. We then compare the values of $Z_k(x)$ with those of $Z_\varepsilon(t_k, x)$. It follows from inequality (3.98) that

$$Z_\varepsilon(t_{n-1}, x)$$
$$\le \inf_{u \in U} \mathbf{E}\{F^a_{[t_{n-1}, T]}(\xi_{t_{n-1}, x}(\cdot), u) + G_{[t_{n-1}, T]}(\xi_{t_{n-1}, x}(\cdot), u)\, h(\xi_{t_{n-1}, x, u}(T))\}$$
$$+ \varepsilon(e^{K \Delta t_n} + \Delta t_n) + \kappa(\Delta t_n)\, \Delta t_n.$$

Therefore,

$$Z_\varepsilon(t_{n-1}, x) \le Z_{n-1}(x) + \varepsilon(e^{K\,\Delta t_n} + \Delta t_n) + \kappa(\Delta t_n)\,\Delta t_n.$$

Analogously,

$$\begin{aligned} &Z_\varepsilon(t_{n-2}, x)\\ &\le \inf_{u\in U} \mathbf{E}\{F^a_{[t_{n-2}, t_{n-1}]}(\xi_{t_{n-2}, x}(\cdot), u)\\ &+ G_{[t_{n-2}, t_{n-1}]}(\xi_{t_{n-2}, x}(\cdot), u) Z_\varepsilon(t_{n-1}, \xi_{t_{n-2}, x}(t_{n-1}))\} + \varepsilon\,\Delta t_{n-1}\\ &+ \kappa(\Delta t_{n-1})\,\Delta t_{n-1}\\ &\le Z_{n-2}(x) + \varepsilon e^{K(T-t_{n-2})} + (\varepsilon + \kappa(|\delta|))(e^{K(t_{n-1}-t_{n-2})}\,\Delta t_n + \Delta t_{n-1}), \end{aligned}$$

where $|\delta| = \max_k \Delta t_k$. Here it was assumed that the function $\kappa(t)$ is monotonically non-decreasing for $t > 0$. An induction argument shows that

$$Z_\varepsilon(t_k, x) \le Z_k(x) + \varepsilon e^{K(T-t_k)} + (\varepsilon + \kappa(|\delta|)) \sum_{j=k+1}^{m} e^{K(t_j - t_k)}\,\Delta t_j.$$

In particular,

$$Z_\varepsilon(0, x) \le Z_0(x) + K_1(\varepsilon + \kappa(|\delta|)), \qquad K_1 = (1 + T)e^{KT}.$$

Thus the function $Z_\varepsilon(0, x)$ provides a lower bound for the optimal cost of controlling.

We shall now describe the manner in which $Z_\varepsilon(0, x)$ can be utilized for constructing ε-optimal controls. This may be accomplished in two ways.

1. Let $\psi_k(x)$ be a Borel function which minimizes for each x the value of the function of u:

$$\mathbf{E}\{F^a_{[t_k, t_{k+1}]}(\xi_{t_k, x}(\cdot), u) + G_{[t_k, t_{k+1}]}(\xi_{t_k, x}(\cdot), u)\; Z_\varepsilon(t_{k+1}, \xi_{t_k, x}(t_{k+1}))\}.$$

We construct a solution of the stochastic differential equation (3.89) using as the control the function $\eta(t) = \psi_k(\xi(t_k))$, $k = 0, 1, \ldots, n-1$. Denote by $\bar{Z}_k(x)$ the cost of controlling during time $[t_k, T]$ under the initial state x at time t_k:

$$\bar{Z}_k(x) = \mathbf{E}F_{t_k}(\xi_{t_k, x, \eta}(\cdot), \eta(\cdot)).$$

It follows from the above that

$$\begin{aligned} \bar{Z}_k(x) = \mathbf{E}\{&F^a_{[t_k, t_{k+1}]}(\xi^{(\psi_k)}_{t_k, x}(\cdot), \psi_k(\cdot))\\ &+ G_{[t_k, t_{k+1}]}(\xi^{(\psi_k)}_{t_k, x}(\cdot), \psi_k(\cdot))\bar{Z}_{k+1}(\xi^{(\psi_k)}_{t_k, x}(t_{k+1}))\}. \end{aligned}$$

Utilizing the definition of function $\psi_k(x)$ we obtain in particular for $k = n-1$

$$\begin{aligned} \bar{Z}_{n-1}(x) &= \mathbf{E}\{F^a_{t_{n-1}}(\xi^{(\psi_{n-1})}_{t_{n-1}, x}(\cdot), \psi_{n-1}(\cdot)) + G_{[t_{n-1}, T]}(\cdot, \cdot)\bar{Z}_n(\xi^{(\psi_{n-1})}_{t_{n-1}, x}(T))\}\\ &= \inf_{u\in U} \mathbf{E}\{F^a_{t_{n-1}}(\xi_{t_{n-1}, x}(\cdot), u) + G_{[t_{n-1}, T]}(\cdot, \cdot) Z_\varepsilon(T, \xi_{t_{n-1}, x}(T))\}. \end{aligned}$$

Consequently,

$$|Z_\varepsilon(t_{n-1}, x) - \bar{Z}_{n-1}(x)| \le \varepsilon\,\Delta t_n + \kappa(\Delta t_n)\,\Delta t_n. \tag{3.99}$$

Furthermore,

$$\begin{aligned}\bar{Z}_k(x) &= \mathbf{E}\{F_{[t_k, t_{k+1}]}(\xi_{t_k, x}(\cdot), \psi_k(\cdot)) + G_{[t_k, t_{k+1}]}(\cdot, \cdot)Z_\varepsilon(t_{k+1}, \xi_{t_k}^{(\psi_k)}(t_{k+1}))\}\\ &\quad + \mathbf{E}\{G_{[t_k, t_{k+1}]}(\xi_{t_k, x}^{(\psi_k)}(\cdot), \psi_k(\cdot))[\bar{Z}_{k+1}(\xi_{t_k, x}^{(\psi_k)}(t_{k+1})) - Z_\varepsilon(t_{k+1}, \xi_{t_k, x}^{(\psi_k)}(t_{k+1}))]\}\end{aligned}$$

so that in view of the definition of function $\psi_k(x)$ and inequality (3.99) inequality

$$\begin{aligned}|Z_\varepsilon(t_k, x) - \bar{Z}_k(x)| &\le \varepsilon\,\Delta t_{k+1} + \kappa(\Delta t_{k+1})\,\Delta t_{k+1}\\ &\quad + e^{K\,\Delta t_{k+1}} \sup_x |Z_\varepsilon(t_{k+1}, x) - \bar{Z}_{k+1}(x)|\end{aligned}$$

holds. Thus,

$$|Z_\varepsilon(t_k, x) - \bar{Z}_k(x)| \le [\varepsilon + \kappa(|\delta|)]e^{K(T-t_k)}(T - t_k).$$

In particular,

$$\begin{aligned}Z_\varepsilon(0, x) - K_1[\varepsilon + \kappa(|\delta|)] \le \bar{Z}_0(x) &\le Z_\varepsilon(0, x)\\ &\quad + K_1[\varepsilon + \kappa(|\delta|)].\end{aligned} \tag{3.100}$$

Observe now that as it easily follows from the definition of the variables $Z_k(x)$ and $\bar{Z}_k(x)$

$$\bar{Z}_k(x) \ge Z_k(x) - \varepsilon e^{K(T-t_k)},$$

provided

$$\bar{Z}_n(x) = Z_\varepsilon(T, x) \ge h(x) - \varepsilon = Z_n(x) - \varepsilon.$$

Utilizing inequality (3.100) we obtain

$$\begin{aligned}Z_\varepsilon(0, x) - K[\varepsilon + \kappa(|\delta|)] &\le Z_0(x) \le \bar{Z}_0(x) + \varepsilon e^{KT}\\ &\le Z_\varepsilon(0, x) + K_1[\varepsilon + \kappa(|\delta|)].\end{aligned}$$

Thus,

$$\bar{Z}_0(x) \le Z_0(x) + 2K_1[\varepsilon + \kappa(|\delta|)].$$

In particular the control $\eta(t)$ ($\eta(t) = \psi_k(\xi(t_k))$) introduced above is an ε_1-optimal control for $t \in [t_k, t_{k+1})$, (here $\varepsilon_1 = 2K_1[\varepsilon + \kappa(|\delta|)]$).

2. The following procedure for constructing an ε-optimal control may turn out to be more convenient. For any (t, x) define the function $\psi = \psi(t, x)$ such that

$$\begin{aligned}&\inf_{u \in U}[L_u Z_\varepsilon(t, x) + g(t, x, u)Z_\varepsilon(t, x) + f(t, x, u)]\\ &\quad = L_\psi Z_\varepsilon(t, x) + g(t, x, \psi)Z_\varepsilon(t, x) + f(t, x, \psi).\end{aligned} \tag{3.101}$$

The same manipulations as in the proof of lemma 3.16 yield the relationship

$$\begin{aligned}\mathbf{E}\{F^a_{[t_1,t_2]}(\xi^{(\psi)}_{t_1,x}(\cdot), \psi(\cdot,\cdot)) + G_{[t_1,t_2]}(\cdot,\cdot)Z_\varepsilon(t_2, \xi^{(\psi)}_{t_1,x}(t_2))\} \\ = \inf_u \{F^a_{[t_1,t_2]}(\xi_{t_1,x}(\cdot), u) + G_{[t_1,t_2]}(\cdot,\cdot)Z_\varepsilon(t_2, \xi_{t_1,x}(t_2))\} \\ + \kappa(t_2 - t_1)(t_2 - t_1),\end{aligned} \quad (3.102)$$

where $\kappa(t) \to 0$ as $t \downarrow 0$ and $\kappa(t)$ is monotonically non-decreasing.

Introduce for each δ the control

$$\zeta(t) = \psi(t_k, \xi(t_k)) \quad \text{for } t \in [t_k, t_{k+1}).$$

Let

$$\tilde{Z}_k(x) = \mathbf{E}F_{t_k}(\xi^{(\zeta)}_{t_k,x}(\cdot), \zeta(\cdot)).$$

The function $\tilde{Z}_k(x)$ satisfies the recurrent relation

$$\tilde{Z}_k(x) = \mathbf{E}\{F^a_{[t_k,t_{k+1}]}(\xi^{(\psi)}_{t_k,x}(\cdot), \psi(\cdot,\cdot)) + G_{[t_k,t_{k+1}]}(\cdot,\cdot)\tilde{Z}_{k+1}(\xi^{(\psi)}_{t_k,x}(t_{k+1})). \quad (3.103)$$

Let

$$|\tilde{Z}_n(x) - Z_\varepsilon(T, x)| = |h(x) - Z_\varepsilon(T, x)| \leq \varepsilon.$$

It easily follows from equations (3.102) and (3.103) that

$$\begin{aligned}|Z_\varepsilon(t_k, x) - \tilde{Z}_k(x)| \leq \varepsilon\, \Delta t_{k+1} + 2\kappa(\Delta t_{k+1})\, \Delta t_{k+1} \\ + e^{K\,\Delta t_{k+1}} \sup_x |Z_\varepsilon(t_{k+1}, x) - \tilde{Z}_{k+1}(x)|.\end{aligned}$$

The last expression yields inequality

$$|Z_\varepsilon(0, x) - \tilde{Z}_0(x)| \leq (\varepsilon + 2\kappa(|\delta|))Te^{KT}$$

so that

$$\tilde{Z}_0(x) \leq Z_0(x) + 2K_1(\varepsilon + 2\kappa(|\delta|))$$

and the control $\zeta(t)$ ($\zeta(t) = \psi(t_k, \zeta(t_k))$ for $t \in [t_k, t_{k+1})$ is also an ε_1-optimal control where $\varepsilon_1 = 2K_1(\varepsilon + 2\kappa(|\delta|))$.

5 Linear Systems without an After-effect

We shall now examine how the generalized Bellman's equation (3.93) can be utilized in problems of optimal control for linear systems.

We shall first discuss systems which are described by *stochastic differential equations linear in u*. We shall write such an equation in the form

$$d\xi = a(t, \xi(t))\, dt + \lambda(\xi(\cdot), dt) + (\tilde{a}(t)\, dt + d\zeta(t))u. \quad (3.104)$$

Here $a(t, x)$, $(t, x) \in [0, T] \times R^d$, is a vector-valued function with values in R^d; $\lambda(x(\cdot), t)$ is a local martingale,

$$\lambda(x(\cdot), t) = \int_0^t \sum_{r=1}^m b^{(r)}(s, x(s))\, dw_r(s)$$

$$+ \int_0^t \int_{R^d} c(s, x(s), y)\tilde{v}(ds, dy);$$

$b^{(r)}(t, x)$ and $c(t, x, y)$ are vector-valued functions,

$$b^{(r)}(t, x) = \{b_i^r(t, x)\}, \qquad c(t, x, y) = \{c_i(t, x)\},$$
$$i = 1, \ldots, d, \quad r = 1, \ldots, m;$$

$\zeta(t)$ is a matrix martingale with independent increments and finite moments of the second order

$$d\zeta(t) = \sum_{r=1}^m \tilde{b}^{(r)}(t)\, dw_r(t) + \int_{R^m} \tilde{c}(t, y)\tilde{v}(dt, dy),$$

$$\tilde{b}^{(r)}(t) = \{\tilde{b}_{ij}^r(t)\}, \qquad \tilde{c}(t, y) = \{\tilde{c}_{ij}(t, y)\}, \qquad \tilde{a}(t) = \{\tilde{a}_{ij}(t)\},$$
$$i = \overline{1, d}, \qquad j = \overline{1, d_1}, \qquad r = \overline{1, m}.$$

Here d is the dimensionality of the system under consideration; d_1 is the dimensionality of the control parameter and m is the dimensionality of the random process acting on the system; $w(t)$ is an m-dimensional standard Wiener process and $v(t, A)$ is a Poisson measure on $\mathfrak{B}_0^m$.

Assume that the conditions for the existence of a solution for equation (3.104) are satisfied. An infinitesimal operator L_u of the Markov process determined by equation (3.104) is of the form

$$L_u V = \frac{\partial V}{\partial t} + (a + \tilde{a}u, \nabla V) + \tfrac{1}{2}\, \mathrm{Sp}(\nabla^2 V[(b + \tilde{b}u)(b + \tilde{b}u)^*])$$

$$+ \int_{R^d} [V(t, x + (c + \tilde{c}u)) - V(t, x) - (c + \tilde{c}u, \nabla V)]q(dy).$$

Here

$$[(b + \tilde{b}u)(b + \tilde{b}u)^*] = \left\{\sum_{r=1}^m \left(b_i^r + \sum_{k=1}^{d_1} \tilde{b}_{ik}^r u_k\right)\left(b_j^r + \sum_{k=1}^{d_1} \tilde{b}_{jk}^r u_k\right)\right\}_{i, j = \overline{1, d}},$$

and ∇V is the vector with components $\partial V/\partial x_k$ and $\nabla^2 V$ denotes the matrix with entries $\partial^2 V/\partial x_i\, \partial x_j$. Set

$$B = \{B_{ij}\}_{i, j = \overline{1, d}}, \qquad B_{ij} = \sum_{r=1}^m b_i^r b_j^r.$$

Define the following linear functions of matrix D of dimension $d_1 \times d$:

$$\breve{B}(D) = \sum_{r=1}^m \tilde{b}^{(r)*} D b^{(r)}, \qquad \tilde{B}(D) = \sum_{r=1}^m \tilde{b}^{(r)*} D \tilde{b}^{(r)}.$$

The first function is a vector-valued function with values in R^{d_1}, the second is a matrix function whose values are matrices of dimensional $d_1 \times d_1$. Using this notation and the symmetry of matrix $\nabla^2 V$ we obtain

$$\operatorname{Sp}(\nabla^2 V[(b+\tilde{b}u)(b+\tilde{b}u)^*]) = \sum_{i,j} \frac{\partial^2 V}{\partial x_i\, \partial x_j} \sum_r b_i^r + \sum_k \tilde{b}_{ik}^r u_k$$

$$\times \left(b_j^r + \sum_k \tilde{b}_{jk}^r u_k\right) = \operatorname{Sp}(B\, \nabla^2 V) + 2(\check{B}(\nabla^2 V), u) + (\tilde{B}(\nabla^2 V)u, u).$$

We shall now introduce a loss function. Assume that this function is of the form

$$F(x(\cdot), u(\cdot)) = f_0(x(T)) + \int_0^T [(F(t)\xi(t), \xi(t)) + (\tilde{F}(t)u(t), u(t))]\, dt,$$

where $f_0(x)$ is a quadratic form, $f_0(x) = (f_0 x, x)$; $F(t)$ and $\tilde{F}(t)$ are non-negative definite symmetric matrices of dimensions $d \times d$ and $d_1 \times d_1$ respectively. In addition we shall require that the matrix $\tilde{F}(t)$ be positive definite uniformly in t:

$$(\tilde{F}(t), u) \geq c\,|u|^2, \qquad \forall t \in [0, T], \quad u \in R^{d_1}.$$

We shall discuss the continuous case only (with $c = \tilde{c} = 0$). In this case equation (3.93) becomes

$$L_0 Z + (F(t)x, x) + \inf_u [(\tilde{a}^* \nabla Z + \check{B}(\nabla^2 Z), u)$$

$$+ \tfrac{1}{2}(\tilde{B}(\nabla^2 Z)u, u) + (\tilde{F}(t)u, u)] = 0, \quad (3.105)$$

where

$$L_0 Z = \frac{\partial Z}{\partial t} + (a(t, x), \nabla Z) + \tfrac{1}{2} \operatorname{Sp}(B\, \nabla^2 Z).$$

This case is special in that the minimum of the corresponding expression in Bellman's equation is explicitly computable. We shall make use of the following lemma.

Lemma 3.17. *If B is a positive definite matrix and $v \in R^{d_1}$ then*

$$\inf_u [(u, v) + \tfrac{1}{2}(Bu, u)] = -\tfrac{1}{2}(B^{-1}v, v),$$

and the infimum is attained for the vector $u_0 = -B^{-1}v$.

Indeed

$$(u, v) + \tfrac{1}{2}(Bu, u) = \tfrac{1}{2}(B(u + B^{-1}v), u + B^{-1}v) - \tfrac{1}{2}(B^{-1}v, v).$$

The assertion of the lemma follows directly from this equality. □

The expression to be minimized in Bellman's equation is of the form

$$K(u) = (\tilde{a}^* \nabla Z + \check{B}(\nabla^2 Z), u) + \tfrac{1}{2}((\tilde{B}(\nabla^2 Z) + 2\tilde{F}(t))u, u).$$

The matrix $\nabla^2 Z$ is symmetric. Assume that this matrix is non-negative definite. In that case the matrix $\tilde{B}(\nabla^2 Z)$ will also possess this property. Indeed

$$(\tilde{B}(\nabla^2 Z)u, u) = \sum_r \sum_{ij} \left(\sum_k \tilde{b}^r_{ik} u_k\right)\left(\sum_l \tilde{b}^r_{jl} u_l\right) \frac{\partial^2 Z}{\partial x_i \, \partial x_j}$$

$$= \sum_{r, i, j} \frac{\partial^2 Z}{\partial x_i \, \partial x_j} \lambda_i^r \lambda_j^r \geq 0 \qquad \left(\lambda_i^r = \sum_k \tilde{b}^r_{ik} u_k\right).$$

Thus assuming that $(\nabla^2 Z u, u) \geq 0$ we obtain that the matrix $C = \tilde{B}(\nabla^2 Z) + 2\tilde{F}(t)$ is symmetric and positive definite. Lemma 3.17 implies that

$$\min_u K(u) = K(u_0) = -\tfrac{1}{2}(C^{-1}(\tilde{a}^* \nabla Z + \check{B}(\nabla^2 Z)), \tilde{a}^* \nabla Z + \check{B}(\nabla^2 Z)),$$

$$u_0 = -C^{-1}(\tilde{a}^* \nabla Z + \check{B}(\nabla^2 Z)).$$

In the case under consideration Bellman's equation becomes

$$L_0 Z + (F(t)x, x) - \tfrac{1}{2}((\tilde{B}(\nabla^2 Z) + 2\tilde{F}(t))^{-1}(\tilde{a}^* \nabla Z + \check{B}(\nabla^2 Z))$$
$$\tilde{a}^* \nabla Z + \check{B}(\nabla^2 Z)) = 0. \quad (3.106)$$

Here it is assumed that the function Z satisfies the boundary condition (cf. (3.94)), $Z(T, x) = \min_u f_0(x, u)$ and the matrix $\nabla^2 Z$ is non-negative definite for all $(t, x) \in [0, T] \times R^d$. The equation obtained is however rather complicated and is a highly non-linear partial differential equation and there are no clear-cut methods for its solution.

If the control is carried out accurately, without interference, then $\tilde{b}^{(r)}(t) = 0$, $\tilde{B}(\nabla^2 Z) = \check{B}(\nabla^2 Z) \equiv 0$ and equation (3.105) becomes

$$L_0 Z + (F(t)x, x) - \tfrac{1}{2}((2\tilde{F})^{-1}\tilde{a}^* \nabla Z, \tilde{a}^* \nabla Z) = 0$$

or

$$\frac{\partial Z}{\partial t} + \sum_{k=1}^d a_k(t, x)\frac{\partial Z}{\partial x_k} + \frac{1}{2}\sum_{i,j=1}^d B_{ij}(t, x)\frac{\partial^2 Z}{\partial x_i \, \partial x_j}$$
$$+ \sum_{i,j=1}^n F_{ij}(t)x_i x_j = \frac{1}{4}\sum_{j,l}\left(\sum_{i,k} \tilde{a}_{ij}(t)\tilde{F}_{ik}^{-1}\tilde{a}_{lk}\right)\frac{\partial Z}{\partial x_j} \cdot \frac{\partial Z}{\partial x_l},$$

where $\tilde{F}_{ik}^{-1}$ are the entries of the matrix $\tilde{F}^{-1}$. The equation obtained is a non-linear parabolic equation involving a quadratic form in the vector ∇Z. In the general case equation (3.106) involves a rational function of partial derivatives of the first and second orders of the unknown function $V(t, x)$.

In the case of a stochastic equation with jumps the elimination of the operation of taking the greatest lower bound in Bellman's equation does not go through. However, if the equation is linear in both ξ and u then Bellman's equation can be substantially simplified by reducing it to solving an ordinary differential equation. Indeed, let the coefficients in equation (3.104) satisfy

$$a(t, x) = a(t)x, \qquad b^{(r)}(t, x) = b^{(r)}(t)x + e^{(r)}(t),$$
$$c(x, t, y) = c(t, y)x + d(t, y).$$

Here $a(t)$, $b^{(r)}(t)$, $c(t, y)$ are matrices of order $d \times d$; $e^{(r)}(t)$ and $d(t, y)$ are vector functions with values in R^d $(r = 1, \ldots, m)$. Equation (3.104) then becomes

$$d\xi = (a(t)\,dt + d\zeta)\xi + (\tilde{a}(t)\,dt + d\tilde{\zeta})u + d\eta, \tag{3.107}$$

where $\zeta(t)$ and $\tilde{\zeta}(t)$ are matrix processes with independent increments while $\eta(t)$ is a vector process of this kind:

$$d\zeta = \sum_{r=1}^{m} b^{(r)}(t)\,dw_r(t) + \int c(t, y)\tilde{\nu}(dt, dy),$$

$$d\tilde{\zeta} = \sum_{r=1}^{m} \tilde{b}^{(r)}(t)\,dw_r(t) + \int \tilde{c}(t, y)\tilde{\nu}(dt, dy),$$

$$d\eta = \sum_{r=1}^{m} e^{(r)}(t)\,dw_r(t) + \int d(t, y)\tilde{\nu}(dt, dy).$$

Equation (3.107) can be interpreted as follows:

Random perturbations acting on the system may be subdivided into 3 categories: (a) *internal random perturbations* inherent in the system which enter additively into the coefficients of the equation at ξ; these are described as a process with independent increments $\zeta(t)$; (b) *random interference in the control mechanism*: these are given by a matrix $(d \times d_1)$ process with independent increments $\tilde{\zeta}(t)$; (c) *additive noise* which can be viewed as the effect of the external environment on the system; these are described by a process with independent increments $\eta(t)$.

It should be noted the processes $\zeta(t)$, $\tilde{\zeta}(t)$ and $\eta(t)$ actually represent arbitrary stochastically continuous processes with independent increments. The only requirement is that they possess finite moments of the second order and that parameters of the distribution of variables $\zeta(t)$, $\tilde{\zeta}(t)$ and $\eta(t)$ be differentiable with respect to time.

It is natural to assume that the external random noise acting on the system is independent of the internal perturbations. We shall accept this premise and assume that

$$d\eta = \sum_{i=1}^{m} b_e^i(t)\,dw_i^e(t) + \int_{R^d} c_e(t, y)\tilde{\nu}_e(dt, dy),$$

where $b_e^i(t)$, $c_e(t, y)$ are vector functions and the processes $\{w_1^e(t), \ldots, w_m^e(t), \tilde{\nu}_e(t, A), A \in \mathfrak{B}_0^d\}$ do not depend on the processes $\{w_1(t), \ldots, w_m(t), \tilde{\nu}(t, A),$

$A \in \mathfrak{B}_0^d\}$. Evidently the system under consideration is a particular case of the system (3.104). To verify this it is sufficient to introduce an $(m + m_1)$-dimensional Wiener process $w'(t) = (w_1(t), \ldots, w_m(t), w_{m+1}^e(t), \ldots, w_{m+m_1}^e(t))$, a $2d$-dimensional centered Poisson measure $\tilde{\nu}'$ such that if A is a cylinder in R^{m+m_1}, over the coordinates $(x_1, \ldots, x_m)$ then $\tilde{\nu}'(t, A) = \tilde{\nu}(t, A)$ and if A_1 is a cylinder over the coordinates $m + 1, \ldots, m + m_1$, then $\tilde{\nu}'(t, A_1) = \tilde{\nu}_e(t, A_1)$.

We shall define more precisely Bellman's equation in the case under consideration. The variance-covariance matrix of the continuous part of the process is of the form

$$\begin{aligned}
\mathbf{E}&\left[\left(\sum_{r=1}^{m} b^{(r)}(t)\, dw_r(t)\right)x + \left(\sum_{r=1}^{m} \tilde{b}^{(r)}(t)\, dw_r(t)\right)u + \sum_{r=1}^{m_1} b_e^r(t)\, dw_r^e(t)\right] \\
&\times \left[\left(\sum_{r=1}^{m} b^{(r)}(t)\, dw_r(t)\right)x + \left(\sum_{r=1}^{m} \tilde{b}^{(r)}(t)\, dw_r(t)\right)u + \sum_{r=1}^{m_1} b_e^r(t)\, dw_r^e(t)\right]^* \\
&= \sum_r b^{(r)}(t)xx^*b^{(r)*} + \sum_r b^{(r)}(t)xu^*\tilde{b}^{(r)*} + \sum_r \tilde{b}^{(r)}(t)ux^*b^{(r)}(t) \\
&\quad + \sum_r \tilde{b}^{(r)}(t)uu^*\tilde{b}^{(r)*} + \sum_r b_e^r b_e^{r*} = B(x, u).
\end{aligned}$$

Furthermore, let

$$\mathbf{E}\nu(t, A) = tq(A), \qquad \mathbf{E}\nu_e(t, A) = tq_e(A),$$

$$\gamma'(t) = \int_{R^m} y\tilde{\nu}(t, dy), \qquad \gamma''(t) = \int_{R^m} z\tilde{\nu}_e(t, dz).$$

Consider the compound process $\gamma(t) = (\gamma'(t); \gamma''(t))$ with values in R^{2m}. Since the processes $\tilde{\nu}(t, \cdot)$ and $\tilde{\nu}_e(t, \cdot)$ are independent and are stochastically continuous the jumps of the function $\gamma(t)$, $\delta_\gamma(t) = \gamma(t) - \gamma(t-)$ are equal with probability 1 equal to $(\delta_{\gamma'}(t); 0)$ or to $(0; \delta_{\gamma''}(t))$. The process $\gamma(t)$ admits representation

$$\gamma(t) = \int_{R^{2m}} (y; z)\tilde{\nu}^*(t, dy \times dz),$$

where the measure $\tilde{\nu}^*(t, \cdot)$ is concentrated on the subspace $R^m \times 0_m$ or the subspace $0_m \times R^m$ (0_m is the null element in R^m). Moreover if $tq^*(A \times B) = \mathbf{E}\tilde{\nu}^*(t, A \times B)$, then $tq^*(A \times 0_m) = tq(A)$, $tq^*(0_m \times A) = tq_e(A)$. The jump component in the stochastic differential equation for $\xi(t)$ can be written as

$$\int C(t, y, z)\tilde{\nu}^*(dt, dy \times dz),$$

where $C(t, y, z) = c(t, y)x + \tilde{c}(t, y)u$ for $z = 0$, $C(t, 0, z) = c_e(t, z)$ and $C(t, y, z) = 0$ outside the set-theoretic sum of the subspaces $y = 0$ and $z = 0$ in R^{2m}.

Thus the infinitesimal operator corresponding to the process $\xi(t)$ for a given u is of the form

$$L_u(Z) = (ax + \tilde{a}u, \nabla Z) + \tfrac{1}{2}\,\mathrm{Sp}[\nabla^2 Z B(x, u)]$$
$$+ \int_{R^{2d}} [Z(t, x + C(t, y, z)) - Z(t, x) - (C(t, y, z), \nabla Z)]q^*(dy, dz).$$

Equation (3.105) can be written in the following manner:

$$L_0(Z) + \inf_u K(t, x, u, Z) = 0, \tag{3.108}$$

where

$$L_0(Z) = \frac{\partial Z}{\partial t} + (ax, \nabla Z) + \frac{1}{2}\sum_r (\nabla^2 Z b^{(r)}x, b^{(r)}x) + (\nabla^2 Z b_e^r, b_e^r)$$
$$+ (F(t)x, x),$$
$$K(t, x, u, Z) = K_1(t, x, u, Z) + K_2(t, x, u, Z),$$
$$K_1(t, x, u, Z) = (\tilde{F}(t)u, u) + (\tilde{a}^* \nabla Z, u)$$
$$+ \sum_r [(\tilde{b}^{(r)} \nabla^2 Z b^{(r)}x, u) + \tfrac{1}{2}(\tilde{b}^{(r)*} \nabla^2 Z \tilde{b}^{(r)}u, u)],$$

$$K_2(t, x, u, Z) = \int_{R^{2d}} [Z(t, x + C(t, y, z)) - Z(t, x)$$
$$- (C(t, y, z), \nabla Z)]q^*(dy, dx).$$

We shall seek the solution of equation (3.108) in the form of a quadratic function $Z(t, x) = p(t) + \frac{1}{2}(P(t)x, x)$ postulating that $P(t)$ be a symmetric non-negative definite matrix. We have $\nabla Z = Px$, $\nabla^2 Z = P$. Substituting these expressions into equation (3.108) we obtain

$$L_0(Z) = p' + \tfrac{1}{2}(P'x, x) + \tfrac{1}{2}(Px, x) + \tfrac{1}{2}(a^*Px, x)$$
$$+ \tfrac{1}{2}(B(P)x, x) + \tfrac{1}{2}b_e(P) + (Fx, x),$$

where
$$B(P) = \sum_r b^{(r)*}Pb^{(r)},$$
$$b_e(P) = \sum_r (Pb_e^r, b_e^{(r)}) = \sum_r b_e^{r*}Pb_e^r.$$

Thus,

$$L_0(Z) = p' + \tfrac{1}{2}b_e(P) + \tfrac{1}{2}([P' + Pa + a^*P + B(P) + 2F]x, x).$$

Furthermore,

$$K_1 = (\tilde{F}u, u) + (\tilde{a}^*Px, u) + (\check{B}(P)x, u) + \tfrac{1}{2}(\tilde{B}(P)u, u),$$

where

$$\breve{B}(P) = \sum_r \bar{b}^{(r)*} P b^{(r)}, \qquad \tilde{B}(P) = \sum_r \bar{b}^{(r)*} P \bar{b}^{(r)},$$

$$K_2 = \frac{1}{2}\int_{R^{2m}} [P(x + C(t, y, z), x + C(t, y, z)) - (Px, x)$$

$$- 2(PC(t, y, z), x)]q^*(dy, dz) = \frac{1}{2}\int_{R^{2m}} (PC, C)q^*(dy, dz).$$

Taking into account the structure of the function $C(t, y, z)$ we arrive at

$$K_2 = \frac{1}{2}\int_{R^m} (P(c(t, y)x + \tilde{c}(t, y)u), c(t, y)x + \tilde{c}(t, y)u)q(dy)$$

$$+ \frac{1}{2}\int_{R^m} (Pc_e(t, y), c_e(t, y))q_e(dy)$$

$$= \frac{1}{2}(C(P)x, x) + (\check{C}(P)x, y) + \frac{1}{2}(\tilde{C}(P)u, u) + \frac{1}{2}c_e(P),$$

where $C(P)$, $\check{C}(P)$ and $\tilde{C}(P)$ are matrix linear functions in P, while $c_e(P)$ is a scalar linear function of the same argument:

$$C(P) = \int_{R^m} c^*(t, y)Pc(t, y)q(dy),$$

$$\check{C}(P) = \int_{R^m} \tilde{c}^*(t, y)Pc(t, y)q(dy),$$

$$\tilde{C}(P) = \int_{R^m} \tilde{c}^*(t, y)P\tilde{c}(t, y)q(dy),$$

$$c_e(P) = \int_{R^m} c_e^*(t, y)Pc_e(t, y)q_e(dy).$$

Hence

$$K(t, x, u, P) = \tfrac{1}{2}(C(P)x, x) + \tfrac{1}{2}c_e(P) + (D(P)x, u)$$
$$+ \tfrac{1}{2}(Q(P)u, u),$$
$$Q(P) = 2\tilde{F} + \tilde{B}(P) + \tilde{C}(P), \qquad D(P) = a^*P + \breve{B}(P) + \check{C}(P).$$

Consequently,

$$\min_u K(t, x, u, P) = \tfrac{1}{2}(C(P)x, x) + \tfrac{1}{2}c_e(P)$$
$$- \tfrac{1}{2}(D^*(P)Q^{-1}(P)D(P)x, x), \tag{3.109}$$

and the minimum is attained at

$$u = u_0 = -Q^{-1}(P)D(P)x. \tag{3.110}$$

Moreover $Q(P)$ and $D(P)$ are linear matrix functions in matrix P and $Q(P)$ is positive definite. Equation (3.108) thus becomes

$$p' + \tfrac{1}{2}b_e(P) + \tfrac{1}{2}c_e(P) + \tfrac{1}{2}([P' + a^*P + Pa + B(P) - D^*(P)Q^{-1}(P)D(P) + 2F]x, x) = 0.$$

This equation will be satisfied if the matrix P is a non-negative definite solution of the ordinary non-linear matrix equation

$$\frac{dP}{dt} + a^*P + Pa + B(P) - D^*(P)Q^{-1}(P)D(P) + 2F = 0, \tag{3.111}$$

and the scalar function $p(t)$ is the solution of equation

$$p'(t) = -\tfrac{1}{2}[b_e(P) + c_e(P)], \qquad p(T) = 0. \tag{3.112}$$

It will be shown below that equation (3.111) indeed possesses a non-negative definite solution. Thus the following result will be verified:

Theorem 3.22. *An optimal control for a solution of stochastic differential equation* (3.107) *is defined by formula* (3.110) *and an optimal cost of control by expression*

$$Z(t, x) = p(t) + \tfrac{1}{2}(\tilde{P}(t)x, x),$$

where $P(t)$ is a solution of equation (3.111) *and $p(t)$ is determined by relation* (3.112).

We shall now prove the existence of a non-negative definite matrix $P(t)$ satisfying equation (3.111). For this purpose we shall use the method of successive approximations.

Choose an arbitrary linear control $u = \varphi_1(t, x)$ ($\varphi_1(t, x)$ is a linear function in argument x). Substitute this expression into equation (3.107). In view of general existence theorems for solutions of stochastic differential equations, equation (3.107) possesses a unique solution $\xi_{tx}^{(1)}(s)$, $s \geq t$, corresponding to control $u = \varphi_1(s, \xi(s))$ and the initial data $\xi_{tx}^{(1)}(t) = x$. Set

$$V_1(t, x) = \int_t^T \mathbf{E}[(F(s)\xi_{tx}^{(1)}(s), \xi_{tx}^{(1)}(s)) + (\tilde{F}(s)\,\varphi_1(s, \xi_{tx}^{(1)}(s)), \varphi_1(s, \xi_{tx}^{(1)}(s)))]\,ds + \mathbf{E}(f_0\,\xi_{tx}^{(1)}(T), \xi_{tx}^{(1)}(T)).$$

In view of known theorems (cf. theorem 3.5) the function $V_1(t, x)$ is twice continuously differentiable with respect to x, once with respect to t and satisfies the boundary condition $V_1(t, x) = (f_0 x, x)$. Utilizing the Markovian property of the process $\xi_{tx}^{(1)}$ we can write

$$\begin{aligned} V_1(t, x) &= \mathbf{E}\int_t^{t+\Delta t} f_1(s, \xi_{tx}^{(1)}(s), \varphi_1)\, ds + \mathbf{E}\left\{\mathbf{E}\left[\int_{t+\Delta t}^{T} f_1(s, \xi_{t+\Delta t, y}^{(1)}(s), \varphi_1)\, ds\right]_{y=\xi_{tx}(t+\Delta t)}\right. \\ &\quad + \mathbf{E}\{[\mathbf{E} f_0(\xi_{t+\Delta t, y}^{(1)}(T))]_{y=\xi_{tx}(t+\Delta t)}\} \\ &= \mathbf{E}\int_t^{t+\Delta t} f_1(s, \xi_{tx}^{(1)}(s), \varphi_1)\, ds + \mathbf{E}V_1(t+\Delta t, \xi_{tx}(t+\Delta t)), \end{aligned}$$

where for brevity we use the notation

$$f_0(x) = (f_0 x, x), \qquad f_1(t, x, u) = (F(t)x, x) + (\tilde{F}(t)u, u).$$

However in view of Itô's formula

$$\mathbf{E}V_1(t+\Delta t, \xi_{tx}(t+\Delta t)) = V_1(t, x) + \int_t^{t+\Delta t} \mathbf{E}\left(\frac{\partial V_1}{\partial t} + L_{\varphi_1} V_1\right)(s, \xi_{tx}(s))\, ds,$$

therefore

$$\int_t^{t+\Delta t} \mathbf{E}\left(\frac{\partial V_1}{\partial t} + L_{\varphi_1} V_1\right)(s, \xi_{tx}(s)) + \mathbf{E}\int_t^{t+\Delta t} f_1(s, \xi_{tx}^{(1)}(s))\, ds = 0.$$

Since the function $\mathbf{E}(\partial V_1/\partial t + L_{\varphi_1} V_1)(s, \xi_{tx}(s))$ is continuous in (t, s), it follows from the last relation that

$$\left(\frac{\partial V_1}{\partial t} + L_{\varphi_1} V_1\right) + f(t, x, \varphi_1(t, x)) = 0, \qquad t \in [0, T],\ V_1(t, x) = f_0(x). \tag{3.113}$$

Observe now that $\xi_{tx}(s)$, $s \geq t$, is a non-homogeneous linear function in x. Indeed if $\xi_{t0}(s)$ is a solution of equation (3.107) for $x = 0$ ($u(s) = \varphi(s, \xi_{tx}(s))$) and $\gamma_{tx}(s)$ is a solution of equation

$$d\gamma = (a(t)\, dt + d\zeta)\gamma + (\tilde{a}(t)\, dt + d\tilde{\zeta})\varphi(t, \xi), \tag{3.114}$$

then $\xi_{tx}(s) = \gamma_{tx}(s) + \xi_{t0}(s)$.

In view of the uniqueness theorem for solutions of stochastic differential equations $\gamma_{tx}(s) = \Gamma(s, t)x$, where $\Gamma(s, t)$ is the fundamental matrix for the solution of equation (3.114) (i.e. the matrix whose i-th column consists of components of the solution $\gamma_{t\,\delta_i}(s)$ of equation (3.113) where δ_t is the vector such that its i-th component equals 1 and all the other components are zero).

Thus $\xi_{tx}(s) = \Gamma(s, t)x + \xi_{t0}(s)$. Substituting this expression in the formula for $V_1(t, x)$ we obtain

$$V_1(t, x) = p_1(t) + (E_1 x, x), \tag{3.115}$$

where E_1 is a non-negative definite matrix. Indeed, in addition to the terms on the r.h.s. of (3.115) the expression for $V_1(t, x)$ will also contain terms of the form

$$\mathbf{E}\int_0^T (G(s)\Gamma(s, t)x, \xi_{t0}(s))\, ds \quad \text{and} \quad \mathbf{E}(G_1\Gamma(T, t)x, \xi_{t0}(T))$$

and analogous summands. We now show that these summands equal zero. Let $\mathfrak{F}'_t$ be the smallest σ-algebra with respect to which $\zeta(\theta)$ and $\tilde{\zeta}(\theta)$ are measurable for $\theta \le t$.

Clearly the quantity $\Gamma(s, t)$ $(t \le s)$ is $\mathfrak{F}'_s$-measurable. We now compute $\mathbf{E}\{\xi_{t0}(s) | \mathfrak{F}'_s\}$. The variable $\xi_{t0}(s)$ is a functional on the values of $\zeta(\theta)$, $\tilde{\zeta}(\theta)$ and $\eta(\theta)$ for $\theta \in [t, s]$, $\xi_{t0}(s) = h(s, \zeta(\cdot), \zeta(\cdot), \eta(\cdot))$. Since $\eta(\cdot)$ does not depend on $\mathfrak{F}'_T$ we obtain that

$$\mathbf{E}\{\xi_{t0}(s) | \mathfrak{F}'_T\} = \mathbf{E}\left\{ h(s, x(\cdot), y(\cdot), \eta(\cdot)) | \mathfrak{F}'_T \right\}\Big|_{\substack{x(\cdot)=\zeta(\cdot)\\ y(\cdot)=\tilde{\zeta}(\cdot)}}$$

$$= \Big[\mathbf{E}h(s, x(\cdot), y(\cdot), \eta(\cdot))\Big]_{\substack{x(\cdot)=\zeta(\cdot)\\ y(\cdot)=\tilde{\zeta}(\cdot)}} = \mathbf{E}\{\xi_{t0}(s) | \mathfrak{F}'_s\} \pmod{P}.$$

Utilizing the equality just established and computing the conditional mathematical expectation with respect to $\mathfrak{F}'_s$ in both sides of the equality

$$\xi_{t0}(s) = \int_t^s (a(\theta)\, d\theta + d\zeta)\xi_{t0}(\theta) + (\tilde{a}(\theta)\, d\theta + d\tilde{\zeta})\varphi(\theta, \xi_{t0}(\theta)) + \eta(s) - \eta(t),$$

we obtain

$$m(s) = \int_t^s (a(\theta)\, d\theta + d\zeta)m_{t0}(\theta) + (\tilde{a}(\theta)\, d\theta + d\tilde{\zeta})\varphi(\theta, m(\theta)),$$

where $m(s) = \mathbf{E}\{\xi_{t0}(s) | \mathfrak{F}'_s\}$. Since $m(t) = 0$ it follows from the obtained equation, in view of the uniqueness of the solution of the stochastic differential equations, that $m(s) = 0$, $s \ge t$. Hence

$$\mathbf{E}\int_t^T (G(s)\Gamma(s, t)x, \xi_{t0}(s))\, ds = \mathbf{E}\int_t^T \mathbf{E}\{(G(s)\Gamma(s, t)x, \xi_{t0}(s)) | \mathfrak{F}'_s\}\, ds$$

$$= \mathbf{E}\int_t^T (G(s)\Gamma(s, t)x, \mathbf{E}\{\xi_{t0}(s) | \mathfrak{F}'_s\})\, ds = 0.$$

Analogously $\mathbf{E}(G_1\Gamma(T, t)x, \xi_{t0}(T)) = 0$. We have thus verified equality (3.115).

Consider the expression

$$L_u(V_1(t, x)) + (F(t)x, x) + (\tilde{F}(t)u, u)$$

and determine the value of u for which this expression attains its minimum. Such a value exists since the preceding expression involves a positive definite quadratic form in u while all the other terms either do not depend on u or depend linearly on u. The corresponding calculations were presented above. The minimum is achieved at $u = \varphi_2(t, x) = -Q^{-1}(E_1)D(E_1)x$. Function $\varphi_2(t, x)$ is linear in x and possesses continuous coefficients.

As a second approximation to the optimal control we choose the Markov control $u = \varphi_2(t, \xi(t))$. Equation (3.113) implies that

$$\frac{\partial V_1}{\partial t} + L_{\varphi_2} V_1 + f(t, x, \varphi_2(t, x)) \le 0. \tag{3.116}$$

Set

$$V_2(t, x) = \int_t^T f(s, \xi_{tx}^{(2)}(s), \varphi_2(s, \xi_{tx}^{(2)}(s)))\, ds + \mathbf{E}(f_0 \xi_{tx}^{(2)}(T), \xi_{tx}^{(2)}(T)),$$

where $\xi_{tx}^{(2)}(s)$ is a solution of equation (3.107) obtained for $u = \varphi_2(s, \xi_{tx}^{(2)}(s))$, $\xi_{tx}^{(2)}(t) = x$. It follows from the above that

$$\frac{\partial V_2}{\partial t} + L_{\varphi_2} V_2 + f(t, x, \varphi_2(t, x)) = 0, \qquad t \le T,$$

$$V_2(T, x) = (f_0 x, x). \tag{3.117}$$

We compare functions V_1 and V_2. Let $\delta(t, x) = V_1(t, x) - V_2(t, x)$. Relations (3.116) and (3.117) yield that

$$\frac{\partial \delta}{\partial t} + L_{\varphi_2} \delta \le 0, \qquad \delta(T, x) = 0,$$

Applying the generalized Itô formula to the function $\delta(s, \xi_{tx}^{(2)}(s))$ we obtain

$$\mathbf{E}\, \delta(T, \xi_{tx}^{(2)}(T)) - \delta(t, x) = \int_t^T \mathbf{E}\left(L_{\varphi_2} \delta(s, \xi_{tx}^{(2)}(s)) + \frac{\partial \delta}{\partial s}\right) ds,$$

whence

$$\delta(t, x) = -\int_t^T \mathbf{E}\left(L_{\varphi_2} \delta(s, \xi_{tx}^{(2)}(s)) + \frac{\partial \delta}{\partial s}\right) ds \ge 0$$

or $V_1(t, x) \ge V_2(t, x)$. From the above we have

$$V_2(t, x) = (E_2(t)x, x) + p_2(t),$$

where $E_2(t)$ is a non-negative definite matrix. Continuing the preceding construction and utilizing an induction argument we obtain a sequence of linear functions $\varphi_1(t, x), \varphi_2(t, x), \ldots, \varphi_n(t, x), \ldots$, defined by equations $\varphi_n(t, x) = -Q^{-1}(E_{n-1})D(E_{n-1})x$ where $E_n(t)$ is a non-negative definite matrix function such that $V_n(t, x) = p_n(t) + (E_n x, x)$, and $V_n(t, x)$ is defined by the relationship

$$V_n(t, x) = \int_t^T \mathbf{E} f(s, \xi_{tx}^{(n)}(s), \varphi_n(s, \xi_{tx}^{(n)}(s))) + \mathbf{E}(f_0 \xi_{tx}^{(n)}(T), \xi_{tx}^{(n)}(T)).$$

Moreover

$$\frac{\partial V_n}{\partial t} + L_{\varphi_n} V_n(t, x) + f(t, x, \varphi_n(t, x)) = 0, \qquad t \le T,$$

$$V_n(t, x) = (f_0 x, x);$$

the function $\varphi_n(t, x)$ coincides with the value of u for which the function $L_u V_{n-1}(t, x) + f(t, x, u)$ achieves its minimum and

$$V_1(t, x) \ge V_2(t, x) \ge \cdots \ge V_n(t, x) \ge \cdots \ge 0.$$

Thus for any $(t, x) \in [0, T] \times R^d$ the limits $\lim V_n(t, x) = V_0(t, x)$, $\lim p_n(t) = p_0(t)$, $\lim E_n(t) = E_0(t)$ exist and $V_0(t, x) = p_0(t) + (E_0(t)x, x)$. The existence of the limit $\lim E_n(t)$ implies the existence of the limit

$$\lim \varphi_n(t, x) = \varphi_0(t, x) = -Q^{-1}(E_0)D(E_0)x.$$

Observe that

$$\left.\begin{aligned} \left(\frac{\partial V_n}{\partial t} + L_{\varphi_n} V_n + f(t, x, \varphi_n)\right) &= \frac{\partial V_n}{\partial t} + (ax + \tilde{a}\varphi_n, \nabla V_n) \\ &+ \frac{1}{2}\sum_{r=1}^{n} \nabla^2 V_n[(b^r x + \tilde{b}^r \varphi_n)(b^r x + \tilde{b}^r \varphi_n)^* + (b^r_e, b^{r*}_e)] \\ &+ \frac{1}{2}\int (\nabla^2 V_n(cx + \tilde{c}\varphi_n, cx + \tilde{c}\varphi_n))\, dq \\ &+ \frac{1}{2}\int (\nabla^2 V_n c_e, c_e)\, dq_e + (Fx, x) + (F\tilde{\varphi}_n, \tilde{\varphi}_n) = 0, \\ V_n = E_n x, \ \nabla^2 V_n &= E_n. \end{aligned}\right\} \qquad (3.118)$$

This equation together with the convergence of E_n and φ_n to their respective limits implies that $\lim \partial V_n/\partial t$ also exists and is a bounded function (in $t \in [0, T]$ for a fixed x). Therefore $\lim \partial E_n/\partial t = E_0'(t)$, $\lim p_n'(t) = p_0'(t)$. Approaching the limit in equation (3.118) as $n \to \infty$ we obtain

$$\frac{\partial V_0}{\partial t} + L_{\varphi_0} V_0 + f(t, x, \varphi_0) = 0.$$

Relation (3.117) shows that $\varphi_0(t, x)$ is chosen in such a manner that

$$\min_u \left(\frac{\partial V_0}{\partial t} + L_u V_0 + f(t, x, u)\right) = \frac{\partial V_0}{\partial t} + L_{\varphi_0} V_0 + f(t, x, \varphi_0) = 0.$$

Thus, the function $V_0(t, x)$ is a solution of the Bellman equation, and the existence of a solution for this equation of the form $V_0(t, x) = p(t) + (E(t)x, x)$, where $E(t)$ is a non-negative definite matrix is verified.

Theorem 3.23. *As $n \to \infty$ the functions $\varphi_n(t, x)$ and $V_n(t, x)$ defined by the method of successive approximations converge uniformly in any bounded region $t \in [0, T]$, $|x| \le a$, to their respective limits $\varphi_0(t, x) = \lim \varphi_n(t, x)$, $V_0(t, x)$ $\lim V_n(t, x)$. Moreover $\varphi_0(t, x)$ is an optimal control and $V_0(t, x)$ is an optimal cost of controlling.*

6 Control Equations with Continuous Noise

In this section control stochastic differential equations with a stochastic component which is continuous and uncontrollable are investigated. The loss function is assumed to be an additive functional and the control is realized on a finite time interval $[0, T]$. Our purpose is to prove the existence of an optimal control and derive the generalized Bellman's equation.

The equations under consideration are of the form

$$d\xi(t) = a(t, \xi(\cdot), u(t))\, dt + b(t, \xi(\cdot))\, dw(t), \qquad \xi(0) = x. \tag{3.119}$$

Here $a(t, x(\cdot), u)$ is a vector-valued function defined on $[0, T] \times C \times U$ with values in R^d; $b(t, x(\cdot))$ is an operator function defined on $[0, T] \times C$ which maps R^d into R^d. As before we shall assume that U is a compact space, $a(t, x(\cdot), u)$ and $b(t, x(\cdot))$ are non-anticipative functions and $b(t, x(\cdot)) \in S(C, L)$. If moreover $a(t, x(\cdot), u) \in S(C, L)$ then equation (3.119) possesses a unique continuous solution with moments of all orders for any measurable process $u(t) \equiv \eta(t)$ adapted to the current of σ-algebras $\mathfrak{F}_t$, $t \ge 0$, with respect to which $w(t)$ is a Wiener process.

We introduce the *control cost*

$$\bar{F}(\eta(\cdot)) = \bar{F}_0(x, \eta(\cdot)) = \mathbf{E} \int_0^T f(s, \xi_{0x}^{(\eta)}(\cdot), \eta(s))\, ds,$$

where $\xi_{0x}^{(\eta)}(\cdot)$ denotes the continuous solution of equation (3.119) corresponding to the given control $\eta(\cdot)$ and the initial condition $\xi(0) = x$. Here it is assumed that x is non-random.

Suppose that the function $f(t, x(\cdot), u)$ satisfies the following conditions:

a. the function is defined for $(t, x(\cdot), u) \in [0, T] \times C \times U$ and is continuous jointly in the variables t, $x(\cdot)$ and u;
b. the function is non-negative and bounded, $0 \le f(t, x(\cdot), u) \le c$;
c. for a fixed (t, u) the function $f(t, x(\cdot), u)$ is $\mathfrak{C}_t$-measurable for all $(t, u) \in [0, T] \times U$.

We define two optimal costs

$$Z(0, x) = \inf_{\eta(\cdot) \in \bar{\mathfrak{U}}} \bar{F}_0(x, \eta(\cdot)),$$

$$Z_1(0, x) = \inf_{\eta(\cdot) \in \mathfrak{U}_1} \bar{F}_0(x, \eta(\cdot)).$$

Here $\bar{\mathfrak{U}}$ is the class of all measurable processes $\eta = \eta(t)$, $t \in [0, T]$, adapted to the current $\{\mathfrak{F}_t, \ t \in [0, T]\}$; $\mathfrak{U}_1$ is the class of all feedback controls $\eta = u(t, x(\cdot))$ for which equation (3.119) possesses a solution. Clearly $\mathfrak{U}_1 \subset \bar{\mathfrak{U}}$. However as it was shown above, the class $\mathfrak{U}_0$ of all step controls is dense in $\bar{\mathfrak{U}}$. Moreover for every step control there exists a feedback control which is at least as good. Thus the class $\mathfrak{U}_1$ is dense in $\bar{\mathfrak{U}}$ and $Z(0, x) = Z_1(0, x)$.

We now impose a new condition on the matrix $b(t, x(\cdot))$ appearing in equation (3.119).

C. The matrix $b(t, x(\cdot))$ is non-degenerate.

This condition may not be plausible in certain problems but will be utilized often in what follows. This condition implies the existence of a solution for equation (3.119) in a wide class of feedback controls. We shall now prove this assertion.

In addition to equation (3.119) we introduce equation

$$d\xi = b(t, \xi(\cdot))\, dw, \qquad \xi(0) = x. \tag{3.120}$$

In view of the assumptions above this equation possesses a unique solution.

Introduce on $\{C, \mathfrak{C}\}$ a measure P_0 generated by the mapping $\omega \to x(t) = \xi(t, \omega)$ where $\xi(t) = \xi(t, \omega)$ is the solution of equation (3.120). Recall that $C = C^d_{[0, T]}$ and $\mathfrak{C}$ is a σ-algebra of Borel sets over C. $\mathfrak{C}$ coincides with the smallest σ-algebra which contains cylinder sets in C. If A is a cylinder set in C of the form $A = \{x(\cdot) : x(t_1), \ldots, x(t_s) \in B\}$ where B is a Borel set in $(R^d)^s$, then

$$P_0(A) = \mathbf{P}\{(\xi(t_1), \xi(t_2), \ldots, \xi(t_s)) \in B\}.$$

Now let the space $\{C, \mathfrak{C}, P_0\}$ be the basic probability space. Denote by $\mathfrak{C}_t$, $t \in [0, T]$, the smallest σ-algebra over C generated by cylinders A with bases over $[0, t]$ (i.e. by the sets A of the form as indicated above such that $0 \le t_1 < t_2 < \cdots < t_s \le t$). A set in $\mathfrak{C}_t$ possesses the following properties: if $B \in \mathfrak{C}_t$ and $x(\cdot) \in B$, then any other continuous function $x'(\cdot)$ such that $x(s) = x'(s)$, $\forall s \in [0, t]$, also belongs to B.

We note that, as it is easy to verify, the current of σ-algebras $\mathfrak{C}_t$, $t \in [0, T]$ is continuous from the left, i.e. $\mathfrak{C}_{t-} = \mathfrak{C}_t$ where $\alpha_{t-} = \bigcup_{s<t} \mathfrak{C}_s$. The random process $\bar{\xi}(t)$ defined on C by equation $\bar{\xi}(t) = \bar{\xi}(t, x(\cdot)) = x(t)$ (in what follows we shall write $x(t)$ in place of $\bar{\xi}(t)$) is a continuous martingale (with respect to measure P_0) with the matrix characteristic $b(t, x(\cdot))b^*(t, x(\cdot))$. The function

$$\tilde{w}(t) = \tilde{w}(t, x(\cdot)) = \int_0^t b^{-1}(t, x(\cdot))\, dx(t) \tag{3.121}$$

is a Wiener process on $\{C, \mathfrak{C}, P_0\}$ and for almost all $x(\cdot) \in C$ (in measure P_0) the equality

$$dx(t) = b(t, x(\cdot))\, d\tilde{w}(t) \tag{3.122}$$

is fulfilled.

It is not difficult to prove these assertions. The process $\zeta(t)$ defined on the initial probability space $\{\Omega, \Gamma, \mathbf{P}\}$ and satisfying equation (3.120) is a continuous martingale possessing moments of the second order. Consequently,

$$\int_A \xi(t)\, d\mathbf{P} = \int_A \xi(t')\, dP, \qquad \forall (t' > t),\ A \in \mathfrak{F}_t.$$

Applying the mapping $f: \Omega \to C$ $(\omega \to x(\cdot) = \xi(\cdot, \omega))$ to the preceding equality we arrive at equality

$$\int_{A^0} x(t)\, dP_0 = \int_{A^0} x(t')\, dP_0,$$

where $A^0 = \{x(\cdot), x(\cdot) = \xi(\cdot, \omega), \omega \in A\}$. In particular we may choose an arbitrary set in $\mathfrak{C}_t$ to serve as A^0. This shows that $x(t)$ is a local $\{\mathfrak{C}_t, P_0\}$-martingale. Furthermore for any subdivision of the interval $[0, t]$ by means of the points $0 = t_0 < t_1 < \cdots < t_n < t_{n+1} = t$ we have

$$\sum_{k=0}^{n} (\Delta x(t_k))(\Delta x(t_k))^* = \sum_{k=0}^{n} (\Delta\xi(t_k, \omega))(\Delta\xi(t_k, \omega))^*,$$

i.e. $\langle x_i, x^k \rangle_t = \langle \xi^i, \xi^k \rangle_t$, $(\xi(\cdot, \omega) \to x(\cdot))$ (mod $\mathbf{P}$), so that the matrix of square variation of the process $x(t)$ becomes

$$\langle x, x \rangle_t = \int_0^t b(s, \xi(\cdot))b(s, \xi(\cdot))^*\, ds = \int_0^t b(s, x(\cdot))b(s, x(\cdot))^*\, ds.$$

It follows from Levy's theorem (Section 1) that the process $\tilde{w}(t)$ defined by formula (3.121) is a Wiener process. Equation (3.122) is a corollary to the expression (3.121).

Below we shall omit the sign $\sim$ in the designation of the Wiener process $\tilde{w}(t)$. Equations (3.121) and (3.122) thus become

$$w(t) = \int_0^t b^{-1}(t, x(\cdot))\, dx(s), \tag{3.123}$$

$$dx(t) = b(t, x(\cdot))\, dw(t); \tag{3.124}$$

$w(t)$ is a $\{\mathfrak{C}_t, P_0\}$-Wiener process defined on the space $\{C, \mathfrak{C}\}$. Integration with respect to measure P_0 on the space C (i.e. the evaluation of the mathematical expectation of the random variable $f(x(\cdot))$ will be designated by the symbol $\mathbf{E}_0$ $(\mathbf{E}_0 f(x(\cdot)))$.

We introduce on the space $[0, T] \times C$ a σ-algebra $\mathscr{G}$ defined in the following manner. The set A is a measurable set in the product $[0, T] \times C$ and belongs to $\mathscr{G}$ if and only if each t-section A_t of the set A $(A_t = \{x(\cdot): (t, x(\cdot)) \in A\})$ belongs to $\mathfrak{C}_t$ and each $x(\cdot)$-section $A^{x(\cdot)}$ of A $(A^{x(\cdot)} = \{t: (t, x(\cdot)) \in A\})$ belongs to $\mathscr{L}^d$. This definition implies that an arbitrary $\mathscr{G}$-measurable function $h(t, x(\cdot))$ is non-anticipative.

We define for any $\mathscr{G}$-measurable function $h(t, x(\cdot))$ with values in R^d

$$\zeta_t^s(h) = \int_s^t [b^{-1}(t', x(\cdot))h(t', x(\cdot))]^* \, dw(t')$$
$$-\frac{1}{2}\int_s^t |b^{-1}(t', x(\cdot))h(t', x(\cdot))|^2 \, dt',$$

$\rho_t^s(h) = \exp\{\zeta_t^s(h)\}$ $(0 \le s < t \le T)$. Observe that

$$\zeta_t^s(h) = \int_s^t (b^{-1}h)^* b^{-1} \, dx(t') - \frac{1}{2}\int_s^t |b^{-1}h|^2 \, dt'.$$

The quantities introduced above are well defined if

$$\int_0^T |b^{-1}(t', x(\cdot))h(t', x(\cdot))|^2 \, dt < \infty \ (\text{mod } P_0).$$

Under this condition the function $\rho_t^s(h)$, $t \ge s$ becomes a positive $\{\mathfrak{C}_t, P_0\}$-supermartingale (cf. Section 1). If the condition

$$\mathbf{E}_0 \rho_T^0(h) = 1 \tag{3.125}$$

is fulfilled then $\{\rho_t^s(h), t \ge s\}$ is a $\{\mathfrak{C}_t, P_0\}$-martingale and in particular $\mathbf{E}\rho_t^s(h) = 1$ (cf. the remark following theorem 3.11).

Assume now that condition (3.125) is fulfilled. Introduce on $\mathfrak{C}$ the measure Q by setting $dQ = \rho_T^0(h) \, dP_0$. In view of Girsanov's theorem (theorem 3.14) the process

$$w'(t) = w(t) - \int_0^t b^{-1}(s, x(\cdot))h(s, x(s)) \, ds$$
$$= \int_0^t b^{-1}(s, x(\cdot)) \, dx(s) - \int_0^t b^{-1}(s, x(\cdot)) \, h(s, x(\cdot)) \, ds$$

is a $\{\mathfrak{C}_t, Q\}$-Wiener process. Thus any function $x(\cdot) \in C$ (mod Q) on the probability space $\{C, \mathfrak{C}_t, Q\}$ satisfies the equation

$$dx = h(t, x(\cdot)) + b(t, x(\cdot)) \, dw'(t), \qquad x(0) = x. \tag{3.126}$$

In our preceding discussion the assumption that $b(t, x(\cdot)) \in S(C, L_N)$ was inessential. It was only important that equation (3.120) possess a solution. In this case equation (3.126) possesses a solution for any function $h(t, x(\cdot))$ for which condition (3.125) is fulfilled. Actual verification of this condition in specific cases may be troublesome. The condition is satisfied if

$$\mathbf{E}_0 \exp\left\{\frac{1}{2}\int_0^T |b^{-1}(t, x(\cdot))h(t, x(\cdot))|^2 \, dt\right\} < \infty \tag{3.127}$$

(cf. [20] or [35]). In particular condition (3.125) is fulfilled for all $\mathscr{G}$-measurable controls provided $|b^{-1}(t, x(\cdot))h(t, x(\cdot))| \le C$ where C is a constant. It is easy to verify that it is also satisfied if the function $b(t, x(\cdot))$ is non-degenerate and $h(t, x(\cdot))$ and $b(t, x(\cdot))$ are linearly bounded. Indeed, let $h_N(t, x(\cdot)) = h(t, x(\cdot))$ for $\|x(\cdot)\|_T \le N$ and $h_N(t, x(\cdot)) = 0$ otherwise. Then $b^{-1}(t, x(\cdot))h_N(t, x(\cdot))$ is bounded and in view of the preceding remark $\mathbf{E}_0\, \rho_T^0(h_N) = 1$. Let Q_N be a measure on $\mathfrak{C}$, $dQ_N = \rho_T^0(h_N)\, dP_0$. Clearly for $N > r$ $Q_N(A \cap S_r)$ does not depend on N where S_r is a sphere of radius r in C: $S_r = \{x(\cdot): \|x(\cdot)\|_T \le r\}$, and $A \subset \mathfrak{C}$. Consequently the limit $\lim_{N\to\infty} Q_N(A \cap S_r) = Q_0(A \cap S_r)$ exists and in view of Fatou's lemma

$$Q_0(A \cap S_r) \le \mathbf{E}_0 \lim \rho_T^0(h_N)\chi(A \cap S_r) = \mathbf{E}_0\, \rho_T^0(h)\chi(A \cap S_r) = Q(A \cap S_r).$$

Q_N is a measure associated with the process $\xi_N(\cdot)$ which satisfies the equation

$$d\xi_N(t) = h_N(t, \xi_N(\cdot))\, dt + b(t, \xi_N(\cdot))\, dw.$$

Therefore

$$Q_N(C \backslash S_r) = \mathbf{P}\{\|\xi_N(\cdot)\|_T > r\} \le \frac{C'(1 + E|\xi_0|^2)}{r^2},$$

where the constant C' does not depend on N (cf. lemma 3.6) so that for any $\varepsilon > 0$ the inequality

$$Q(S_r) \ge \lim_{N\to\infty} Q_N(S_r) \ge 1 - \varepsilon$$

is satisfied for r sufficiently large. Thus $Q(C) = \mathbf{E}_0\, \rho_T^0(h) \ge 1$. However, as it was pointed out above (cf. Section 1) we always have $\mathbf{E}_0\, \rho_T^0(h) \le 1$. Consequently $\mathbf{E}_0\, \rho_T^0(h) = 1$.

We apply the result obtained to the controlled stochastic differential equation (3.119) and obtain

Theorem 3.24. *Let* a) $b(t, x(\cdot)) \in S(C, L_N)$ *and the matrix* $b(t, x(\cdot))$ *be non-degenerate*; b) *the function* $a(t, x(\cdot), u)$ *be* $\mathcal{G}$*-measurable and*

$$|a(t, x(\cdot), u)| \le C(1 + \|x(\cdot)\|_t).$$

Then for any $\mathcal{G}$*-measurable control* $u = u(t, x(\cdot))$ *a Wiener process* $\{w(t), \mathfrak{C}_t, P_u\}$ *can be found such that*

$$x(t) = x(0) + \int_0^t a(s, x(\cdot), u(s, x(\cdot))\, ds$$

$$+ \int_0^t b(s, x(\cdot))\, dw(s), \qquad \forall t \in [0, T],$$

for all $x(\cdot) \in C$ (mod P_u) *where* P_u *is a probability measure on* $\{C, \mathfrak{C}, P_u\}$, $dP_u = \rho_T^0(a^u(\cdot))\, dP_0$ *and* $a^u(t) = a(t, x(\cdot), u(t, x(\cdot)))$.

The function $\rho_T^0(a^u(\cdot))$ is a density of the measure P_u with respect to measure P_0:

$$\frac{dP_u}{dP_0} = \rho_T^0(a^u(\cdot)).$$

We shall now consider the question of weak compactness of the family of densities $\rho_T^0(a^u(\cdot))$.

Lemma 3.18. *If the function $b(t, x(\cdot))$ is bounded, $|b(t, x)| \le C$ and*

$$|b^{-1}(t, x(\cdot))a(t, x(\cdot), u)|^2 \le C(1 + \|x(\cdot)\|_t^2),$$

then for some $p > 1$

$$\sup_u \mathbf{E}_0 \exp(p\zeta_T^0(a^u(\cdot)) \le k.$$

Proof. We have

$$\exp(p\zeta_T^0(a^u(\cdot))) = \rho(\zeta_T^0(pa^u(\cdot)))\exp\left\{\frac{p^2 - p}{2}\int_0^T |b^{-1}a^u|^2\, dt\right\}$$
$$\le \rho(\zeta_T^0(pa^u(\cdot)))\exp\left\{\frac{p^2 - p}{2} CT(1 + \|x\|_T^2)\right\}.$$

It was shown above that $\mathbf{E}_0\,\rho(\zeta_T^0(pa^u(\cdot))) = 1$. We introduce on $\mathfrak{C}$ a measure Q by setting $dQ = \rho(\zeta_T^0(pa^u(\cdot)))\, dP_0$. With respect to this measure $x(\cdot)$ can be viewed as a solution of the stochastic differential equation

$$dx = pa(t, x(\cdot), u)\, dt + b(t, x(\cdot))\, dw.$$

Set $\mu(t) = \int_0^t b(t, x(\cdot))\, dw$. Then $\{\mu(t), \mathfrak{C}_t, Q\}$ is a continuous square integrable martingale. We have

$$x(t) = x_0 + \int_0^t pa(s, x(\cdot), u)\, ds + \mu(t),$$

whence

$$\|x(\cdot)\|_t \le 3\left(|x_0|^2 + Tp^2\int_0^t |a(s, x(\cdot), u)|^2\, ds + \|\mu(\cdot)\|_t^2\right)$$
$$\le 3(|x_0|^2 + CT^2p^2 + \|\mu(\cdot)\|_T^2) + 3CTp^2\int_0^t \|x(\cdot)\|_s^2\, ds.$$

Utilizing lemma 3.5 we obtain

$$\|x(\cdot)\|_t^2 \le 3(|x_0|^2 + CT^2p^2 + \|\mu(\cdot)\|_t^2)e^{3Ct^2p^2}.$$

Thus

$$\mathbf{E}_0 \exp\{p\zeta_T^0(a^u(\cdot))\} \le \mathbf{E}_Q \exp\left\{\frac{p^2-p}{2}(A_0 + A_1\|\mu(\cdot)\|_T^2)\right\},$$

where $A_0 = CT[1 + 3(|x_0|^2 + CT^2p^2)e^{3CT^2p^2}]$, $A_1 = 3CTe^{3CT^2p^2)}$.

Since the function $\exp A|\mu(t)|^2$ is a submartingale, we have

$$\mathbf{E}_Q \exp\left\{\frac{p^2-p}{2}A_1\|\mu(\cdot)\|_T^2\right\} = \mathbf{E}_Q \sup_{0\le t\le T} \exp\left\{\frac{p^2-p}{2}A_1|\mu(t)|^2\right\}$$

$$\le 4\mathbf{E}_Q \exp\left\{\frac{p^2-p}{2}A_1|\mu(T)|^2\right\}.$$

For any vector α and any positive numbers k and λ the equality

$$Q((\alpha, \mu(T)) > k) = Q\left\{\lambda(\alpha; \mu(T)) - \frac{\lambda^2}{2}(\langle\mu\rangle_T\alpha, \alpha) > \lambda k - \frac{\lambda^2}{2}(\langle\mu\rangle_T\alpha, \alpha)\right\}$$

is valid. Here $\langle\mu\rangle_T$ is the matrix characteristic of martingale μ, $\langle\mu\rangle_T = \{\int_0^T b(t, x(\cdot))b^*(t, x(\cdot))\,dt\}$. Since the function $b(\cdot, \cdot)$ is bounded, a constant C can be found such that $(\langle\mu\rangle_T\alpha, \alpha) \le C^2|\alpha|^2$. If $|\alpha| = 1$ we have

$$Q\{(\alpha, \mu(T)) > k\} \le Q\{e^{\lambda(\alpha, \mu(T)) - (\lambda^2/2)(\langle\mu\rangle_T\alpha, \alpha)} > e^{(\lambda k - \lambda^2C^2/2)}\}.$$

However,

$$\mathbf{E}_Q e^{\lambda(\alpha, \mu(T)) - (\lambda^2/2)(\langle\mu\rangle_T\alpha, \alpha)} = 1$$

(this follows from the condition (3.127) in which $b^{-1}h$ should be replaced by $\lambda\alpha b$ and from the condition $|b| \le C$). Utilizing Chebyshev's inequality we obtain

$$Q\{(\alpha, \mu(T)) > k\} \le \frac{1}{\exp\left\{\lambda k - \dfrac{\lambda^2C^2}{2}\right\}}.$$

Setting $\lambda = k/C^2$ we have $Q\{(\alpha, \mu(T)) > k\} \le e^{-k^2/2C^2}$. In particular for each component $\mu_j(T)$ of the vector $\mu(T)$ the inequality $Q\{\mu_j(t) > k\} \le \exp(-k^2/2C^2)$ is valid. Since the preceding bounds are applicable to the martingale $\mu(t)$, we have $Q\{|\mu_j(t)| > k\} \le 2\exp(-k^2/2C^2)$. Now for $\alpha > 0$ we obtain

$$\mathbf{E}_Q e^{\alpha|\mu(T)|^2} = \int_1^\infty Q(e^{\alpha|\mu(T)|^2} > t)\,dt$$

$$= \int_1^\infty Q\left(|\mu(T)| \ge \sqrt{\frac{\ln t}{\alpha}}\right)dt \le 2d\int_1^\infty \exp\left\{-\frac{\ln t}{2\alpha C^2}\right\}dt$$

$$= 2d\int_1^\infty \left(\frac{1}{t}\right)^{1/2\alpha C^2} dt.$$

This expression is finite provided $\alpha > 0$ is sufficiently small. The lemma is thus proved. □

Corollary 1. *If the conditions of lemma* 3.18 *are satisfied then the family of densities* $p_T^0(a^u)$, $u \in \mathscr{G}$, *is uniformly integrable.*

Corollary 2. *If the conditions of lemma* 3.18 *are satisfied then the family of densities* $\{\rho_T^0(a^u),\ u \in \mathfrak{U}_1\}$ *is a relatively weak compact family (cf.* [13]) *and limit density is represented by* (3.127).

This property means that given an arbitrary sequence of controls $u^{(n)}(\cdot)$ one can select a subsequence $u^{(n_k)}(\cdot)$ (which will again be denoted by $u^{(n)}(\cdot)$) and an integrable function $\rho = \rho(x(\cdot))$, $x(\cdot) \in C$, can be found such that

$$\int_C f(x(\cdot))\rho_T^0(a^{u_n}(\cdot))\, dP_0 \to \int_C f(x(\cdot))\rho\, dP_0$$

as $n \to \infty$ for any bounded $\mathfrak{C}$-measurable function $f(x(\cdot))$. This implies in particular that $\rho \ge 0$ and $\int_C \rho\, dP_0 = 1$. Let $A = \{x(\cdot) : \rho = 0\}$. Then $\rho_T^0(a^{u_n}) \to 0$ as $n \to \infty$ on the set A (mod P_0) and

$$\zeta(a^{u_n}) = \int_0^T (b^{-1}a^{u_n})^*\, dw - \frac{1}{2}\int_0^T |b^{-1}a^{u_n}|^2\, dt \to -\infty \quad \text{on } A \ (\text{mod } P_0).$$

However, as was mentioned above, the quantity $\int_0^T |b^{-1}a^{u_n}|^2\, dt$ is stochastically bounded, so that $\int_0^T (b^{-1}a^{u_n})^*\, dw \to -\infty$, $x(\cdot) \in A$ (mod P_0). However

$$\mathbf{E}\chi_A \varliminf \left|\int_0^T (b^{-1}a^{u_n})^*\, dw\right|^2 \le \varliminf_{n\to\infty} \mathbf{E}\left|\int_0^T (b^{-1}a^{u_n})\, dw\right|^2$$

$$= \varliminf_{n\to\infty} \mathbf{E}\int_0^T |b^{-1}a^{u_n}|^2\, dt < \infty.$$

Thus $P_v(A) = 0$ and $\rho > 0$ (mod P_0).

Set $\rho_t = E\{\rho \mid \mathfrak{C}_t\}$. Then $\{\rho_t, \mathfrak{C}_t, t \in [0, T]\}$ becomes a continuous positive martingale. Hence in view of theorem 3.12 it admits the representation of the form

$$\rho_t = \exp\{\mu(t) - \tfrac{1}{2}\langle \mu, \mu\rangle_t\},$$

where $\mu(t)$ is a continuous local martingale adapted to $\{\mathfrak{C}_t, t \in [0, T]\}$. Since the σ-algebra $\mathfrak{C}_t$ is generated by the process $w(s)$, $s \in [0, t]$, the representation

$$\mu(t) = \int_0^t (c(s), dw(s))$$

is valid and

$$\rho_t = \exp\left\{\int_0^t (c(s), dw(s)) - \frac{1}{2}\int_0^t |c(s)|^2\, ds\right\}.$$

We now return to the optimal cost function. Let $\mathfrak{U}_1$ denote the class of all $\mathscr{G}$-measurable controls $u = u(t, x(\cdot))$. Taking into account the above stated fact that the functions $Z(t, x(\cdot))$ and $Z_1(t, x(\cdot))$ coincide one can formulate the optimality principle in the case under consideration as follows:

$$Z(t, x(\cdot)) \le \mathbf{E}\left[\int_t^s f(t', \xi^{(u)}_{t, x(\cdot)}(\cdot), u(t'))\, dt' + Z(s, \xi^{(u)}_{t, x(\cdot)}(\cdot))\right], \tag{3.128}$$

where the equality sign is attained if and only if the control $u(t, x(\cdot))$ is optimal. Here $\xi^{(u)}_{t, x(\cdot)}(s)$ is the solution of equation (3.119) in the region $s \ge t$ under the control $u = u(t) = u(t, x(\cdot))$ and for the given "past" $\xi^{(u)}_{t, x(\cdot)} = x(s)$ for $s \le t$.

Observe that theorem 3.19 was originally stated for the case of cost functionals of a somewhat different form. However its proof is fully applicable to the case under consideration since only the continuity of functionals F^a_{ts}, G_{ts} and F_s in $x(\cdot)$ is used in that proof.

Taking into account that the conditional mathematical expectation of a random variable in $\{C, \mathfrak{C}\}$ with respect to $\mathfrak{C}_t$ is a $\mathfrak{C}_t$-measurable function of $x(\cdot)$, we can rewrite the inequality (3.128) also in the form

$$Z(t, x(\cdot)) \le \mathbf{E}_u\left[\int_t^s f[t', y(\cdot), u(t', y(\cdot))]\, dt' + Z(s, y(\cdot)) \,|\, \mathfrak{C}_t\right], \tag{3.129}$$

where $\mathbf{E}_u$ is the symbol of integration in $\{C, \mathfrak{C}\}$ with respect to argument $y(\cdot)$ and measure P_u. To simplify the notation we set $f_t^{(u)} = f(t, y(\cdot), u(t, y(\cdot)))$. We then have the following expression for the cost function $\bar{F}_t(u(\cdot))$:

$$\bar{F}_t(u(\cdot)) = \mathbf{E}_u\left\{\int_t^T f_s^{(u)}\, ds \,|\, \mathfrak{C}_t\right\}.$$

Lemma 3.19. *A control $u^*(\cdot) \in \mathfrak{U}_1$ is optimal if and only if for any $u(\cdot) \in \mathfrak{U}_1$ there exists an integrable process $\alpha_u(t)$ adapted to $\{\mathfrak{C}_t, t \in [0, T]\}$ and satisfying the conditions*

$$\mathbf{E}_u\left\{\int_0^T \alpha_u(s)\, ds \,|\, \mathfrak{C}_0\right\} = F^*,$$

$$\inf_{u(\cdot)}\{f_t^{(u)} - \alpha_u(t)\} = f_t^{(u^*)} - \alpha_{u^*}(t) = 0,$$

where F^* is a constant independent of $u(\cdot)$.

Proof. Assume that u^* is optimal. Let $F^* = \bar{F}_0(u^*)$, $k(u(\cdot)) = F^*[\bar{F}_0(u)]^{-1}$. Then $k(u(\cdot)) \le 1$ and $k(u(\cdot)) = 1$ if and only if control $u(\cdot)$ is optimal. Set $\alpha_u(t) = k(u(\cdot))f_t^{(u)}$. Then

$$\mathbf{E}_u\left\{\int_0^T \alpha_u(t)\, at \,|\, \mathfrak{C}_0\right\} = k(u(\cdot))\bar{F}_0(u(\cdot)) = F^*$$

does not depend on $u(\cdot)$ and $f_t^{(u)} - \alpha_u(t) = (1 - k(u(\cdot)))f_t^{(u)} \ge f_t^{(u^*)} - \alpha_{u^*}(t) \equiv 0$. Thus the necessity of the lemma's conditions is verified.

Let the lemma's conditions be fulfilled. Set

$$V_t(u(\cdot)) = \mathbf{E}_u\{\alpha_u(T)\,|\,\mathfrak{C}_t\} - (\alpha_u(t), \alpha_u(t)) = \int_0^t \alpha_u(s)\, ds.$$

Then $V_0(u(\cdot)) = F^*$ does not depend on $u(\cdot)$. We have

$$\bar{F}_t(u(\cdot)) - V_t(u(\cdot)) = \mathbf{E}_u\left\{\int_t^T [f_s^{(u)} - \alpha_u(s)]\, ds \,|\, \mathfrak{C}_t\right\} \geq 0,$$

and the equality sign is attained if $u(\cdot) = u^*(\cdot)$. Hence $\bar{F}_t(u(\cdot)) \geq V_t(u(\cdot))$, $\bar{F}_0(u(\cdot)) \geq V_0(u(\cdot)) = F^*$; the equality sign is attained for $u(\cdot) = u^*(\cdot)$. Then $F^* = \min_{u(\cdot)} \bar{F}_0(u(\cdot)) = V(0, x(\cdot))$. The lemma is thus proved. □

Remark. If the conditions of the lemma are fulfilled for all $u(\cdot)$ belonging to a class $\mathfrak{U}'$ of admissible controls then u^* is an optimal control in the class $\mathfrak{U}'$.

We shall call a control $u = u(t, x(\cdot))$ *monotone* if for all $(t, t+h) \in [0, T] \times [0, T]$, $h > 0$,

$$Z(t, x(\cdot)) \geq \mathbf{E}_u\{Z(t+h, \xi_{tx}^{(u)}(\cdot))\,|\,\mathfrak{C}_t\},$$

i.e. if $Z(t, x(\cdot))$ is a $\{\mathfrak{C}_t, P_u\}$-supermartingale. Clearly if an optimal control $u^*(t, x(\cdot))$ exists, then such a control is monotone. For brevity we denote $Z_t = Z(t, x(\cdot))$.

Lemma 3.20. *Let a control $u(\cdot) \in \mathfrak{U}_1$ be monotone. Then processes $L_u Z_t$ and $\Lambda_u Z_t$ exist with values in R and R^d respectively adapted to $\{\mathfrak{C}_t, t \in [0, T]\}$ and such that*

$$\mathbf{E}_u \int_0^T |L_u Z_t|\, dt < \infty,$$

$$\int_0^T |\Lambda_u Z_t|^2\, dt < \infty \pmod{P_u}$$

and

$$Z_t = F^* + \int_0^t L_u Z_s\, ds + \int_0^t (\Lambda_u Z_s, dx(s)) \pmod{P_u}. \tag{3.130}$$

Proof. Since the function Z_t is a supermartingale and the function f is bounded it follows from the optimality principle that

$$0 \leq \mathbf{E}_u(Z_t - Z_{t+h}) \leq \mathbf{E}_u \int_t^{t+h} f_s^{(u)}\, ds \leq ch,$$

where c is a constant. Hence $\mathbf{E}_u Z_t$ is a continuous function. It is easy to verify that the current of σ-algebras $\mathfrak{C}_t$ is continuous. Thus the supermartingale Z_t admits modification for which the sample functions are continuous from the right with probability 1. Moreover this process is regular, i.e. $\lim \mathbf{E}Z_{\tau_n} = \mathbf{E}Z_\tau$ for any monotonically non-decreasing sequence of random times τ_n on $\mathfrak{C}_t$, $t \in [0, T]$, where $\tau = \lim \tau_n$. This also follows from the inequality presented above with $t = \tau_n$ and $t + h = \tau$.

In view of Meyer's theorem [36, or 20, vol III] the supermartingale Z_t can be represented in the form

$$Z_t = \mathbf{E}_u\{A_T \,|\, \mathfrak{C}_t\} - A_t,$$

where A_t is a continuous monotonically non-decreasing process adapted to the current $\{\mathfrak{C}_t\}$ and $A_0 = 0$. In view of a theorem in the theory of martingales (cf. [20], vol III, Chapter 1, Section 1) relation

$$A_t = \mathbf{P}\lim_{|\delta| \to 0} \sum_{k=0}^{n} \mathbf{E}_u\{-\Delta Z_{t_k} \,|\, \mathfrak{C}_{t_k}\}$$

is valid for process A_t (here δ denotes the subdivision of the interval $[0, t]$ generated by the points $0 = t_0 < t_1 < \cdots < t_n < t_{n+1} = t$ and $\Delta Z_{t_k} = Z_{t_{k+1}} - Z_{t_k}$. Moreover

$$0 \in \mathbf{E}_u\{-\Delta Z_{t_k} \,|\, \mathfrak{C}_{t_k}\} \le \mathbf{E}_u\left\{\int_{t_k}^{t_{k+1}} f_s^{(u)}\, ds \,\middle|\, \mathfrak{C}_{t_k}\right\} \le c(t_{k+1} - t_k),$$

which implies that $A_t - A_s \le c(t - s)$, $dt \times dP_u$-almost for all $(t, x(\cdot))$. Therefore the process A_t is with probability 1 absolutely continuous in t and

$$A_t = \int_0^t \alpha(s, x(\cdot))\, ds,$$

where α_t is a process adapted to the current $\mathfrak{C}_t$. Proceding to process $\eta_t = \mathbf{E}_u\{A_T \,|\, \mathfrak{C}_t\}$ we note that this process is a $\{\mathfrak{C}_t, P_u\}$-martingale, and since $x(t)$ is a non-anticipative functional of $w(t)$ it follows that $\eta_t = \varphi(t, w(\cdot))$. Here $\varphi(t, w(\cdot))$ is a non-anticipative function of $w(t)$ and $w(t)$ is a $\{\mathfrak{C}_t, P_u\}$-Wiener process. In such a case (cf. [20] vol III, Chapter 3, Section 1) there exists a vector process $\psi(t)$ adapted to $\mathfrak{C}_t$ such that

$$\eta_t = \gamma + \int_0^t (\psi(s), dw(s)), \qquad \int_0^T |\psi(s)|^2\, ds < \infty.$$

Setting $t = 0$ we obtain $\gamma = \eta_0 = Z_0$. Since

$$dw(t) = b^{-1}(t, x(\cdot))[dx(t) - a(t, x(\cdot), u(t, x(\cdot))\, dt],$$

we arrive at the representation (3.130) such that

$$L_u Z_t = -(\alpha(t, x(\cdot)) + (\psi, b^{-1}(t, x(\cdot))a[t, x(\cdot), u(t, x(\cdot))]),$$
$$\Lambda_u Z_t = b^{-1*}(t, x(\cdot))\psi(t).$$

The lemma is thus proved. □

Theorem 3.25. *Let there exist an optimal control. Then the control* $u^* = u^*(t, x(\cdot))$ *is optimal if and only if there exists a constant* F^* *and for every monotone control there exist processes* $\gamma^{(u)}(t)$, $\beta^{(u)}(t)$ *with values in* R *and* R^d *respectively adapted to* $\{\mathfrak{C}_t, t \in [0, T]\}$ *such that*

a. $\int_0^T |\beta^{(u)}(t)|^2\, dt < \infty (\text{mod } P_u)$, *and* $\mathbf{E} \int_0^T \beta^{(u)}(s)\, dx(s) = 0$;

b. $V^{(u)}(T) = 0$, *where*

$$V^{(u)}(t) = F^* + \int_0^t \gamma^{(u)}(s)\, ds + \int_0^t (\beta^{(u)}(s), dx(s)); \tag{3.130}$$

c.

$$\inf_u \{\gamma^{(u)}(t) + (\beta^{(u)}(t), a^{(u)}(t)) + f_t^{(u)}\} = \gamma^{(u^*)}(t) + (\beta^{(u^*)}(t), a^{(u^*)}(t)) + f_t^{(u^*)} = 0; \tag{3.131}$$

$dt \times dP_u$*-almost for all* (t, x).

If the conditions of the theorem are fulfilled then $V^{(u^*)}(t) = \bar{F}_t(u^*) = Z_t$; $F^* = \bar{F}(u^*) = Z_0$ *is an optimal cost of control. Here* $a^{(u)}(t) = a(t) = a[t, x(\cdot), u(t, x(\cdot))]$ *and* $f_t^{(u)} = f[t, x(\cdot), u(t, x(\cdot))]$.

(Equation (3.131) should be considered as an analog of Bellman's equation; we shall verify this fact below.)

Proof. Let $u = u(t, x(\cdot))$ be a monotone control. In view of lemma 3.20 we have

$$\begin{aligned} Z_t &= F^* + \int_0^t L_u Z_s\, ds + \int_0^t (\Lambda_u Z_s, dx(s)) \\ &= F^* + \int_0^t (L_u Z_s + (\Lambda_u Z_s, a^{(u)}(s)))\, ds \\ &\quad + \int_0^t (\Lambda_u Z_s)^* b(s, x(\cdot))\, dw(s). \end{aligned}$$

Bellman's optimality principle and lemma 3.20 yield that

$$\begin{aligned} \mathbf{E}_u\left\{\int_t^{t+h} f_s^{(u)}\, ds \,\middle|\, \mathfrak{C}_t\right\} &\geq Z_t - \mathbf{E}_u\{Z_{t+h} \mid \mathfrak{C}_t\} \\ &= -\mathbf{E}_u\left\{\int_t^{t+h} (L_u Z_s + (\Lambda_u Z_s, a^{(u)}(s)))\, ds \,\middle|\, \mathfrak{C}_t\right\}. \end{aligned}$$

whence

$$L_u Z_t + (\Lambda_u Z_t, a^{(u)}(t)) + f_t^{(u)} \geq 0 \text{ mod}(dt \times dP_u), \tag{3.132}$$

and the equality sign is valid only for an optimal control (provided it exists). Thus the assertions of the theorem are fulfilled with $\gamma^{(u)}(t) = L_u Z_t$, $\beta^{(u)}(t) = \Lambda_u Z_s$, $F^* = Z_0$.

Now let $u^*(\cdot, x(\cdot))$, F^*, $\gamma^{(u)}(t)$, $\beta^{(u)}(t)$ exist and satisfy the conditions of the theorem. Consider a monotone control $u = u(t, x(\cdot)) \in \mathfrak{U}_1$. We have

$$V^{(u)}(t) = F^* + \int_0^t (\gamma^{(u)}(s) + (\beta^{(u)}(s), a^{(u)}(s)))\, ds + \int_0^t (\beta^{(u)}(s), b(s, x(\cdot))\, dw(s))$$

Set $\alpha^{(u)}(t) = -(\gamma^{(u)}(t) + (\beta^{(u)}(t), a^{(u)}(t)))$. Then

$$\mathbf{E}_u\left\{\int_0^T \alpha^{(u)}(s)\, ds \,\middle|\, \mathfrak{C}_0\right\} = F^*, \qquad f^{(u)}(t) - \alpha^{(u)}(t) \geq 0,$$

in view of the conditions of the theorem. Moreover if $u = u^*$ then the equality is valid in the last inequality. It follows from lemma 3.19 that u^* is an optimal control in the class of all monotone controls. Since an optimal control is itself monotone it coincides with u^*. Furthermore

$$\begin{aligned}
&\mathbf{E}_{u^*}\left\{\int_t^T f^{(u^*)}(s)\, ds \,\middle|\, \mathfrak{C}_t\right\} \\
&= \mathbf{E}_{u^*}\left\{\int_t^T (-\gamma^{(u^*)}(s) - (\beta^{(u^*)}(s), a^{(u^*)}(s)))\, ds - \int_t^T (\beta^{(u^*)}(s), b(s, x(\cdot))\, dw(s)) \,\middle|\, \mathfrak{C}_t\right\} \\
&= \mathbf{E}_{u^*}\left\{-\int_t^T \gamma^{(u^*)}(s)\, ds - \int_t^T (\beta^{(u^*)}(s), dx(s)) \,\middle|\, \mathfrak{C}_t\right\}.
\end{aligned}$$

In view of condition b) of theorem 3.25 we have

$$\begin{aligned}
&\mathbf{E}_{u^*}\left\{\int_t^T f^{(u^*)}(s)\, ds \,\middle|\, \mathfrak{C}_t\right\} \\
&= F^* + \mathbf{E}_{u^*}\left\{\int_0^t \gamma^{(u^*)}(s)\, ds + \int_0^t (\beta^{(u^*)}(s), dx(s)) \,\middle|\, \mathfrak{C}_t\right\} \\
&= F^* + \int_0^t \gamma^{(u^*)}(s)\, ds + \int_0^t (\beta^{(u^*)}(s), dx(s)),
\end{aligned}$$

so that $Z_t = V^{(u^*)}(t)$. The theorem is proved. □

Since we have not established the existence of an optimal control we don't know yet whether the random process $Z_t = Z(t, x(\cdot))$ admits representation of the form (3.130). We shall now show that this is indeed the case.

Lemma 3.21. *Let the conditions of lemma 3.18 be satisfied. Then there exists a process $h(t, x(\cdot))$ such that $\{Z_t, \mathfrak{C}_t, P^*\}$ is a supermartingale. Here*

$$d\mathbf{P}^* = \exp\{\zeta_T^0(h)\}\, dP_0.$$

PROOF. Choose a sequence $u_n(\cdot) \in \mathfrak{U}_1$ such that $\bar{F}_0(u_n(\cdot)) \downarrow Z_0$. Since the set of densities is weakly compact there exists a subsequence of densities (to be denoted by $\rho_T^0(u_n(\cdot)))$ such that $\rho_T^0(u_n(\cdot)) \to \rho^*$ weakly in $L_1(P_0)$ where ρ^* admits the representation

$$\rho^* = \exp\left\{\int_0^T (c(s), dw(s)) - \frac{1}{2}\int_0^T |c(s)|^2 \, ds\right\}.$$

Set $\theta_t = \int_0^t (c(s), dw(s)) - \frac{1}{2}\int_0^t |c(s)|^2 \, ds$. It is easy to verify that $\mathbf{E}\{\rho^* \mid \mathfrak{C}_t\} = \exp\{\theta_t\}$. The weak convergence of $\rho_T^0(u_n(\cdot))$ to ρ^* implies that

$$\rho_t^0(u_n(\cdot)) = \mathbf{E}_0\{\rho_T^0(u_n(\cdot) \mid \mathfrak{C}_t\} \to \mathbf{E}_0\{\rho^* \mid \mathfrak{C}_t\} = \exp\{\theta_t\}.$$

We introduce in $\{C, \mathfrak{C}_t\}$ a measure $P^* : dP^* = \rho^* \, dP_0$. We show that for any $t \in [0, T)$, $\Delta > 0$ and $A \in \mathfrak{C}_t$ the inequality

$$\int_A (Z_{t+\Delta} - Z_t) \, dP^* \leq 0 \tag{3.133}$$

is satisfied. This will imply the assertion of the lemma. For brevity of notation we set $\rho_* = \exp\{\theta_{t+\Delta}\}$, $\rho_n = \rho_{t+\Delta}^0(u_n(\cdot))$. We then have

$$\begin{aligned}
&\int_A (Z_{t+\Delta} - Z_t) \, dP^* \\
&= \int_A (\rho_* - \rho_n)[Z_{t+\Delta} - Z_t] dP_0 + \int_A \rho_n[\bar{F}_t(u_n(\cdot)) - Z_t] \, dP_0 \\
&+ \int_A \rho_n[Z_{t+\Delta} - \bar{F}_{t+\Delta}(u_n(\cdot))] \, dP_0 + \int_A \rho_n[\bar{F}_{t+\Delta}(u_n(\cdot)) - \bar{F}_t(u_n(\cdot))] \, dP_0.
\end{aligned}$$

Since $Z_{t+\Delta} \leq \bar{F}_{t+\Delta}(u_n(\cdot))$ the third summand on the r.h.s. is non-positive while the fourth is non-positive in view of the fact that the relation

$$\bar{F}_t(u_n(\cdot)) = \mathbf{E}\left\{\int_t^{t+\Delta} f_s^{(u)} \, ds \mid \mathfrak{C}_t\right\} + \mathbf{E}\{\bar{F}_{t+\Delta}(u_n(\cdot)) \mid \mathfrak{C}_t\}$$

implies that $\bar{F}_t(u_n(\cdot))$ is a supermartingale with respect to P_{u_n}. Let $\varepsilon > 0$ be fixed and $\bar{F}_0(u_n(\cdot)) < Z_0 + \varepsilon$ for $n \geq n_0$. Then $\mathbf{E}_{u_n}[\bar{F}_t(u_n(\cdot)) - Z_t] < \varepsilon$ (since relation (3.128) implies that $Z_0 \leq \bar{F}_0(u_n) - F_t(u_n) + E_{u_n}\{Z_t \mid \mathfrak{C}_0\}$ or $\mathbf{E}(\bar{F}_t(u_n) - Z(t)) \leq \mathbf{E}(\bar{F}_0(u_n) - Z(0))$. Furthermore $(Z_{t+\Delta} - Z_t)$ is a bounded variable therefore there exist n' such that for all $n > n'$ the inequality

$$\int_A (\rho_* - \rho_n)(Z_{t+\Delta} - Z_t) \, dP < \varepsilon$$

is valid. We thus have for $n \geq \max(n_0, n')$, $\int_A \rho_*(Z_{t+\Delta} - Z_t) \, d\mathbf{P} < 2\varepsilon$. Since ε is arbitrary inequality (3.133) follows. □

Theorem 3.26. *Let the conditions of lemma 3.18 be satisfied. Then there exist processes $\{LZ_t\}$ and $\{\Lambda Z_t\}$ with values in R and R^d, respectively, adapted to $\{\mathfrak{C}_t\}$ and such that*

$$\int_0^T |\Lambda Z_t|^2 \, dt < \infty, \qquad \mathbf{E} \int_0^T |LZ_t| \, dt < \infty,$$

$$Z_t = F^* + \int_0^t LZ_s \, ds + \int_0^t (\Lambda Z_s, dx(s)) \ (\text{mod } P^*), \qquad \forall t \in [0, T]. \tag{3.134}$$

PROOF. We utilize the notation and constructions used in the proof of the preceding lemma.

Let the sequence u_n and measure P^* be chosen as described above. Then

$$|\mathbf{E}^*(Z_{t+h} - Z_t)| \le |\mathbf{E}_0\{(\rho^* - \rho(u_n(\cdot)))\}(Z_{t+h} - Z_t)| + |\mathbf{E}_0 \rho(u_n(\cdot))(Z_{t+h} - Z_t)|.$$

The first summand on the r.h.s. of the inequality tends to zero in view of the weak convergence of $\rho(u_u)$ to ρ^*. Furthermore

$$|\mathbf{E}_0 \rho(u_n)(Z_{t+h} - Z_t)| \le |\mathbf{E}_0 \rho(u_n(\cdot))(Z_{t+h} - \bar{F}_{t+h}(u_n(\cdot)))| + |\mathbf{E}_0 \rho(u_n)(Z_t - \bar{F}_t(u_n))| + |\mathbf{E}_0 \rho(u_n)(\bar{F}_{t+h}(u_n) - \bar{F}_t(u_n))|.$$

As it was shown in the course of the proof of the preceding lemma the first two summands tend to zero as $n \to \infty$ and the last summand is at most ch. Thus $|\mathbf{E}^*[Z_{t+h} - Z_t]| \le kh$. This implies that the supermartingale $\{Z_t, \mathfrak{C}_t, P^*\}$ can be written in the form $Z_t = \mathbf{E}^*\{A_T | \mathfrak{C}_t\} - A_t$, where A_t is an absolutely continuous monotone $\mathfrak{C}_t$-process. The rest of the proof of the theorem is carried out analogously to the proof of lemma 3.19. □

Remark. In the same manner as inequality (3.132) was obtained one can verify that

$$LZ_t + (\Lambda Z_t, a_t^{(u)}(t)) + f_t^{(u)} \ge 0, \quad \text{mod}(dt \times dP_0), \tag{3.135}$$

for any $u(t, x(\cdot))$; moreover the equality sign mod$(dt \times dP_0)$ is valid only in the case when the control is optimal.

The result just obtained can be utilized to prove the existence of an optimal feedback control. For this purpose we introduce the *Hamiltonian* $H(t, x, p, u)$ of the control system by setting

$$H(t, x, p, u) = (p, a(t, x, u)) + f(t, x, u),$$

where p is an arbitrary vector in R^d, $(t, x, u) \in [0, T] \times C \times U$.

We shall also introduce the function

$$H(t, x, p) = \min_u H(t, x, p, u). \tag{3.136}$$

This minimum exists since $a(t, x, u)$ and $f(t, x, u)$ are continuous and U is compact. Consequently one can define a Borel function $y^*(t, x, p)$, $(t, x, p) \in [0, T] \times C \times R^d$ for which the minimum is attained in formula (3.136), i.e. this function satisfies equality

$$H(t, x, p) = H(t, x, p, y^*(t, x, p)).$$

Theorem 3.27. *Let $(a, b) \in S(C, L_N)$, $b(t, x)$ be a non-degenerate bounded matrix and function $b^{-1}a(t, x)$ be linearly bounded. Then an optimal control exists and is given by the function*

$$u^*(t, x) = y^*(t, x, \Lambda Z(t, x(\cdot))). \tag{3.137}$$

PROOF. The function $u^*(t, x)$ is clearly $\mathscr{G}$-measurable. It follows from the remark following theorem 3.26 that to prove the optimality of control u^* it is sufficient to show that equality

$$LZ_t + (\Lambda Z_t, a^{(u^*)}(t)) + f_t^{(u^*)} = 0 \quad \text{mod}(dt \times dP_0) \tag{3.138}$$

is fulfilled. We shall now utilize equation (3.134). Replacing dx in it by $a^{(u)}(t)\, dt + b dw$, we obtain

$$Z_t = F^* + \int_0^t [LZ_s + (\Lambda Z_s, a^{(u)}(s))]\, ds + \int_0^t \Lambda Z_s b\, dw(s),$$

whence

$$F^* + \mathbf{E}_u \int_0^T [LZ_s + (\Lambda Z_s, a^{(u)}(s))]\, ds = 0.$$

Furthermore in view of the definition of function u^* $(\Lambda Z_t, a^{(u)}(t)) + f_t^{(u)} \geq (\Lambda Z_t, a_t^{(u^*)}) + f_t^{(u^*)}$, so that

$$F^* \leq \mathbf{E}_u \int_0^T f_s^{(u)}\, ds - \mathbf{E}_u \int_0^T [LZ_s + \alpha]\, ds,$$

where $\alpha = f_s^{(u^*)} + (\Lambda Z_s, a^{(u^*)}(s))$. Set $\xi = \int_0^T (LZ_s + \alpha)\, ds$. Then $\xi \geq 0$ (cf. (3.135)). For any n one can find a u_n such that

$$\mathbf{E}_{u_n} \int_0^T f_s^{(u_n)}\, ds < F^* + \frac{1}{n}.$$

Thus $\mathbf{E}_{u_n} \xi < 1/n$. If $\xi^{(N)} = (\xi \wedge N)$ then all the more $\mathbf{E}_{u_n} \xi^{(N)} < 1/n$. Let $\rho_n = \rho_T^0(a^{(u_n)}(\cdot))$. In view of the above ρ_n is weakly compact in L_1. Therefore there exists a density ρ_0 in C^d and a subsequence ρ_n such that

$$\lim_{k \to \infty} \mathbf{E}_{u_{n_k}} \xi^{(N)} = \lim_{k \to \infty} \mathbf{E}_0\, \rho_{n_k} \xi^{(N)} = \mathbf{E}_0\, \rho_0\, \xi^{(N)} = 0;$$

$\rho_0\, \xi^{(N)} = 0$ (mod P_0), but $\rho_0 > 0$ (mod P_0). Therefore $\xi^{(N)} = 0$ and $\xi = 0$ (mod P_0). This implies (3.138) and the theorem is proved. □

7 Controlled Diffusion Processes

Assume that functions $a(t, \xi(\cdot), u)$ and $b(t, x(\cdot))$ are of the form

$$a(t, x(\cdot), u) = a(t, x(t), u), \qquad b(t, x(\cdot)) = b(t, x(t)).$$

If the control $u = u(t)$ is non-random or $u = u(t, x(t))$, then the solution of equation

$$d\xi = a(t, \xi(t), u(t, \xi(t)))\,dt + b(t, \xi(t))\,dw(t), \tag{3.139}$$

provided it exists and is unique, defines a Markov process. The control $u = u(t, x(t))$, where $u(t, x)$ is a Borel function of arguments $(t, x) \in [0, T] \times R^d$ is called *Markovian.*

Let the control cost be of the form

$$\bar{F}(u) = \mathbf{E}\int_0^T f(t, \xi^{(u)}(t), u(t, \xi^{(u)}(t)))\,dt,$$

where $f(t, x, u)$ is a continuous non-negative function, $(t, x, u) \in [0, T] \times R^d \times U$, $\xi^{(u)}(t)$ be the solution of equation (3.139) corresponding to the control $u = u(t, \xi^{(u)}(t))$. Denote $f_t^{(u)} = f(t, \xi^{(u)}(t), u(t, \xi^{(u)}(t)))$. Let the conditions of the preceding section be satisfied. Namely:

a. $(a, b) \in S(C, L_N)$;
b. the matrix $b(t, x)$ is non-degenerate and bounded;
c. the function $b^{-1}(t, x)\,a(t, x, u)$ is linearly bounded.

We shall assume as before that the space $\{C, \mathfrak{C}, P_0\}$ where P_0 is the measure associated with the solution of the stochastic differential equation

$$dx(t) = b(t, x(t))\,dw(t)) \tag{3.140}$$

is the basic probability space.

Under these conditions equation (3.139) possesses a solution for any control $u = u(t, \xi(t))$, where $u(t, x)$ is a Borel function. This class of controls will be denoted by $\mathfrak{U}_2$. It turns out that the class of controls $\mathfrak{U}_2$ is dense in the class $\mathfrak{U}$ which is the class of all feedback controls.

Theorem 3.28. *Let the conditions* a-c *be fulfilled. Then for any* $\varepsilon > 0$ *there exists an* ε*-optimal Markov control.*

This theorem follows from theorem 3.29 proved below. □

As it was shown above

$$\inf_{\eta(\cdot)\in\mathfrak{U}} F(\xi^{(\eta)}(\cdot), \eta(\cdot)) = \lim_{|\delta|\to 0}\ \inf_{\eta(\cdot)\in\mathfrak{U}_\delta} \mathbf{E}F^\delta(\xi^{(\eta)}(\cdot), \eta(\cdot)). \tag{3.141}$$

In the case under consideration the functional $F(x(\cdot), u(\cdot))$ is of the form

$$F(x(\cdot), u(\cdot)) = \int_0^T f(s, x(s), u(s))\, ds,$$

and

$$F^\delta(x(\cdot), u(\cdot)) = \sum_{k=0}^{n} \int_{t_k}^{t_{k+1}} f(s, x(t_k), u(t_k))\, ds,$$

$0 = t_0 < t_1 < \cdots < t_{n+1} = T$. Since the function $f(t, x, u)$ is continuous, equality (3.141) remains valid if we presuppose that

$$F^\delta(x(\cdot), u(\cdot)) = \sum_{k=0}^{n} f(t_k, x(t_k), u(t_k))\, \Delta t_k. \tag{3.142}$$

In what follows we shall assume that the functional F^δ is given by formula (3.142). As it was shown above an optimal control for the solution of equation (3.139) with the cost functional (3.142) can be determined in the following manner. Construct a sequence of functions $Z_k(x)$, $Z_k^*(x, u)$, $k = n+1, n, \ldots, 0$ by setting $Z_k(x) = \min_u Z_k^*(x, u)$,

$$Z_k^*(x, u) = f(t_k, x, u)\, \Delta t_k + \mathbf{E} Z_{k+1}(\xi_{t_{k+1}}(t_k, x, u)),\ Z_{n+1}^*(x, u) = 0,$$

where

$$\xi_t(t_k, x, u) = x + \int_{t_k}^{t} a(s, x, u)\, ds + \int_{t_k}^{t} b(s, x, u)\, dw(s).$$

Denote by $g_k(x)$ a Borel function such that $Z_k(x) = Z_k^*(x, g_k(x))$. The control $u = g_\delta(t, x) = g_k(x)$ for $t \in [t_k, t_{k+1})$, $k = \overline{0, n}$, is an ε-optimal control for $|\delta| = \max_k |t_{k+1} - t_k|$ sufficiently small.

Now let $\eta(t) = \varphi(t, \xi(t))$ be a Markov control for the solution of equation (3.139) such that for all t the inequality

$$\mathbf{E} f[t_k, \xi^\varphi_{t_{k-1}x}(t_k), \eta(t_k)] \le \inf_u f(t_k, \xi_{t_k}(t_{k-1}, x, u), u) \tag{3.143}$$

is satisfied. Then

$$\begin{aligned} &\mathbf{E} f(t_k, \xi^\varphi_{0x}(t_k), \eta(t_k)) \\ &\quad = \mathbf{E}\{\mathbf{E}[f(t_k, \xi^\varphi_{t_{k-1}y}(t_k), \varphi(t_k, \xi_{t_{k-1}y}(t_k))]\,|_{y=\xi_{0x}(t_{k-1})}\} \\ &\quad \le \mathbf{E} f[t_k, \xi_{t_k}(t_{k-1}, \xi_{0x}(t_{k-1}), u), u], \end{aligned}$$

and using an induction argument we obtain

$$\mathbf{E} f[t_k, \xi^\varphi_{0x}(t_k), \eta(t_k)] \le \mathbf{E} f[t_k, \xi_k, g_k(\xi_k)],$$

where $\xi_0, \xi_1, \ldots, \xi_n, \xi_{n+1}$ is a Markov sequence defined by equations

$$\xi_0 = x,$$

$$\xi_{k+1} = \xi_{t_{k+1}}(t_k, \xi_{t_k}, g_k(\xi_{t_k})), \qquad k = 0, 1, \ldots, n.$$

Therefore to construct an ε-optimal Markovian control it is sufficient to be able to construct a Markovian control such that the inequality (3.143) is satisfied. Clearly, for this purpose it is sufficient to construct for each interval $[t_k, t_{k+1}]$ the corresponding function $\varphi(t, x)$. It is also evident that for the solution of the problem the following simplifying assumptions can be made:

a. in place of the interval $[t_k, t_{k+1}]$ the interval $[0, T]$ can be considered;
b. it may be assumed that the set U consists of a finite number of points $u_1, \ldots, u_N$;
c. functions $a(t, x, u)$ and $b(t, x)$ are twice continuously differentiable with respect to x;
d. the function $f(t_k, x, u) = h(x, u)$ are twice continuously differentiable with respect to x and continuous in u.

Since U is compact, when constructing an ε_1-net in U we observe that if some ε-optimal control with values in U is constructed, then for an ε_1 sufficiently small one can construct a 2ε-optimal control whose values are points of a ε_1-net on the set U. This justifies assumption b. It is also easy to show that if the functions $a_n(t, x, u)$ and $b_n(t, x, u)$ satisfy the conditions of linear boundedness and that of the Lipschitz condition with constants independent of n and $|a - a_n| + |b - b_n| \le \rho_n(1 + |x|)$, where $\rho_n \to 0$, then a solution of equation (3.139) with coefficients (a_n, b_n) will converge uniformly in u to a solution of a stochastic differential equation with coefficients (a, b).

Thus in the problem under consideration it may be assumed that functions $a(t, x, u)$ and $b(t, x)$ are sufficiently smooth, for example, satisfying condition c. Clearly condition d is also acceptable.

The required control can be constructed in the following manner:
Set

$$Z(t, x) = \min_{u \in U} \mathbf{E}h(\xi_T(t, x, u); u).$$

Let $\varphi(t, x)$ satisfy equation

$$Z(t, x) = \mathbf{E}h[\xi_T(t, x, \varphi(t, x)); \varphi(t, x)].$$

For a given subdivision δ of the interval $[0, T]$ using points $0 = t_0 < t_1 \cdots < t_n < t_{n+1} = T$, we denote by $\eta(t)$ a step control such that $\eta(t_k) = \varphi(t_k, \xi(t_k))$ and let $\check{\xi}_{t_k x}(t)$ be a solution of equation (3.139) on the interval $[t_k, T]$ corresponding to the control $\eta(t)$. Then

$$\begin{aligned} Z(t_n, x) &= \mathbf{E}h[\xi_T(t_n, x, \varphi(t_n, x)), \varphi(t_n, x)] \\ &= \mathbf{E}h[\check{\xi}_{t_n x}(T), \varphi(t_n, \check{\xi}_{t_n x}(t_n))], \\ Z(t_{n-1}, x) &= \mathbf{E}h[\xi_T(t_{n-1}, x, \varphi(t_{n-1}, x)), \varphi(t_{n-1}, x)] \\ &= \mathbf{E}h[\xi_T(t_n, \check{\xi}_{t_{n-1} x}(t_n), \varphi(t_{n-1}, x)), \varphi(t_{n-1}, x)] \\ &\ge \mathbf{E} \inf_u \mathbf{E}\{h[\xi_T(t_n, \check{\xi}_{t_{n-1} x}(t_n), u), u] \check{\xi}_{t_{n-1} x}(t_n)\} \\ &= \mathbf{E}\mathbf{E}\{h[\xi_T(t_n, \check{\xi}_{t_{n-1} x}(t_n)), \varphi(t_n, \check{\xi}_{t_{n-1} x}(t_n)), \end{aligned}$$

$$\varphi(t_n, \check{\xi}_{t_{n-1}x}(t_n))] \mid \check{\xi}_{t_{n-1}x}(t_n)\}$$
$$= \mathbf{E}h[\xi_T(t_n, \check{\xi}_{t_{n-1}x}(t_n)), \varphi(t_n, \check{\xi}_{t_{n-1}x}(t_n)), \varphi(t_n, \check{\xi}_{t_{n-1}x}(t_n))]$$
$$= \mathbf{E}h[\check{\xi}_{t_{n-1}x}(T), \varphi(t_n, \check{\xi}_{t_{n-1}x}(t_n))].$$

Continuing in this manner we obtain

$$Z(t_k, x) \geq \mathbf{E}h[\check{\xi}_{t_kx}(T), \varphi(t_n, \check{\xi}_{t_kx}(t_n))].$$

In particular the last inequality implies

$$\mathbf{E}h[\check{\xi}_{0x}(T), \eta(T)] \leq \inf_u h[\xi_T(0, x, u), u].$$

If the finite-dimensional (marginal) distributions of processes $\check{\xi}_{0x}(t)$ and $\eta(t)$ for $|\delta| \to 0$ converge to the finite-dimensional (marginal) distributions of some processes $\xi_{0x}(t)$ and $\eta(t)$, then

$$\mathbf{E}h[\xi_{0x}(T), \eta(T)] \leq \inf_u \mathbf{E}h[\xi_T(0, x, u), u].$$

This will establish the existence of ε-optimal Markovian controls under the condition that $\xi_{0x}(t)$ is a solution of the equation

$$d\xi_{0x} = a(t, \xi_{0x}(t), \eta(t))\, dt + b(t, \xi_{0x}(t))\, dw(t) \tag{3.144}$$

and $\eta(t) = \varphi(t, \xi(t))$ for almost all t with probability 1.

Denote by Δ_k the set

$$\Delta_k = \{(t, x) : \varphi(t, x) = u_k\}.$$

Consider the process $\xi^\delta(t)$ satisfying the equation

$$\xi^\delta(t) = \xi^\delta(t_k) + \sum_{i=1}^{N} \chi_{\Delta_i}(t_k, \xi^\delta(t_k)) \int_{t_k}^{t} a_i(s, \xi^\delta(s))\, ds$$
$$+ \int_{t_k}^{t} b(s, \xi^\delta(s))\, dw(s), \qquad t \in [t_k, t_{k+1}],$$

where $a_i(t, x) = a(t, x, u_i)$ and $\chi_{\Delta_i}(t, x)$ is the indicator function of the set Δ_i. Denote by $\mathbf{P}_\delta$ the measure on $\{C, \mathfrak{C}\}$ corresponding to the process $\xi^\delta(t)$ and by $\mathbf{P}_0$ the measure on the same space generated by the solution of equation

$$dx = b(t, x(t))\, dw(t), \qquad x(0) = x_0.$$

As it was mentioned above measures $\mathbf{P}_\delta$ are absolutely continuous with respect to $\mathbf{P}_0$ and

$$\rho_\delta(T) = \frac{d\mathbf{P}_\delta}{d\mathbf{P}_0}(x(\cdot)) = \exp\left\{\int_0^T (b^{-1}(s, u(s))A_\delta(s, x(\cdot)), dx(s))\right.$$
$$\left. - \frac{1}{2}\int_0^T |b^{-1}(s, x(s))A_\delta(s, x(\cdot))|^2\, ds\right\},$$

where $A_\delta(t, x(\cdot)) = \sum_{i=1}^N \chi_{\Delta_i}(t_k, x(t_k))\, a_i(t, x(t))$ for $t \in [t_k, t_{k+1}]$. We show that under the condition

$$|b^{-1}(t, x)a_i(t, x)| \le c \tag{3.145}$$

the family of densities $\rho_\delta(T)$ is uniformly integrable (with respect to $\mathbf{P}_0$). We have

$$\int_C \rho_\delta^2(T)\, d\mathbf{P}_0 = \int_C \exp\left\{2\int_0^T (b^{-1}A_\delta, dx) - 2|b^{-1}A_\delta|^2\, ds\right\}$$

$$\times \exp\left\{\int_0^T |b^{-1}A_\delta|^2\, ds\right\} dP_0 \le C_1 \int_C \exp\left\{2\int_0^T (b^{-1}A_\delta, dx)\right.$$

$$\left. - 2\int_0^T |b^{-1}A_\delta|^2\, ds\right\} dP_0 \le C_1,$$

since under the condition $|b^{-1}A_\delta| \le c$

$$\int_C \exp\left\{2\int_0^T (b^{-1}A_\delta, dx) - 2\int_0^T |b^{-1}A_\delta|^2\, ds\right\} d\mathbf{P}_0 \le 1.$$

Assume, furthermore, that

$$|b(t, x) - b(t', x')| \le c(|t - t'| + |x - x'|), \tag{3.146}$$

$$\sup_{(t,x)}[|a(t, x, u)| + |b(t, x)| + |b^{-1}(t, x)|] < \infty. \tag{3.147}$$

As it was mentioned above the additional conditions (3.146) and (3.147) are not essential under the conditions of theorem 3.26. These conditions imply that the process $\{\xi(t), \mathfrak{C}_t, \mathbf{P}_0\}$ is a Markov process and that there exists for this process a transition probability density $p(s, x, t, y)$ ($0 \le s < t \le T$, $x, y \in R^d$) satisfying the inequality

$$p(s, x, t, y) \le L(t - s)^{-d/2} \exp\left\{-\frac{\alpha|y - x|^2}{t - s}\right\}, \tag{3.148}$$

where L and α are positive constants (cf. for example, [20], vol III, Chapter 3, Section 3).

If $S_r(x)$ is a sphere of radius r in R^d centered at x, then it is easy to verify that

$$\int_{R^d \setminus S_r(x)} p(s, x, t, y)\, dy \le c \exp\left\{-\frac{\gamma r^2}{t - s}\right\}, \tag{3.149}$$

where c and γ are constants.

Formula (3.148) yields that for any $\beta \in (1, 1 + 2/m)$

$$\int_0^T \int_{R^d} [p(s, x, t, y)]^\beta\, ds\, dy \le L_\beta$$

for all $s \in [0, T]$ and $x \in R^d$. Utilizing the Hölder inequality we easily obtain the bound

$$\iint_D p(s, x, t, y)\, dy\, ds \le L_\beta^{1/\beta}[m_L(D)]^{1/\beta'}, \qquad \frac{1}{\beta} + \frac{1}{\beta'} = 1 \tag{3.150}$$

for any measurable set $D \subset [0, T] \times R^d$. Here $m_L(D)$ is the Lebesgue measure of the set D.

Lemma 3.22. *Let conditions* (3.146) *and* (3.147) *be satisfied. Then*

$$\lim_{|\delta| \to 0} \mathbf{E}_0 \int_0^T |A_\delta(t, x(\cdot)) - a(t, x(t))|^2\, dt \to 0.$$

PROOF. We have

$$\mathbf{E}_0 \int_0^T |A_\delta(t, x(\cdot)) - a(t, x(t))|^2\, dt$$

$$= \sum_{k=0}^n \mathbf{E}_0 \int_{t_k}^{t_{k+1}} \left| \sum_{j=1}^N (\chi_{\Delta_j}(t_k, x(t_k)) - \chi_{\Delta_j}(t, x(t)))a_j(t, x(t)) \right|^2$$

$$\le c_N \sum_{j=1}^N \sum_{k=0}^n \mathbf{E}_0 \int_{t_k}^{t_{k+1}} (\chi_{\Delta_j}(t_k, x(t_k)) - \chi_{\Delta_j}(t, x(t)))^2\, dt.$$

Thus it is sufficient to show that

$$B_j^\delta = \sum_{k=0}^n \mathbf{E}_0 \int_{t_k}^{t_{k+1}} (\chi_{\Delta_j}(t_k, x(t_k)) - \chi_{\Delta_j}(t, x(t)))^2\, dt$$

$$= \sum_{k=0}^n \int_{t_k}^{t_{k+1}} dt \int_{R^d} p(0, x, t_k, y)\, dy \int_{R^d} (\chi_{\Delta_j}(t_k, y) - \chi_{\Delta_j}(t, z))^2$$

$$\times\, p(t_k, y, t, z)\, dz$$

approaches 0 for all j as $|\delta| \to 0$. Since the set Δ_j is a Borel set, one can find for any $\varepsilon > 0$ two closed sets F_i, $i = 1, 2$, such that $F_1 \subset \Delta_j$, $F_2 \subset R^d \backslash \Delta_j$ and $m_l(\bar{F}_2 \backslash F_1) < \varepsilon$ where $\bar{F}_2 = R^d \backslash F_2$.

Let $\psi(t, x)$ be a continuous function, $0 \le \psi(t, x) \le 1$ and $\psi(t, x) = 0$ for $(t, x) \in F_2$, $\psi(t, x) = 1$ for $(t, x) \in F_1$. We have

$$B_j^\delta \le 3 \sum_{k=0}^n (I_k^{(1)} + I_k^{(2)} + I_k^{(3)}),$$

where

$$I_k^{(1)} = \int_{t_k}^{t_{k+1}} dt \int_{R^d} p(0, x, t_k, y)\, dy \int_{R^d} (\chi(t_k, y) - \psi(t, z))^2 p(t_k, y, t, z)\, dz, \tag{3.151}$$

and $I_k^{(2)}$ and $I_k^{(3)}$ are analogous to $I_k^{(1)}$ and are obtained if one replaces $\chi(t_k, y) - \psi(t, z)$ by $\chi_{\Delta_j}(t_k, y) - \psi_k(t, y)$ and $\chi_{\Delta_j}(t, z) - \psi(t, z)$ respectively in the r.h.s. of equality (3.151).

As it is easy to verify the continuity of the functions and inequality (3.149) imply that $\sum_{k=1}^n I_k^{(1)} \to 0$ for $|\delta| \to 0$ and an arbitrary $\varepsilon > 0$.

We now bound the sum $\sum_{k=1}^n I_k^{(2)}$. We have

$$I_k^{(2)} = \Delta t_k \int_{R^d} (\chi_{\Delta_j}(t_k, y) - \psi(t_k, y))^2 p(0, x, t_k, y)\, dy.$$

Construct a continuous function $g(t, x) = g^\varepsilon(t, x)$ such that $0 \le g(t, x) \le 1$, $g(t, x) = 1$ for $(t, x) \in \bar{F}_2 \setminus F_1$ and $m_L\{(t, x) : g(t, x) > 0\} < 2\varepsilon$. Then firstly $(\chi_{\Delta_j}(t_k, y) - \psi(t_k, y))^2 \le g(t, x)$ and furthermore

$$\begin{aligned}
\overline{\lim_{|\delta|\to 0}} \sum_{k=0}^{n} I_k^{(2)} &\le \overline{\lim_{|\delta|\to 0}} \sum_{0}^{n-1} \Delta t_k \int_{R^d} g(t_k, y) p(0, x, t_k, y)\, dy \\
&= \int_0^T dt \int_{R^d} g(s, y) p(0, x, s, y)\, dy \le L_\beta(m_L\{(t, x) : g(t, x) > 0\})^{1/\beta} \\
&\le L_\beta(2\varepsilon)^{1/\beta}.
\end{aligned}$$

We now turn to the quantities $I_k^{(3)}$. We have

$$I_k^{(3)} \le \int_{t_k}^{t_{k+1}} dt \int_{R^d} p(0, x, t_k, y)\, dy \int_{R^d} g(t, z) p(t_k, y, t, z)\, dz = \mathscr{T}_k^{(1)} + \mathscr{T}_k^{(2)},$$

where

$$\mathscr{T}_k^{(1)} = \int_{t_k}^{t_{k+1}} dt \int_{R^d} p(0, x, t_k, y)\, dy \int_{R^d} [g(t, z) - g(t, y)] p(t_k, y, t, z)\, dz,$$

$$\mathscr{T}_k^{(2)} = \int_{t_k}^{t_{k+1}} dt \int_{R^d} g(t, y) p(0, x, t_k, y)\, dy.$$

For the quantities $\mathscr{T}_k^{(2)}$ we obtain an expression analogous to a preceding one:

$$\overline{\lim_{|\delta|\to 0}} \sum_{k=0}^{n} \mathscr{T}_k^{(2)} \le L_\beta(2\varepsilon)^{1/\beta}.$$

Furthermore,

$$\begin{aligned}
\sum_{k=0}^{n} \mathscr{T}_k^{(1)} \le \sum_{k=0}^{n} \int_{t_k}^{t_{k+1}} dt \int_{R^d} p(0, x, t_k, y)\, dy \left(\int_{|z-y| \le \varepsilon_1} + \int_{|z-y| > \varepsilon_1} \right) \\
\times [g(t, z) - g(t, y)] p(t_k, y, t, z).
\end{aligned}$$

The continuity of the function $g(t, z)$ and inequality (3.149) yield

$$\sum_{k=0}^{n} \mathscr{T}_k^{(1)} \le \kappa(\varepsilon_1, \varepsilon) + \sum_{k=0}^{n} 2 \int_{t_k}^{t_{k+1}} dt\, c \exp\left\{ -\frac{\gamma \varepsilon_1^2}{t - t_k} \right\} \int_{R^d} p(0, x, t_k, y)\, dy,$$

where $\kappa(\varepsilon_1, \varepsilon) \to 0$ as $\varepsilon_1 \to 0$ for any $\varepsilon > 0$.

Thus, for any $\varepsilon > 0$ and $\varepsilon_1 > 0$

$$\overline{\lim_{|\delta|\to 0}} \sum_{k=0}^{n} I_k^{(3)} \le L_\beta(2\varepsilon)^{1/\beta} + \kappa(\varepsilon_1, \varepsilon).$$

We note that

$$\overline{\lim_{|\delta|\to 0}} B_j^{(\delta)} \le 2L_\beta(2\varepsilon)^{1/\beta} + \kappa(\varepsilon_1, \varepsilon)$$

for any $\varepsilon > 0$ and $\varepsilon_1 > 0$. Hence $\lim_{|\delta|\to 0} B_j^{(\delta)} = 0$. □

Theorem 3.29. *Let the coefficients $a_j(t, x)$, $j = 1, \ldots, N$, and $b(t, x)$ satisfy a local Lipschitz condition and conditions* (3.146) *and* (3.147). *Then the finite-dimensional (marginal) distributions of processes $\xi^{(\delta)}(t)$ weakly converge as $|\delta| \to 0$ to the corresponding distributions of the process satisfying equation*

$$d\xi = a(t, \xi(t))\, dt + b(t, \xi(t))\, dw(t), \qquad a(t, x) = \sum_{j=1}^{N} \chi_{\Delta_j}(t, x) a_j(t, x). \tag{3.152}$$

Proof. It follows from lemma 3.22 that

$$\mathbf{E}\left[\int_0^T (b^{-2}(t, x)A_\delta, dx(s)) - \int_0^T (b^{-1}(t, x), a, dx(s))\right]^2$$
$$\le c^2 \mathbf{E} \int_0^T |A_\delta(s, x(\cdot)) - a(s, x(s))|^2\, ds \to 0$$

as $|\delta| \to 0$. Analogously,

$$\mathbf{E}\left|\int_0^T |b^{-1}A_\delta|^2\, ds - \int_0^T |b^{-1}a|^2\, ds\right| \le \mathbf{E}\left|\int_0^T (b^{-2}A_\delta, A_\delta - a)\, ds\right|$$
$$+ \mathbf{E}\left|\int_0^T (b^{-2}(A_\delta - a), a)\, ds\right| \le 2c^2 T\left(\mathbf{E}\int_0^T |A_\delta - a^2|\, ds\right)^{1/2} \to 0$$

as $|\delta| \to 0$. Therefore $\lim_{|\delta|\to 0} \rho_\delta(T) = \rho(T)$, where

$$\rho(T) = \exp\left\{\int_0^T (b^{-1}a, dx(s)) - \frac{1}{2}\int_0^T |b^{-1}a|^2\, ds\right\}.$$

Furthermore, for any bounded and continuous on C function $f(x(\cdot))$ in view of the uniform integrability of the family of densities $\{\rho_\delta(T)\}$ we have

$$\lim_{|\delta|\to 0} \int_C f(x(\cdot))\, d\mathbf{P}_\delta = \lim_{|\delta|\to 0} \int_C f(x(\cdot))\rho_\delta(T)\, d\mathbf{P}_0 = \int_C f(x(\cdot))\rho(T)\, d\mathbf{P}_0.$$

Thus the measure $\mathbf{P}_\delta$ weakly converges as $|\delta| \to 0$ to the measure $\mathbf{P}$, which is absolutely continuous with respect to $\mathbf{P}_0$ and possesses the density $d\mathbf{P}/d\mathbf{P}_0 = \rho(T)$. However, the measure $\mathbf{P}$ corresponds in C to the solution of equation (3.152). The theorem is thus proved. □

As it was mentioned above, the theorem just proved implies theorem 3.28 so that

$$Z(t, x) = \inf_{u \in \mathfrak{U}_2} \mathbf{E}\left\{\int_t^T f(s, \xi^{(u)}(s), u(s))\, ds \,|\, \mathfrak{C}_t\right\}$$

$$= \inf_{u \in \mathfrak{U}_2} \mathbf{E} \int_t^T f(s, \xi_{t_x}^{(u)}(s), u(\cdot))\, ds.$$

The existence of ε-optimal Markov controls allows us to prove the existence of optimal Markov controls and to determine their form.

Lemma 3.23. *Under the conditions* a, b *and* c *(see the beginning of this section) there exist measurable functions* $LZ: [0, T] \times R^d \to R^d$ *and* $\Lambda Z: [0, T] \times R^d \to R$ *such that*

$$\mathbf{E} \int_0^T |LZ(t, x(t))|\, dt < \infty,$$

$$\int_0^T |\Lambda Z(t, x(t))|^2\, dt < \infty,$$

$$Z(t, x(t)) = F^* + \int_0^T LZ(s, x(s))\, ds + \int_0^T \Lambda Z(s, x(s))\, dx(s),$$

where $F^* = \inf Z(0, x)$.

PROOF. It was shown above (cf. theorem (3.24)) that

$$Z(t, x(t)) = F^* + \int_0^t \eta_s\, ds + \int_0^t \zeta_s\, dw(s). \tag{3.153}$$

It remains to verify that η_s and ζ_s are functions of $x(s)$. We have

$$\int_0^t \zeta_s \zeta_s^*\, ds = [Z, Z]_t,$$

where $[Z, Z]_t$ is the matrix quadratic variation of the process $Z(t, x(t))$. This implies that the function $[Z, Z]_t$ is absolutely continuous with probability 1 and that

$$[Z, Z]_t = \int_0^t \psi_s\, ds,$$

where ψ_s is a function of $x(s)$ and $\zeta_t \zeta_t^* = \psi_t$. Consequently, the vector ζ_t can be defined in such a manner that ζ_t will be a function of $x(t)$. In turn it follows from that proven above and formula (3.153) that $\eta(t)$ is also a function of $x(t)$. □

Remark. Let the function $Z(t, x)$ possess continuous derivatives with respect to t of the first order and that of the second order with respect to x. Then

$$LZ = \frac{\partial Z}{\partial t} + \frac{1}{2} \sum_{kj} \frac{\partial^2 Z}{\partial x_k \, \partial x_j} (bb^*)_{kj}, \tag{3.154}$$

$$\Lambda Z = \nabla Z. \tag{3.155}$$

PROOF. Since $dx(t) = b(t, x(t))\, dw(t)$ (mod $\mathbf{P}_0$) we have, in view of Itô's formula

$$dZ(t, x(t)) = \left(\frac{\partial Z}{\partial t} + \frac{1}{2} \sum_{kj} \frac{\partial^2 Z}{\partial x_k \, \partial x_j} (bb^*)_{kj} \right) dt + (\nabla Z, dx).$$

Comparing the last formula with formula (3.153) we obtain equations (3.154) and (3.155) due to the uniqueness of the form of a stochastic differential.

Analogously to theorem 3.23, the following theorem can be proved.

Theorem 3.30. *A Markov control $u^* \in \mathfrak{U}_2$ is optimal if and only if there exist a constant F^* and measurable functions $\gamma(t, x)$ and $\beta(t, x)$ $[(t, x) \in [0, T] \times R^d]$ with values in R and R^d respectively such that*

a. $\int_0^T |\beta(t, x(t))|^2 \, dt < \infty$ (mod $\mathbf{P}_0$), $\quad \mathbf{E} \int_0^T \beta(t, x(t)) \, dx(t) = 0$;

b. $V(T) = 0$, *where*

$$V(t) = F^* + \int_0^t \gamma(s, x(s)) \, ds + \int_0^t \beta(s, x(s)) \, dx(s);$$

c.

$$\gamma(t, x(t)) + (\beta(t, x(t)), a_t^{(u)}) + f_t^{(u)} \geq 0 \text{ (mod } \mathbf{P}_0);$$
$$\gamma(t, x(t)) + (\beta(t, x(t)), a_t^{(u^*)}) + f_t^{(u^*)} = 0 \text{ (mod } \mathbf{P}_0),$$

where

$$a_t^{(u)} = a[t, x(t), u(t, x(t))],$$
$$f_t^{(u)} = f[t, x(t), u(t, x(t))].$$

The proof of this theorem is the same as that of theorem 3.25 but, instead of lemma 3.20, lemma 3.23 is utilized. Observe that if the transition probability $P(s, x, t, A)$ corresponding to the Markov process which is determined by equation (3.140) possesses density $p(s, x, t, y)$ then condition c is equivalent to the following:

$$\gamma(t, x) + \min_{u \in U} [(\beta(t, x), a(t, x, u) + f(t, x, u)] = 0,$$

and the optimal control is characterized by the property: the variable $(\Lambda Z(t, x), a(t, x, u)) + c(t, u, x)$ is minimized at $u = \varphi^*(t, x)$. This implies, in particular, the existence of an optimal Markov control.

The following theorem which is analogous to theorem 3.30 is valid.

Theorem 3.31. *Let*

$$H(t, x, p) = \min_u H(t, x, p, u),$$

$$H(t, x, p, u) = (p, a(t, x, u)) + f(t, x, u),$$

and the coefficients of equation (3.26) *satisfy conditions* a–c. *Then the Markov control* $u^*(t, x)$ *defined by relation*

$$u^*(t, x) = y^*(t, x, \Lambda Z(t, x)),$$

where $y^*(t, x, p)$ *is the solution of equation*

$$H(t, x, p) = H(t, x, p, y^*(t, x, p)),$$

is an optimal control.

Bibliography

1. Aoki, M. *Optimization of Stochastic Systems*. New York, London: Academic Press, 1967.
2. Astrem, K. J. *Introduction to Stochastic Control Theory*. New York: Academic Press, 1970.
3. Bellman, R. A Markov decision process. *J. Math. Mech.*, **6**, 679–684, 1957.
4. Bellman, R. *Dynamic Programming*. Princeton: Princeton University Press, 1957.
5. Benes, V. E. Existence of optimal strategies based on specified information, for a class of stochastic decision problems. *SIAM J. Control*, **2**, 179–188, 1970.
6. Benes, V. E. Existence of optimal stochastic control laws. *SIAM J. Control*, **9**, 446–472, 1971.
7. Blackwell, D. Discrete dynamic programming. *Ann. Math. Stat.*, **33**, 719–726, 1962.
8. Blackwell, D. Positive dynamic programming. Proc. of the 5th Berkeley Symp., vol. 3, 415–428, 1965.
9. Blackwell, D. On stationary policies, Proc. of the 37th Session of the Intern. Statist. Institute, London, Sept. 1969. (Russian translation in Matematika **14**, No. 2, 155–159, 1970.)
10. Davis, M. H. A. On the existence of optimal policies in stochastic control. *SIAM J. Control*, **11**, 587–994, 1973.
11. Davis, M. H. A., Varaiya, P. P. Dynamic programming conditions for partially observable stochastic integrals. *SIAM J. Control*, **11**, 226–261, 1973.
12. Duncan, T. E., Varaiya, P. P. On the solution of a stochastic control system. *SIAM J. Control*, **9**, 354–371, 1971.
13. Dunford, N., Schwartz, J. T. *Linear Operators, General Theory*. New York: Interscience Publishers, 1958.
14. Dynkin, E. B., Yushkevich, A. A. *Controlled Markov Processes and their Applications*. Moscow: Nauka, 1975 (in Russian).
15. Fleming, W. Some Markovian optimization problems. *J. Math. and Mech.*, **12**, 131–140, 1963.
16. Fleming, W. Duality and a priori estimates in Markovian optimization problems. *J. Math. Anal. Appl.*, **16**, 254–279, 1966. Erratum, *ibid.* **19** (1966) p. 204.
17. Fleming, W. Optimal continuous parameter stochastic control. *SIAM Rev.*, **11**, 470–509, 1969.

18. Gihman, I. I. On a weak compactness of a set of measures corresponding to solutions of stochastic differential equations. In: *Mathematical Physics* **7**, Kiev: Naukova Dumka, 49–65, 1970 (in Russian).
19. Gihman, I. I., Skorohod, A. V. *Stochastic Differential Equations*. Berlin, Heidelberg, New York: Springer-Verlag, 1972.
20. Gihman, I. I. and Skorohod, A. V. *The Theory of Stochastic Processes*, Volumes I, II, III. Berlin, Heidelberg, New York: Springer-Verlag, 1974, 1975, 1979.
21. Girsanov, I. V. On transforming a certain class of stochastic processes by absolutely continuous substitution of measures. *Theory Probab. and Applic.* **5**, 285–301, 1960.
22. Gubenko, L. G. Controlled Markov and semi-Markov models and some specific problems of optimization of the stochastic system. Abstract of candidate's thesis, Kiev State University, Kiev, 1972 (in Russian).
23. Gubenko, L. G. and Shtatland, E. S. On controlled discrete time Markov decision processes. *Teor. Veroyatn. i Matemat. Statist.*, **7**, 51–64, 1972. English translation **7**, 47–62, 1975.
24. Hausdorff, F. *Grundzüge der Mengenlehre*, Leipzig: Teubner 1914.
25. Hinderer, K. *Foundation of Nonstationary Dynamic Programming with Discrete Time Parameter*. Berlin, Heidelberg, New York: Springer-Verlag, 1970.
26. Howard, R. A. *Dynamic Programming and Markov Processes*. Cambridge: Technology Press, and New York: J. Wiley, 1960.
27. Howard, R. A. *Dynamic Probabilistic Systems, V.2. Semi-Markov and Decision Processes*. New York: J. Wiley, 1971.
28. Krylov, N. W. On the existence of ε-optimal homogeneous Markov strategies for controlled chains. *Dokl. Akad Nauk SSSR*, **155**, 747–750, 1964 (in Russian).
29. Krylov, N. W. The construction of an optimal strategy for a finite controlled chain. *Theory Probab. and Applic.*, **10**, 45–54, 1965.
30. Krylov, N. W. Control of a solution of a stochastic integral equation. *Theory Probab. and Applic.*, **17**, 114–131, 1972.
31. Krylov, N. W. On controlling a solution of a stochastic integral equation in the presence of degeneracy. *Izvest. Akad. Nauk SSSR, Ser. Matemat.* **36**, 248–261 (in Russian) 1972.
32. Krylov, N. W. On Bellman's equation. In the volume: *Trudy Shkoly-Seminara po Teorii Sluchainyh Protsessov* (*Proc. of the Seminar on the Theory of Stochastic Processes*). Izdat. Akad. Nauk Lit SSSR, Vil'nyus, 1975.
33. Kushner, H. J. *Stochastic Stability and Control*. New York, London: Academic Press, 1967.
34. Kushner, H. J. *Introduction to Stochastic Control*. New York: Holt, 1971.
35. Lipcer, R. Sh. and Shiraev, A. N. *Stochastics of Random Processes*. Berlin, Heidelberg, New York: Springer-Verlag, 1976.
36. Dellacherie, C. and Meyer, P. A. *Probabilities and Potential*. New York: Elsevier, 1977.
37. Portenko, N. I. and Skorohod, A. V. On the existence of ε-optimal Markov strategies for controlled diffusion processes. In: *Voprosy Statistiki i Upravleniya Sluchaĭnymi Protsessami* (*Problems of Statistics and Control of Random Processes*). Izdat. IM Akad. Nauk USSR, Kiev, 204–227, 1974.
38. Pragarauskas, G. On a control theory for discontinuous random processes. In: *Trudy Shkoly-Seminara po Teorii Sluchaĭnych Protsessov* (*Proc. of the Seminar on the theory of stochastic processes*), Izdat. Akad. Nauk LitSSR, Vil'nyus, 1975.
39. Rishel, R. Necessary and sufficient dynamic programming conditions for continuous-time stochastic optimal control. *SIAM J. Control*, **8**, 559–571, 1970.
40. Shiryaev, A. N. *Statistical Sequential Analysis*. Providence: Amer. Math. Soc., 1973 (Russian edition, Nauka 1969).
41. Skorohod, A. V. *Studies in the Theory of Random Processes*. Reading: Addison Wesley, 1965.

42. Skorohod, A. V. and Slobodenyuk, N. P. *Limit Theorems for Random Walks.* Kiev: Naukova Dumka, 1970.
43. Wald, A. *Statistical Decision Function.* New York: J. Wiley, 1950.
44. Wonham, W. M. Stochastic problems in optimal control. Tech. report, Baltimore, 63–74, 1963.

Additional References for the English Edition

45. Wonham, W. M. Random differential equations in control theory. In A. T. Bharucha-Reid ed., *Probabilistic Methods in Applied Math.*, vol 2, New York, London: Academic Press, 131–212, 1970.
46. Chow, Y. S., Robbins, H., Siegmund, D. *Great Expectations: The Theory of Optimal Stopping.* Boston: Houghton Mifflin, 1971.
47. Fleming, W. H., Richel, R. W. *Deterministic and Stochastic Control.* Berlin, Heidelberg, New York: Springer-Verlag, 1975.
48. Gatun, A. P., Gihman, I. I. Controlled stochastic differential equations without an after-effect. Theory of stoch. processes, **5**, 14–21, Kiev: Naukova Dumka, 1977 (in Russian).
49. Krylov, N. W. *Controlled Processes of the Diffusion Type.* Moscow: Nauka 1977 (in Russian).
50. Skorohod, A. W. *On One General Scheme of Controlled Stochastic Processes. Controlled Stochastics Processes and Systems.* Kiev: Naukova Dumka, 1972 (in Russian).
51. Snell, J. L. Application of martingale system theorems. *Trans. Amer. Math. Soc.*, **73**, 293–312, 1952.
52. Stone, L. D. Necessary and sufficient conditions for optimal control of semi-Markov jump process. *SIAM J. Control*, **11**, 187–201, 1973.
53. Strauch, R. E. Negative dynamic programming. *Ann. Math. Stat.* **37**, 871–890, 1966.

Historical and Bibliographical Remarks

Chapter 1

The basic ideas of optimal control theory are linked to sequential analysis and the theory of statistical decision functions which originated and were developed by A. Wald (see, e.g. [43]). R. Bellman developed this theory for more specific cases under the name of dynamic programming (see [3], [4]). Further development of multi-staged Markov processes was stimulated by Howard [26], [27] who also studied the problem of the controlled Markov chain. The general results on optimal controlled Markov chains with arbitrary sets of states were obtained by D. Blackwell [7], [8], [9], R. E. Strauch [53], and N. W. Krylov [28], [29]. A detailed exposition of the theory of controlled Markov chains is contained in Kushner's monograph [34] and a more general theory may be found in E. B. Dynkin and A. A. Yushkevich's memoir [14].

The general description of controlled stochastic objects presented in the first chapter of the present book and the constructions of optimal and ε-optimal controls are based on A. V. Skorohod's paper [50].

The bibliography of works devoted to the problem of optimal stopping is quite substantial. A general solution of this problem is due to Snell [51]. A more detailed exposition of optimal stopping theory is presented in A. N. Shiryaev's [40] and in Y. S. Chow *et al.*'s monographs [46].

Chapter 2

The definition and construction of arbitrary controlled (in general non-Markov) objects with continuous time given in the second chapter is along the lines of A. V. Skorohod's paper [50] cited above as well as N. I. Portenko and A. V. Skorohod's work [37]; however, our definition is wider in scope and more far reaching. This construction of the representation of a controlled object appears in the literature presumably for the first time.

Optimal controlled semi-Markov processes with continuous time are discussed by L. D. Stone [52].

Chapter 3

The theory of stochastic differential equations is studied in more detail in the following books: [19] and [20] vol 3, for example.

Initially the development of the theory of optimal control for solutions of stochastic differential equations was concerned solely with linear systems with a square cost of control and with problems related to the theory of "Bellman's equation".

Section 5 deals with the problem of control for a solution of a linear system of stochastic differential equations with complete observations of a somewhat more general type than usually considered in the literature, namely the random perturbations affecting the system may be general continuous processes with independent increments and finite moments of the second order.

A review of the results dealing with the theory of linear systems is presented for instance in A. W. Wonham's paper [45].

In connection with Bellman's equation and its justification for stochastic processes with continuous time, the reader is referred to the works of H. J. Kushner [33], W. Fleming [16], [17], and N. W. Krylov [30–32], where the results of the theory of parabolic equations with partial derivatives are substantially utilized. The most far-reaching study of controlled diffusion processes based on a direct probabilistic approach to Bellman's equation is presented in N. W. Krylov's monograph [49]; see also W. H. Fleming and R. W. Rishel's memoir [47].

The convergence of finite-difference approximations in the problem of optimal control for solutions of stochastic differential equations was considered in the papers of A. V. Skorohod [50], A. P. Gatun and I. I. Gihman [48].

The material in Sections 6 and 7 is based on M. H. A. Davis and P. P. Varaiya's [11] and M. H. A. Davis' contributions [10] (see also V. E. Benes [5–6].)

In this monograph only the case of complete observations is studied. A more general case of controlling incomplete observations is discussed in [11]. The construction of ε-optimal Markov controls is based on N. I. Portenko and A. V. Skorohod's exposition [37].

Index

D

E

F

G

H

I

J

K

L

M

N

O

P

R

S

V

W